The Total Survey Error Approach

The Total Survey Error Approach

A GUIDE TO THE NEW SCIENCE OF SURVEY RESEARCH

Herbert F. Weisberg

THE UNIVERSITY OF CHICAGO PRESS · CHICAGO AND LONDON

Herbert F. Weisberg is professor of political science and director of the Center for Survey Research at Ohio State University. He is a coauthor of the widely used text *Introduction to Survey Research, Polling, and Data Analysis*, 3d ed. (1996, with Jon A. Krosnick and Bruce D. Bowen).

The University of Chicago Press, Chicago 60637
The University of Chicago Press, Ltd., London
© 2005 by The University of Chicago
All rights reserved. Published 2005
Printed in the United States of America

14 13 12 11 10 09 08 07 06 05   1 2 3 4 5

ISBN: 0-226-89127-5 (cloth)
ISBN: 0-226-89128-3 (paper)

Library of Congress Cataloging-in-Publication Data

Weisberg, Herbert F.
    The total survey error approach : a guide to the new science of survey research / Herbert F. Weisberg.
        p.   cm.
    Includes bibliographical references and index.
    ISBN 0-226-89127-5 (cloth : alk. paper) — ISBN 0-226-89128-3 (pbk. : alk. paper)
    1. Social surveys. 2. Social sciences—Research. I. Title.
HM538.W45 2005
300.72′3—dc22
                                                           2005007816

♾ The paper used in this publication meets the minimum requirements of the American National Standard for Information Sciences—Permanence of Paper for Printed Library Materials, ANSI Z39.48-1992.

Contents

Preface

In recent decades, the survey field has been revolutionized with increased attention to both its basis in social psychology and its statistical underpinnings. The "total survey error" approach provides a new paradigm by which to understand and study this field, so it is time to proclaim the true arrival of survey research as a new science. This book uses the total survey error approach to present a unified approach to understanding the components of good survey research.

Survey research is an interesting research approach because the trade-offs between theory and practical considerations are quite direct. This has always been evident in survey sampling, where sampling statisticians learned early on to balance statistical theory with the difficulty of sampling in the real world. The initial knowledge of question writing was developed from practical interviewing, but in recent years the understanding of this topic has become more theoretically informed through insights from social psychology. The chapters in this book reflect the current balance between theory and practice, statistics and cognitive psychology.

The total survey error approach has become crucial to understanding survey research. The usual presentation of this approach focuses on the trade-offs between survey error and costs. This book extends this approach by generalizing the notion of survey constraints to include time and ethical considerations as well as monetary cost constraints. Additionally, it broadens the coverage by emphasizing the role of survey-related effects, effects that can be studied but cannot be minimized. Thus, this book argues that a full statement of the total survey error approach requires consideration of survey errors, survey constraints, and survey-related effects.

The total survey error framework is introduced in the opening three chapters, after which separate chapters detail each of the sources of survey error: measurement error (due to interviewers and respondents), nonresponse (at the item and unit levels), coverage error, and sampling error. Additional chapters deal with postsurvey error, mode differences, and survey ethics. The treatment of these topics is generally nonmathematical, presenting findings from the most recent journal articles and edited volumes, including extensive treatment of Internet polls.

The main chapters are organized around the different types of survey error. Each chapter begins with theoretical elements, pulling together insights from different disciplines, such as the social psychology of interviewing and relevant statistical theories. This is followed by detailed treatment of the specifics of the type of error and potential solutions. The chapters end with discussion of the constraints on error minimization due to costs, time, and ethics. The topic of measurement error due to respondents is so large that I have broken it into two separate chapters, one on question wording and one on questionnaire construction.

The organization of this book is unconventional, particularly by covering sampling error after rather than before the other major types of error. However, the point of the total survey error approach is that there is more to survey error than just the usual emphasis on sampling error. Beginning with sampling error would be to continue that overemphasis, whereas treating it last serves to emphasize the need to deal directly with the other sources of error. Still, the chapters on specific types of error—chapters 4 through 10—are written so that they can be read in any order; thus the readers can follow their own preferences regarding topic ordering.

The main focus of this book is cross-section surveys of the mass public, with coverage of face-to-face, telephone, and Internet surveys in addition to self-administered questionnaires. Some discussion of panel surveys and establishment surveys is also provided. I have attempted to cover a wide variety of problems with surveys, giving evidence on both sides when there is no consensus yet as to how to deal with those problems.

This book was written because of two sets of indirect forces. First, Paul Beck, Kathleen Carr, Jerry Kosicki, Jon Krosnick, Paul Lavrakas, Kathleen McGraw, Tom Nelson, Randy Olsen, Toby Parcel, Randall Ripley, and Elizabeth Stasny all, not for this purpose, allowed me to learn from observing and participating in Ohio State University's high-quality survey operations and political psychology program. Second, unexpected personnel changes in spring 2001 led to my teaching a new practicum course in OSU's fledgling Graduate Interdisciplinary Specialization in Survey Research, giving me the not-necessarily-desired opportunity to see how much the survey research field had changed since the early 1990s as the total survey error approach has become dominant. I am also very appreciative of those who contributed more directly to this project: Steve Mockabee for his pounding into me the importance of the total survey error approach; Margaret Williams for her intensive journal review; Ron Rapoport, Andy Farrell, Elizabeth Stasny, Mike Traugott, Delton Daigle, and anonymous readers for their comments on the manuscript; Erin McAdams for her careful indexing; and the people at the University of Chicago Press for their fine work on this book.

The Total Survey Error Approach

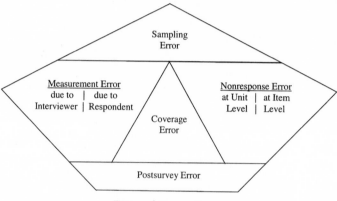

Types of Survey Error

PART I

Survey Error Theory

1

Scientific Survey Research
THE DEVELOPMENT OF A DISCIPLINE

Around 1939 George Gallup's American Institute of Public Opinion published a pamphlet entitled *The New Science of Public Opinion Measurement*. That was a heady era in the survey world, with researchers having learned how to measure attitudes and how to sample the mass public. As the pamphlet indicated, this new science would use the "sampling referendum" (as it termed what we now call the sample survey), based on understanding the sample size required for confidence in results and on devising methods for obtaining representative cross-sections of respondents. Yet in retrospect it is clear that survey research was still in its infancy back then. Both the statistical underpinnings of survey methods and the underlying social psychology were primitive. It was appropriate for Gallup's shop to exult in the early successes of their survey approach, but much more research was needed before it would be time to call this a science. Today, more than sixty years later, it finally is time to proclaim *the new science of survey research*. In the last few decades, survey research has accumulated a solid basis in both statistical methods and social psychology. The "total survey error" approach now provides a paradigm by which to understand and study this new science.[1]

The purpose of this book is to explain the survey research process in the context of this new approach to the field. The first three chapters establish the background by reviewing the development of the survey field, outlining and extending the total survey error approach, and providing a common understanding of the different survey modes. The middle of the book follows the total survey error approach's emphasis on the different sources of survey errors, discussing the different elements of surveys through a focus on the sources of error. The final chapters turn to broader and summary issues.

The notion of a scientific survey will be explained carefully in chapter 2, but it is useful to give a quick definition here. Surveys have people answer questions so as to develop quantitative descriptions; usually a sample of people are interviewed, with the intention of generalizing the findings to a larger population.

The Development of Modern Survey Research

It is appropriate to begin with a brief recounting of the development of survey research as a scientific field. This section reviews the early institutionalization of survey research, later changes due to evolving survey technology, and the recent professionalization of the field.

The Early Institutionalization of Survey Research

Polling can be dated back to censuses taken several millennia ago and to straw polls in early U.S. elections, but surveys did not become commonplace until the twentieth century. Jean Converse (1987) identifies three more proximate sets of ancestors: social reformers, science, and business. Social reformers around 1900 sought to document the extent of poverty in cities in England (Booth 1889; Rowntree 1901) and the United States (Residents of Hull House 1895; DuBois 1899) through "social surveys" (Bulmer, Bales, and Sklar 1991) and to detail living patterns in American rural areas through the Country Life movement (Bailey 1906; Bowers 1974). Sampling techniques were also developed during the first half of the twentieth century, particularly for agricultural research. A second independent source was the world of social science, resulting from the work in the 1920s on attitude measurement in psychology and sociology (Thurstone 1928). This led to some early academic political surveys (e.g., Gosnell 1927; Rice 1928).

Converse argues, however, that the third source was the most direct ancestor of modern survey research: the world of business as represented by market research and opinion polling in journalism. Consumer research developed in the decade of the 1910s in several large industries. Such researchers as George Gallup, Elmo Roper, and Archibald Crossley pioneered in straw-vote journalism with their correct prediction that Franklin Delano Roosevelt would win reelection in 1936.[2] This led to innovation in sampling procedures, attention to interviewing techniques, and developing formats for questions. By the 1935–40 period, linkages were being established between those doing applied research for business and government and academic survey researchers. This was typified by the founding of the *Public Opinion Quarterly* journal in 1937 and the establishment of survey organizations in the government and at universities. Converse documents the early history of three influential survey organizations: Rensis Likert's Division of Program Surveys in the Department of Agriculture; Hadley Cantril's Office of Public Opinion Research at Princeton University, where Gallup was headquartered; and Paul Lazarsfeld's Office of Radio Research (to develop measures of listenership so that radio stations could charge for ads) at Columbia University, which became the Bureau of Applied Social Research.

What is notable is how many of the basic understandings that underlie survey research were in place by 1940. It was recognized that people could be studied and could give useful information and that different people would give different information so that it was important to study multiple people. Statisticians realized that they could generalize from a sample without studying the entire population, and they had developed probability sampling methods to permit generalizations (Wright 2001). Psychologists learned that attitudes could be measured. Panel surveys in which the same respondents were reinterviewed over time were devised. Each of these understandings is taken for granted today, but each was an important advance for survey research.

But the events that were most important for the development of the survey field occurred just after this time. Many social scientists went to Washington, D.C., during World War II to work for the federal government in the war effort. That led to significant development of polling technology, as evidenced in the publication of the major work *The American Soldier* (Stouffer et al. 1949), based on surveys of soldiers to deal with wartime morale and issues that might affect their return to civilian society. After the war, the social scientists themselves went back to academia, often in teams, such as Likert's Department of Agriculture team, which went to Ann Arbor to found their Institute of Social Research (ISR) with its important Survey Research Center. Lazarsfeld's Bureau of Applied Research at Columbia University did several single community surveys in the 1940s as well as contributing to the intellectual understanding of survey research and data analysis. The National Opinion Research Center (NORC), a group that had been founded just before the war at the University of Denver, moved to the University of Chicago in 1947 and did national face-to-face interviews for social scientists. Thus by 1950 several large academic survey organizations were in place, and on the commercial side of polling, George Gallup and Elmo Roper, among others, had their own national survey operations.

The late 1940s was a period of building the institutions of survey research. This is best exemplified by the founding of a professional organization, the American Association for Public Opinion Research (AAPOR), in 1947 after an initial meeting in 1946, with *Public Opinion Quarterly* becoming its official journal in 1948. The *International Journal of Opinion and Attitude Research* was also started in 1947. The Survey Research Center at the University of Michigan began its summer training program in survey research methods in 1948, attracting students from across the United States and around the world. In the 1950s, textbooks on survey research methods were written (Parten 1950; Hyman 1955), the range of topics studied through surveys was expanded, and survey research spread to more countries.

The federal government also had a major involvement in the development of survey research through its Census Bureau. Originally commissioned to conduct only a single census every ten years, the Census Bureau became increasingly involved in the conduct of surveys and the study of survey accuracy in order to assess the precision of its own census efforts. Additionally, the federal government's need to generate statistics about such diverse topics as employment, crime, and health as well as a desire to be able to evaluate the results of government programs led to the development of excellent survey research capabilities in other federal government agencies. Indeed, some of the best work on methodological issues involving surveys is still done in the federal statistical agencies, including the Bureau of the Census, the Bureau of Labor Statistics, and the National Center for Health Statistics.

Both fact-based and attitude surveys became common by the 1950s. Fact-based surveys, which include most surveys undertaken directly by the federal government, generally focus on past behavioral experiences, such as asking people about their medical history or asking businesses about their sales in the past month. Attitude surveys ask people how they feel toward particular objects, from political candidates in election surveys to household products in marketing research. Attitudes themselves are the positive or negative orientations toward these objects. Along with attitudes, these surveys typically measure preferences between objects, based on comparisons of attitudes toward objects, such as whether the person prefers the Republican or Democratic candidate or prefers one brand of toothpaste or another. Attitude surveys also often include questions about beliefs—the respondents' opinions about the objective state of the world. These are measured in attitude surveys because people's attitudes will depend on what they think is true of the world. Thus public views about the health of the economy are likely to affect voting plans as well as purchasing plans.

The first surveys were used mainly for simple reporting of the results of individual questions in order to assess the frequency of particular facts or attitudes, often with comparisons of different subgroups of the population. By the 1960s repeated surveys became common, so that trends over time in both facts and attitudes could be traced. Examples of large-scale continuing surveys include the National Election Studies in political science, the General Social Survey in sociology, and the National Longitudinal Study of Labor Market Experience. Social science researchers increasingly used these surveys to construct causal explanations of behaviors and attitudes by examining the relationships between those dependent variables and predictor variables that are appropriate in the relevant fields of study.

Looking back at this period, Jean Converse (1987) argues that the early survey researchers made significant contributions in addition to founding survey organizations. She calls particular attention to the development of theory, especially attitude theory, to the creation of exemplars of quality research, and to progress in methods, including sampling, interviewing, and question wording. Yet she concludes that even by 1960 survey research was not itself a discipline: "while it was no garden-variety tool, it was still something less than a discipline" (390–91).

The Impact of Evolving Survey Technology

While the survey research field had begun to broaden its scope by the 1960s, it growth was being limited by technology. The understanding at the time was that quality surveys could be obtained only through face-to-face interviewing in dwelling units, and the expense of that mode of polling restricted its expansion potential.

By the mid-1970s the field was rethinking the optimal way to conduct surveys. Whereas telephone surveys were originally frowned upon because many households did not have phones, telephones had finally become nearly universal in the United States by the 1970s, with a coverage rate nationally of about 90%. Furthermore, the costs of telephoning across the nation began to fall around the same time that the gasoline crisis of that era increased the transportation costs involved in personal interviewing. Procedures for sampling telephone numbers were developed and refined. The shift from room-size computers to desktop computers in the 1980s made it possible for even small telephone survey shops to own their own computers to process survey results. By the 1980s telephone interviewing had become the predominant mode of conducting surveys in the United States.

The shift to telephone technology permitted widespread growth in the field of survey research. It became possible to conduct a national survey from just a small room equipped with a dozen telephones. The simultaneous development of computer-assisted telephone-interviewing (CATI) technology further facilitated this switch and provided the added benefit of being able to generate results virtually instantaneously. Led by Berkeley's Survey Research Center, most academic survey centers and the established commercial pollsters switched to phone interviewing. The institutionalization of the survey field continued as more universities established survey research centers, additional commercial polling companies entered the field, and several important research firms were founded in the Washington, D.C., area to consult for federal agencies that used survey data. The survey research industry blossomed and experienced robust growth after the nation emerged from the recessions of the early 1990s, with the federal and state governments commissioning many more surveys.

Further technological changes occurred in the 1990s, continuing into the first decade of the 2000s. Just as telephone survey methods were being refined, changes in telephone usage patterns began to make phone interviews more problematic. The chances of getting through to a residential phone when calling a phone number decreased with the popularity of cellular phones and with people putting answering machines, fax machines, and computer modems on their telephone lines. Thus it may be that the high point of telephone surveys is already in the past. Meanwhile, the development of the Internet permits new ways of contacting respondents through e-mail and through Web surveys. The different modes of survey research will be discussed in more detail in chapters 3 and 12.

Professionalization of the Field

The survey field has professionalized as it has institutionalized and evolved. One aspect of its professionalization has been the development of professional standards. This includes disclosure standards (such as what information should be provided in press releases on surveys, as discussed in chapter 13), technical standards (particularly standard ways of computing response rates, as will be described in chapter 8), and ethics codes that are incorporated in statements of professional conduct (see chapter 14).

Several professional associations have been formed. AAPOR brings together survey researchers from academia, government, and commercial firms. The Survey Research Methods Section of the American Statistical Association initially focused on sampling methods but subsequently broadened to cover the full range of survey-related methods. The Council of American Survey Research Organizations (CASRO) is a trade association for commercial survey firms, the Council for Marketing and Opinion Research (CMOR) emphasizes market research, and the National Council of Public Polls (NCPP) focuses on media reporting of polls.

Another aspect of professionalization is that survey research has become truly international. The World Association for Public Opinion Research (WAPOR) has been in existence since the late 1940s, while the International Association for Survey Statisticians was formed in 1973. Excellent research on the survey process is being conducted at universities and government statistical agencies in a wide variety of countries. There are frequent international conferences on various aspects of survey research, yielding major compendia of research reports, such as *Survey Nonresponse* (Groves et al. 2002), and *Cross-Cultural Survey Methods* (Harkness, van de Vijver, and Mohler 2003). The internationalization of the field is reflected in the articles published in journals such as *Public Opinion Quarterly*, *Survey Methodology* (published by Statistics Canada), the *Journal of Official Statistics* (published by Statistics Sweden), the *Survey Statistician* (published by the International Association for Survey Statisticians), and the *International Journal of Public Opinion Research* (published in England for WAPOR).

The professionalization of survey research is also evident in the establishment in the 1990s of several university-based degree programs. The Joint Program in Survey Methodology of the Universities of Michigan and Maryland in conjunction with Westat is a major training program for federal government employees in the Washington, D.C., area. Additionally, certification, master's, and/or doctoral programs in survey research were instituted at the Universities of Cincinnati, Connecticut, Maryland, Michigan, Nebraska, and Ohio State University. If survey research was not yet a separate discipline in 1960, it was approaching that status by 2000.

One distinguishing feature of the survey research field is that it is inherently multidisciplinary. It is a social science field since it deals with human behavior, but it is based on sampling procedures developed in statistics. Many of the most important insights come from psychology, as it is essential to understand how people react to survey questions as well as how people form their attitudes. The scientific study of communication is also relevant since interviews are a communication process. Many important developments in the field also come from political science because of the importance of election polls and from market research because of extensive polling of consumers.

Approaches to Survey Research

As the institutionalization, technology, and professionalization of survey research have evolved, approaches to the field have also changed. The original approach to the field was based on the lore that accumulated from early surveys, assembling the experiences of survey researchers into lists of recommended do's and don'ts. The procedures that seemed to work best became the accepted practices in the field. Some procedures were based on the social psychology of the time, such as structuring the relationship between a survey interviewer and respondent on the basis of the relationship between a psychological therapist and patient, but most of this early work was atheoretical.

The survey field gradually began to emphasize *standardization*. The more each interview proceeded in the same way, the more scientific the polling seemed to be. Interviewers were instructed to proceed in a standardized fashion. The best questions were thought to be those that were the standards in the field. This approach was still atheoretical, but it seemed more systematic than just codifying experiences of survey researchers.

The survey research field underwent *scientization* in the 1990s, applying theories from other fields to the survey research field. In particular, the cognitive revolution in psychology has led to new understanding of the survey response process (see chapter 5), and developments in statistics have provided new imputation techniques for dealing with missing data (chapter 7).

As part of this scientization, a new approach to survey research began to predominate by the 1990s: the "total survey error approach." It was always recognized that there are several types of error involved in a survey, but the greatest emphasis was on the type that is easiest to estimate statistically: sampling error—the error that occurs in a survey when one surveys a sample of the population rather than the entire

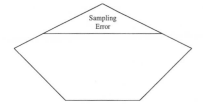

Figure 1.1 Sampling error as the tip of the iceberg

population. However, sampling error is just the tip of the iceberg in surveys (see fig. 1.1).

Early depictions often represented the error in surveys using a right triangle, with sampling error as one leg of the triangle, nonsampling error as the other leg, and the hypotenuse as the total survey error (fig. 1.2). This representation reminded people that the total survey error is much greater than just sampling error.[3] In practice, however, the emphasis in the field was mainly on sampling error because there are elegant mathematical formulas to estimate its magnitude, while nonsampling error was seen as too difficult to estimate. Furthermore, sampling error could be lowered if the sample size was increased, whereas the other sources of error were seen as more difficult to minimize.

The "total survey error" phrase first appeared in a book title involving health surveys (Andersen, Kasper, Frankel, and associates 1979). Applying ideas developed by Hansen, Hurwitz, and Madow (1953) and Kish (1965), the approach took off with Robert Groves's book *Survey Errors and Survey Costs* (1989). The Groves book was the first systematic treatment of all the different considerations involved in this approach, detailing the several sources of survey error and analyzing the costs involved in minimizing each. The total survey error approach has subsequently dominated research in the field, becoming the prevailing paradigm for understanding surveys. Chapter 2 will elaborate on this approach, including extensions of its usual formulation.

Figure 1.2 Sampling error as one component of total error

It must be admitted, however, that the recent explosion of research on surveys has led to reappraisals of several longstanding assumptions. As an example, a review by Jon Krosnick (1999) challenges much of the conventional wisdom about survey methodology. Additionally, there are some other frameworks that have been proposed for surveys, particularly Dillman's (2000) "tailored design method" for non–interviewer assisted surveys such as mail and Internet surveys and self-administered questionnaires.[4] Thus the move to a new paradigm is not fully smooth, reminding us that paradigms in science are fragile and the development of a paradigm is not necessarily decisive.

Survey Limitations

Survey research has become an important means of data collection that is useful for assessing the frequency of attitudes and human behaviors as well as measuring changes in those attitudes and behaviors over time. The underlying assumption is that people will respond candidly, sharing their views and experiences with researchers. Survey research will be important as long as governments, businesses, and social scientists value these measures. Yet as with any research approach, it is also necessary to recognize the limitations of surveys. There are several sources of error in surveys to consider, as will be detailed throughout this book. Furthermore, there are cost, time, and ethical issues that impinge on survey research (see chapter 2).

The most fundamental challenge would be widespread public disillusionment with surveys, leading to diminished cooperation and lack of faith in the results. George Gallup justified public opinion polling by emphasizing its usefulness in a democracy, allowing officials to know what the public is thinking. However, as will be seen in chapter 8, the rate of cooperation with surveys has already decreased considerably. There always has been some public suspicion of surveys, especially when results differ from the person's own views. Surveying becomes even more problematic when public figures encourage people to lie to pollsters on their vote intentions, though few real people are likely to go to that extreme. An even more negative argument against polls would be that the public does not know enough and is not intelligent enough for its views to be given credence. Gallup instead accepted Charles-Maurice de Talleyrand's assertion that "the only thing wiser than anybody is everybody," because democracy requires acceptance of the primacy of the public.

These limitations do not diminish as survey research becomes more scientific. Indeed, many people are skeptical of claims of the scientific status of surveys. Still, it is important for survey researchers to keep developing its scientific basis. The development of the total survey error approach is best seen as a stepping-stone in the continuing progress of the field.

2

Survey Error
THE ESTABLISHMENT OF A PARADIGM

The total survey error approach is based on analyzing the several different sources of error in surveys and considering how to minimize them in the context of such practical constraints as available money. This chapter introduces the total survey error approach by summarizing the different sources of error that will be treated in more detail in later chapters and by explaining the statistical implications of error. Additionally, this chapter extends previous statements of the total survey error approach by generalizing the notion of survey constraints to include time and ethics considerations and by recognizing the limitations imposed by survey-related effects.

The Theory of Survey Error

The total survey error approach is based on the fact that error can occur at every stage of a survey. Table 2.1 outlines these stages and indicates the chapter in this book in which each is discussed. A preliminary step is to decide on one's research objectives and whether a survey is the best way to meet them. As part of that decision, it is necessary to select a target population, such as residents of a particular city or senior citizens across the country.

The next step is to decide how to administer the survey: whether to do a face-to-face survey, a phone survey, a mail survey, an Internet survey, or some other type. This is known as the survey *mode*. Along with that decision is the choice of a survey *design*: whether it is a one-time *cross-section*

Table 2.1 **Steps of the Survey Process**

Stage	Chapter
Decide research objectives	
Determine target population	
Choose survey mode and design	3, 12
Choose sampling frame	9
Select sampling method	10
Write questions	5
Pretest questionnaire	6
Recruit respondents	8
Ask questions	4
Process data	11
Analyze results	

survey or a *longitudinal* survey that involves repeated interviews of the same people to look at changes in their attitudes and/or behavior over time.

Then comes deciding on how to sample respondents. Part of this decision involves moving from the target population to a realistic *sampling frame*—the set of units from which the sample will be drawn. A sampling frame for senior citizens, for example, might consist of single- and multiple-family dwelling units but would necessarily exclude hospitals, thereby missing part of the target population. Another part of this decision is the choice of sampling method—whether people will be selected randomly or some other sampling technique will be used.

The questions to be asked must be written and assembled into a questionnaire (sometimes termed the *survey instrument*). It is best to "pretest" the questionnaire by trying it out on some people before the actual interviewing so as to make sure that it works as intended.

The actual interviewing comes next. That includes recruiting actual respondents and asking them the questions. Afterward, the data must be processed and the results analyzed.

While these steps are listed in order in Table 2.1, the process is more fluid than has been described so far. Often, for example, researchers decide on the questions they want to ask people before deciding on a sampling method. But the basic steps are the same, regardless of their order.

Types of Error

While sample surveys draw responses from individual people, the analysis is usually based on aggregating the results to obtain statistics for the sample, statistics that hopefully estimate the corresponding population values (known as population parameters). The survey may be designed to measure some abstract constructs (such as abortion attitudes). It does so by a process of measurement, asking specific questions intended to measure the construct. The ultimate intention is to infer population values from the sample results.

The term *error* is usually thought of as a synonym for *mistake*, but in the survey context it refers to the difference between an obtained values and the true value, usually the true value for the larger population of interest. Several sources of error are distinguished in the total survey error approach.

The first three types of error are related to selecting respondents for a survey. The best known is *sampling error*. As stated in chapter 1, sampling error is the error that occurs when a sample of the population rather than the entire population is surveyed. There can be systematic bias when nonprobability sampling is employed, whereas probability sampling has the advantage of permitting mathematical computation of the sampling error.

Another type of respondent-selection error is *coverage error*, the error that occurs when the list from which the sample is taken (known as the *sampling frame*) does not correspond to the population of interest. This is the type of error that concerned researchers before the 1970s, when they avoided phone surveys: telephone coverage was not universal, so samples obtained through phoning people would not have covered the full adult population.

The other category of error related to respondent selection is *unit nonresponse error*. This error occurs when the designated respondent does not participate in the survey, thereby limiting how representative the actual respondents are of the population of interest. This form of nonresponse error calls attention to a key question: Are high response rates essential in surveys?

Three additional types of survey error involve the accuracy of responses. First is *item nonresponse error*—the error that occurs when the respondent participates but skips some questions. "Don't know" responses are an important item nonresponse problem, making it necessary for researchers to decide how to deal with missing item data in the data analysis.

Response accuracy issues also arise because of *measurement error*, the error that occurs when the measure obtained is not an accurate measure of what was to be measured. *Measurement error due to the respondent* involves whether the respondent gave an accurate answer to the question, which is often really a matter of how well the researcher worded the survey question. *Measurement error due to the interviewer* occurs when effects associated with the interviewers lead to inaccurate measures.

The final type of error is *postsurvey error*, the error that occurs in processing and analyzing survey data. Some accounts of the total survey error approach include postsurvey error, though others do not since postsurvey error is really not error from the survey process itself.

Another survey administration issue has to do with *mode effects*, the effects that the choice between face-to-face interviewing, phone surveys, mail questionnaires, and other survey modes has on the results that are obtained. The final survey administration issue involves *comparability effects*, the differences between survey results obtained by different survey organizations or in different nations or at different points in time.

Figure 2.1 takes the iceberg depiction of survey error and breaks it into three tiers in accord with the above discussion. Sampling error is still shown as the tip of the iceberg, with other respondent-selection issues also near the top of the iceberg. Response accuracy issues are further below the surface. Survey administration issues are shown in the bottom tier, detachable from the rest, as postsurvey error, mode effects, and house effects are usually not included in discussions of survey error. The relative sizes of the different areas in figure 2.1 should not be taken literally, though it is true that two of the larger areas, those for sampling error and measurement error due to respondents, both depict major topics.

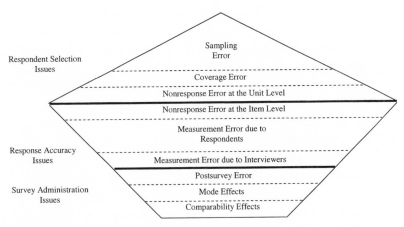

Figure 2.1 Types of survey error

The chapters that follow discuss these several potential sources of error in detail. These are called *potential sources of error* because they are not necessarily errors. For example, survey nonresponse does not lead to error if the people who do not participate in the survey are similar to those who do participate. Yet we generally cannot tell whether they are similar, since we lack data on the people who do not participate; this makes it important to understand nonparticipation as a potential source of error.

Scientific Sample Surveys

Since this book is dealing with surveys, it is appropriate to consider what exactly is a scientific survey. The American Association for Public Opinion Research (AAPOR) attempted in 2001–2 to define a "scientific sample survey" as a means of helping the media and, through them, the public to distinguish between high-quality surveys and ones that should not be taken seriously. As could be expected, there was not full agreement on the initial draft, so in the end AAPOR did not adopt an official definition. Still, it is useful to examine how the AAPOR committee of distinguished survey researchers initially attempted to define the term (AAPOR 2001, 476–77).

First, they emphasized that surveys and polls provide a means to study people's "attitudes, behaviors, and characteristics." Their focus was on sample surveys, not censuses or other surveys that interview the entire population of interest.

Their list of four characteristics of a scientific sample survey or poll emphasizes the points listed in the previous section: coverage, sampling, nonresponse, and measurement.

1. Coverage: A scientific survey "samples members of the defined population in a way such that each member has a known nonzero probability of selection. Unless this criterion is adhered to, there exists no scientific basis for attempting to generalize results beyond those individuals who completed the survey."

2. Sampling: A scientific survey "collects data from a sufficient number of sampled units in the population to allow conclusions to be drawn about the prevalence of the characteristic in the entire study population with desired precision (for example, + or − 5%) at a stated level of confidence (e.g. 95%)."

3. Nonresponse: A scientific survey "uses reasonable tested methods to reduce and account for unit and item nonresponse error (differences between characteristics of respondents and

nonrespondents) by employing appropriate procedures for increasing unit and item response rates and/or making appropriate statistical adjustments."

4. Measurement: A scientific survey "uses reasonable tested methods to reduce and account for errors of measurement that may arise from question wording, the order of questions and categories, the behavior of interviewers and of respondents, data entry, and the mode of administration of the survey."

At the same time, the AAPOR preliminary statement admits that "in practice all sample surveys fall short of perfection in meeting one or more of the above criteria." Because of that lack of perfection, its definition emphasizes the importance of "disclosure of the exact procedures by which a survey is conducted, including the sample size and sampling methods, coverage characteristics, the questionnaire itself, data collection methods, response rates, and information about the characteristics of respondents as well as nonrespondents, [since this] makes it possible for others to evaluate likely survey errors and reach an informed judgment on the resultant quality of the findings."

Finally, the AAPOR committee gave a few examples of what leads to a survey's "failing to meet the test of being scientific." Letting volunteers fill out a questionnaire rather than using random selection is nonscientific, for generalization beyond the people who filled out the questionnaire is not warranted. A low response rate can make the survey unscientific: "In some surveys, a relatively small percentage of people selected in a sample actually complete a questionnaire, and the respondents differ significantly from non-respondents on a characteristic of relevance to the survey. If this fact is ignored in the reporting of results, the survey fails to meet an important criterion of being scientific." The final example is that a survey is not scientific if the question's wording produces biased answers.

This definition of a scientific sample survey has some ambiguities, which shows the complexity of defining a scientific survey. Take the matter of response rates. The nonresponse paragraph seems to indicate that a survey is scientific so long as it tries to minimize and/or adjust for nonresponse, but the later paragraph on nonscientific surveys seems to say that nonresponse can invalidate a survey. This inconsistency may properly reflect the nature of response rate as a problem—it can render a survey nonscientific but does not necessarily do so—but it also shows that there always will be ambiguities in working through what makes a survey scientific.

The Statistical Impact of Error

While it would be best if all of the types of error described above could be eliminated, that is not possible. The goals instead are to keep them at minimal levels and to measure them so that their statistical impact can be assessed. Survey error affects the statistical analysis of survey data. Statisticians make two distinctions between types of error that have important implications for statistical analysis.

The first key distinction is between systematic and random error. *Systematic error* is usually associated with a patterned error in the measurement. This patterned error is also known as *bias*. An example of systematic bias would be measuring the average age from a survey if older people were less willing to be interviewed. Another example is when people who are dissatisfied with a product are more likely to choose to participate in a customer satisfaction survey, rendering estimates of the extent of dissatisfaction wrong. Thus systematic error directly affects the average value of the variable (its *mean*), depressing it in these two examples. In statistical terminology, the sample mean is a biased estimator of the population mean in the presence of systematic error.[1] The *validity* of a measure is whether it measures the construct of interest. Variables that are measured with systematic bias are not valid. If the systematic error is just a constant, the variance of the variable will still be correct. Also, adding a constant to a variable does not affect correlations or regression coefficients, so those statistics can still be correct, depending on the nature of the systematic error.

By contrast, *random error* is error that is not systematic. For example, if people do not think hard enough about a question to answer it correctly, some may err on one side and others may err in the opposite direction, but without a systematic tendency in either direction. Random error would have a mean of zero, so random error does not affect the mean of a variable. However, random error does increase the variance of a variable.[2] The *reliability* of a measure has to do with the amount of random error it contains. The reliability of a measure can be thought of as the extent to which repeated applications would obtain the same result. An unreliable measure has a large amount of response variance. Random measurement error reduces the reliability of a measure, as when asking a question in different tones of voice yields different answers. Because of the increased variance, random error directly affects correlations with other variables, which are reduced in magnitude (an effect known as *attenuation*). Regression coefficients are similarly attenuated by random error in the independent variable, and hypothesis tests have less statistical power (making it harder to obtain significant results) than if measures were perfectly reliable.

One way to put systematic and random error together is to think in terms of the *mean squared error* on a variable—the average of the squared differences between survey respondents' values on the variable and the true population mean. The mean squared error is the sum of the squared bias and the observed variance, respectively reflecting validity and reliability concerns (Biemer and Lyberg 2003, 56). This formulation is of only theoretical interest, since of course the true population mean is unknown, but it usefully shows how validity and reliability concerns fit together in survey measurement.

A second key distinction is between uncorrelated and correlated measurement error. The discussion of systematic and random error above was based on *uncorrelated error*. Uncorrelated error occurs when the errors for different individuals are unrelated, as when an interviewer happens to ask a question in one tone one time and in a different tone of voice another time, with no particular pattern as to when each tone is used. This sounds benign, but, as seen in the above paragraphs, even uncorrelated random measurement error does affect statistics. One solution (Fuller 1991) would be to reinterview some respondents to obtain estimates of the reliability of measures in order to adjust the statistics, but researchers generally would prefer spending money to interview more respondents rather than to take more interviews with the same respondents.

Correlated error occurs when the errors for different respondents are related. If an interviewer systematically asks questions in a particular wrong way, for example, all of the interviews that person conducts will be slanted in the same way. Correlated error can occur when interviewers take multiple interviews and when coders code open-ended questions (those that respondents answer in their own words, rather than choosing between response categories that are part of the question) in multiple interviews. As shown in table 2.2, correlated error is more serious than uncorrelated error because it directly multiplies the variance of the variable, which makes it harder to get significant results. The mathematics of correlated error will be discussed in chapter 4 as part of the treatment of measurement error due to interviewers.

Survey-Related Effects

While the standard total survey error approach naturally emphasizes survey errors, it is important to extend this approach to take explicit account

Table 2.2 The Effects of Error on Variance and Estimates

	Random Error	Systematic Error
Uncorrelated Errors	Unbiased estimates; increased random variance	Biased estimates
Correlated Errors	Unbiased estimates; variance multiplied	Biased estimates

of *survey-related effects*. These effects are recognized in the current literature, but they have not been previously incorporated into the total survey error approach. They should, however, be considered explicitly, because they limit the conclusions that can be drawn from survey evidence even if they do not constitute errors.

As an example, one set of survey-related effects is question-related. No single question wording is "correct," but, as will be seen in chapter 5, different question wordings can yield somewhat different results. Question order is known to affect results, but question order effects cannot be eliminated, since it is not possible to have every question come first in a survey. Similarly the order in which response options are offered affects answers; the *recency effect*, for example, occurs when phone respondents tend to choose the last answer on a list because they remember it better than those read to them earlier. What distinguishes these various effects is that there is no way to minimize them. Different ways to structure the question yield somewhat different results, but that does not mean that a particular question wording is more valid than any other wording. One way to think about this is that attitudes are always context dependent, so it would be chimerical to try to eliminate context. It is better to conceptualize true answers as a range and then view alternative question wordings (including question orders and response-option orderings) as permitting researchers to examine that range. According to this perspective, it is better to experiment with multiple question wordings than to seek a single optimal wording.

As already mentioned above, there are also mode effects associated with the choice between face-to-face, phone, mail, Internet, and other modes of administering surveys, and there are comparability effects involved with comparisons between surveys. Similarly, the term "interviewer effects" is sometimes used rather than "interviewer error," since some of the effects associated with interviewers are not mistakes but are simply the inevitable results of interviewer gender, race, or tone of voice— factors that cannot be totally eliminated so long as interviewers are used. Again, such effects cannot be eliminated; instead it is important to understand such effects and study how much they can affect survey results.

The statistical effects of survey-related effects vary considerably. They can have either random or systematic effects and can be uncorrelated or correlated. Fortunately, it is possible to study these effects, such as by testing the effects of different question orderings or different response-option orders.

Survey Error and Survey Constraints

Survey researchers want to minimize error in their work, but the degree to which error can be reduced is limited by practical constraints. Groves's (1989) classical exposition of the total survey error approach focuses on the balance between survey errors and one important constraint: costs. However, it is important to generalize this treatment by recognizing that time requirements and ethics also impose important constraints.

The *cost* side is relevant in every research endeavor, but survey research involves particularly large costs, especially when face-to-face or telephone interviewing is used. Some survey costs are fixed, but most are variable, depending on the number of interviews and the average length of the interviews. Survey organizations typically generate cost proposals to potential clients based on those considerations, taking the number of interviews and average interview length into account and adding a term to cover fixed costs. For example, if it is estimated that the interviewer will, on average, complete one interview per hour for a study, and if 1,200 interviews are to be taken, that multiplies out to 1,200 interview hours. If the cost per interview hour were $15 (covering administrative costs as well as interviewer wages), variable costs would multiply out to $18,000. Additionally, the fixed costs associated with taking on the project must be added on to cover the development of the questionnaire, programming it if the survey is computer assisted, and data management. Survey cost estimates also take into account the number of callbacks requested, the amount of screening required (screening is expensive since it means more calls are required to obtain a completed interview), and how complicated the sampling is.

In practice, the cost side is more complicated because of the competitive nature of the survey field. Since researchers always want to get the most they can with their limited funds, they typically will seek bids from different survey organizations. As a result, survey organizations must price their product at a competitive level to win survey contracts. Of course, there are also nonprice bases for such competition, as when survey organizations emphasize the quality of their research and argue that high-quality research is more expensive. For example, it is common for university-based telephone survey operations to justify their charges on the basis of making many callback phone calls to find an intended respondent at home, rather than just going on to the next phone number as some commercial survey firms do.

Time is also an important consideration in survey research. Surveys are often conducted on a tight time schedule. Sometimes this is due to external constraints that make the information less valuable if it is not available by a particular time. Preelection polling, for example, must by definition end before the election actually occurs. Also, some attitudes and facts will change over time, so it is necessary to measure them in a timely fashion. When commercial airlines hire polling operations to phone recent customers and ask about their satisfaction, that information must be gathered soon enough after a flight that the passengers remember it and before they have been on subsequent flights. However, there are also drawbacks to instantaneous polling. Some survey organizations conduct tracking surveys every evening to look for changes in attitudes, but that research design limits the results since it overrepresents the types of people who are at home on particular evenings.

A third relevant constraint on survey research involves *ethics*. As explained in chapter 14, there are a number of ethical guidelines for research that affect surveys, and it is important that researchers think through some of these ethical issues before conducting surveys. Indeed, preclearance is necessary to conduct human subjects research at universities and other organizations that receive funds from the federal government. Ethical considerations can be considered to be a constraint on research, as the rules restrict certain types of surveys. Surveys on children, for example, are tightly regulated. The review boards that consider proposals for human subjects research may not approve questions that would cause stress for respondents. Having to justify these types of questions to a review panel has the side benefit of discouraging such questions unless they are essential to the survey. Thus ethical considerations can constrain survey activities just as survey costs and time needs do.

The Total Survey Error Approach

The total survey error approach is often discussed in terms of a trade-off between survey errors and costs. However, a broader statement of this approach emphasizes the several different types of survey error, inevitable survey-related effects, and the full set of cost, time, and ethical considerations that constrain research. Research always involves compromises, and here the compromise is between trying to minimize survey error and recognizing that survey constraints may require accepting more error, while also taking into account the impact of survey-related effects.

This is not meant as an argument for tolerating large amounts of error in surveys but rather as an argument for finding the best survey design within the researcher's financial and time constraints as well as applicable ethical factors. For example, many specialists would view face-to-face interviews as having less error than most other types of surveys because of their higher response rates and the greater rapport that interviewers can achieve with respondents, but they are also much more costly and take much longer to complete. At the opposite end of the continuum, Internet polls can be fairly inexpensive and can give fast results, even if they have lower response rates and interviewers cannot probe "don't know" responses. What is important is to understand the trade-offs involved in the choices, since it is not possible to maximize all benefits at once.

The total survey error approach is used in many ways in the survey research field. Some survey organizations use it to structure their discussions with clients, making them aware of the different sources of error and working with them to achieve a research design that deals with those error sources within their budget. A survey can be thought of as no better than its weakest link (Fowler 2002, 8), which makes it important that clients understand the extent to which each source of error affects their survey. The natural inclination is to maximize the sample size and/or the response rate, but careful consideration of the different sources of error may suggest that it is more important to take steps to work on coverage or other errors. There are inevitably trade-offs in research, and the total survey error approach helps make those trade-offs more explicit.

Some survey organizations also use the total survey error approach in training interviewers and their supervisors, making sure that all employees understand how their performance affects the quality of the product delivered to clients. The total survey error approach is relevant after the data are collected as well. In particular, reports on the survey should discuss the relevant sources of errors, detailing the decisions in the data collection stage that affect interpretation of the results.

Additionally, the total survey error approach is used to structure some academic training programs in the field, including the Joint Program in Survey Methodology (JPSM) of the University of Maryland and the University of Michigan in collaboration with Westat.

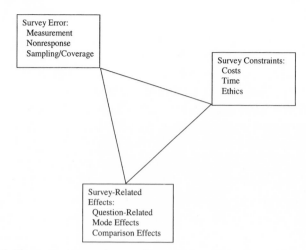

Figure 2.2 The survey research triangle: Survey error, constraints, and effects

There is, of course, never a perfect paradigm for understanding everything in a field, and so there are other perspectives that can be used in approaching survey research as a field of study. However, the total survey error approach is certainly the dominant approach today, and part of its advantage is that some alternative approaches can be treated as part of its perspective. For example, one recent aspect of the scientific approach to surveys has been the application of increased understanding of the human thought process (the so-called cognitive revolution in psychology) to understanding the survey research process. Although this could be considered an alternative perspective, it can also be subsumed under the total survey error approach, as part of increasing our understanding of the errors in the measurement process associated with the respondent.

Finally, the total survey error approach provides a useful way to structure this book. It reminds us of the importance of identifying, discussing, measuring, and seeking ways to minimize all the factors that lead to errors in surveys. It emphasizes the need to balance error minimization against survey costs, time, and ethical considerations. Additionally, it recognizes that there are a number of survey-related effects that will always affect survey results. Figure 2.2 summarizes the main elements of these three factors, while implying the difficulty of balancing between them. Chapters 4 through 14 will focus on the various sources of survey error, survey constraints, and some survey effects.

3

Survey Modes
RESPONSES TO EMERGING TECHNOLOGIES

There are several different modes through which surveys can be administered: face-to-face, telephone, mail, Internet, self-administered, and so on. Before the sources of survey error are examined in detail, it is useful to give preliminary consideration to these different modes of survey administration. Two basic distinctions between survey modes will be developed in this chapter, after which some related research approaches are discussed. Finally, computer-assisted survey information collection and Internet surveys will be described in more detail.

Survey Modes

A variety of different survey modes have been used over the years, with much of the change reflecting changes in available technology. As will be seen, the different modes vary in their costs and time requirements, factors that must be considered when one is choosing an interviewing mode.

Two Important Distinctions

There are two important distinctions to make about survey modes (see table 3.1). One is how personal they are in terms of the extent of interviewer involvement: whether the survey is *self-administered* or *interviewer-administered*. Face-to-face and telephone surveys are generally interviewer administered. Face-to-face interviews are usually in the respondents' natural surroundings, as when interviewers go to people's homes or offices. Skillful interviewers can usually get a high response rate and can conduct

Table 3.1   Survey Modes by Administration Approaches

	Self-Administered	Interviewer-Administered
Not computerized (static)	SAQ: Self-administered questionnaires Mail surveys Static e-mail and Internet surveys	Telephone surveys Face-to-face surveys
Computerized (interactive)	Interactive Internet surveys CSAQ: computerized self-administered questionnaires IVR: interactive voice response	CATI: computer-assisted telephone interviewing CAPI: computer-assisted personal interviewing

very high-quality interviews. It is harder to exercise quality control because these interviews are usually not monitored, though even monitoring is possible with modern recording technology. The response rate and interview quality are generally lower in phone interviews. Telephone surveys can be conducted out of a central telephone facility (a "calling center"), which improves quality control since a supervisor can oversee the process. The data can be collected much faster than with mail surveys or face-to-face interviews, which is important to clients who want daily updates.

Interviewers provide several advantages in surveys. They can probe answers that are unclear or inadequate. Interviewers are essential if respondents' skills in reading, writing, and/or computer usage are too low to permit them to fill out self-administered surveys. Also, interviewers are especially useful in establishment surveys of business firms, since they can try to interview the desired respondent in a business rather than letting someone in the organization decide who will fill out a questionnaire. However, interviewers also add a source of error to surveys, as will be detailed in chapter 4, and these errors are eliminated when no interviewer is involved.

Survey modes that do not involve interviewers include self-administered questionnaires, mail surveys, Internet surveys, and interactive voice response surveys. These surveys afford respondents greater privacy, as no one can overhear answers and the respondents will not be embarrassed by their answers to an interviewer. Self-administered questionnaires, such as those given in classes, are often given to volunteer subjects and/or captive groups.[1] Unfortunately, that means that the respondents are not necessarily representative of the larger group of interest, except when questionnaires are given to people as they undergo an experience (such as having people fill out a questionnaire as they complete a training program). Still, self-administered questionnaires are particularly useful when respondents need to check their records in order to answer questions, as when surveys of businesses have to answer detailed personnel and financial questions. Mail surveys can be sent out to a representative sample of the population of interest. Large numbers of mail questionnaires are usually sent out, in the hope that people will fill them out and mail them back. However, because of the impersonality of mail surveys, the response rate can be very low, and people who return them may not be representative of the larger population. There has been a considerable increase in self-administered surveys (Dillman 2000, 7–8). Many federal government surveys are now self-administered, for reasons such as lower cost and the ability to conduct them in-house without hiring a professional firm to conduct interviewer-administered surveys.

The other important distinction involves whether or not the survey is *computerized in an interactive sense*. Computerizing permits control of the flow of the questionnaire. For one thing, researchers often want the answer to one question to determine which question is asked next (for example, questions about the education of one's children can be skipped for people who do not have children). Computer-based surveys can show people the next question that is appropriate given their answer to the last question, whereas directions in a written questionnaire as to how the next question depends on answers to the last question can be confusing. As described below, computerized surveys also permit "survey experiments" in which random subsamples are asked different versions of the same question or the same questions in different orders to test the effects associated with question wording and question order.

As shown in table 3.1, computerized surveys can be interviewer administered or self-administered. Interviewer-administered computerized surveys can be conducted either on the telephone (computer-assisted telephone interviewing, CATI) or face to face with interviewers using personal or handheld computers (computer-assisted personal interviewing, CAPI). Computer-assisted surveys combine the advantages of an interviewer and of computerized control of the question flow.

Internet surveys have become the best-known form of self-administered computerized surveys. "Virtual interviewing" resembles self-administered questionnaires in that there is no interviewer. However, some of the less expensive Internet surveys involve only a static questionnaire posted on the Web, not taking advantage of the potential for controlling the question flow in an interactive manner. It is also possible to conduct self-administered computerized surveys without using the Internet, either through automated phone surveys (now often using interactive voice response technology) or by having respondents fill out a questionnaire at a computer without going onto the Internet (known as CSAQ technology, for computerized self-administered questionnaires).

A third possible distinction between different survey modes involves the "channel of communication"—whether visual or aural communication is invoked. Self-administered questionnaires use only the visual channel, whereas telephone surveys just use the aural channel. Internet surveys are primarily visual, though some use voice as well. Face-to-face interviews are primarily aural but can include visual stimuli (as well as nonverbal communication from the interviewer). Surveys that are presented only visually are problematic for respondents who have difficulty reading, while surveys that are presented only aurally can challenge the respondents' listening abilities. Some computerized surveys have people put on earphones to hear the survey while they read it on the computer screen, as a means of providing respondents both types of input.

The choice of survey mode can affect survey results. These effects are discussed in chapter 12, after survey procedures are discussed in detail in intervening chapters.

Measuring Change

In addition to the selection of a survey mode, an important design aspect of surveys involves whether and how to measure change over time. A *cross-section survey* takes a single one-time sample. Many surveys are simply single cross-sectional surveys with no intended time perspective, but survey projects are increasingly interested in assessing the amount of change that occurs in facts or attitudes.

There are several types of *longitudinal surveys* (Rose 2000a; Menard 2002), with the more complex designs permitting more complete assessment of change. Repeated cross-section surveys involve taking a series of separate surveys from the same sampling frame without reinterviewing the same people; this strategy allows an estimate of net change at the aggregate level but cannot show which people change. By contrast, a longitudinal *panel survey* involves reinterviews with the same people each time, a design that excels in looking at individual-level change but cannot take changes in the composition of the population into account.[2] The U.S. National Election Studies began a panel survey in 1990, with the first interview (known as the first "wave") after the midterm election that year.

Reinterviews were conducted in 1991 after the first Gulf War and again before and after the 1992 presidential election.

These two approaches can be mixed in various ways. *Overlapping panels* bring new people into the study periodically, whereas *rotating panels* rotate people out of the survey after a certain number of interviews. A *revolving panel* drops subsamples for one period and then includes them again in a later period. Other options include following people when they move and maintaining records at the individual level across panel waves. A *split panel* combines a panel study with either a repeated or a rotating panel survey, though the combination limits the resources that could otherwise be used to buy a higher sample size for one of its components (Kalton and Citro 2000). The most complete form is the *longitudinal survey with rotation*, bringing in new cases periodically, following people when they move, and maintaining records for individuals across waves (Bailar 1989). Minimally, augmenting panel surveys with fresh cross-sections is useful for obtaining estimates of biases resulting from panel surveys (Bartels 2000). This strategy was followed in the 1992–97 NES panel study, in which some respondents were interviewed as many as seven times, but the panel was "refreshed" in 1994 and 1996 with the addition of new cross-sections to make up for the inevitable loss of respondents over repeated interviewing. The sources of bias in panel studies will be discussed in later chapters of this book.

The choice between cross-sectional and longitudinal designs has implications for the choice of survey mode. Long-term panel surveys need to track people when they move, and such tracking is facilitated when at least the initial contact with the respondent is through an interviewer. As will be seen later in this chapter, there are also some interesting experiments in maintaining ongoing short-term panel surveys over the Internet. In any case, when selecting a survey mode, researchers need to think about how best to obtain change measures if they are required.

Related Research Approaches

Several other research modes share some features of surveys. Indeed, participants in some of these modes are often chosen on the basis of a short phone survey. *Focus groups* select a small group of people to come together to answer questions posed by a discussion facilitator. The group discussion is meant to resemble natural decision processes more than surveys, though a person with a strong view on the topic being discussed can dominate the conversation and discourage others from expressing contrary opinions. The results are looked at qualitatively, not quantitatively as with surveys, such as listening to see which ways to frame an advertising pitch for a product or political issue work best. *Deliberative polls* get together a large group of people to be given information about a policy problem and discuss possible solutions with each other. Deliberative polls are rare, but they are interesting because they allow people to discuss ideas before finalizing their opinions as people do in real life. *Audience reaction research* entails having people record their reactions to a presentation—for example, turning a dial to record how positive or negative their reactions are to each moment of a political advertisement.

Diaries have respondents keep track themselves of their behavior over a period of time. Prior to the advent of cable television, this was one of the main methods of measuring television viewing: people were asked to write down what they watched. Diaries have also been used to have people keep track of their food intake, their time use, and their household purchases. Diaries would be highly accurate if people filled them out as they performed these acts, but cooperation usually declines over time, with people recording fewer activities. Further, some people change their behavior because of the data collection (such as watching the TV shows they feel they should watch instead of those they would normally watch).

The sampling methods used in surveys are often employed in these research approaches, and surveys are often conducted in conjunction with them. For example, the 1996 National Issues Convention deliberative poll was based on interviews taken nationally by NORC with 915 people, with a completion rate of 72%. In the end, 50% of those respondents accepted paid trips to Austin, Texas, for three and a half days to discuss the public issues of that election year, and they filled out questionnaires at the beginning and end of those sessions so that opinion change could be measured (Merkle 1996).

Two additional research modes employ special forms of contacting survey respondents. *Exit polls* ask people leaving election polling places how they voted (Mitofsky 1991), while *intercept surveys* ask questions of people in malls or other settings. Exit polls and intercept surveys use specific sampling approaches in deciding which people to interview (see the discussion of systematic sampling in chapter 10).

Some survey modes are not considered scientifically valid. As an example, the *phone-in poll* has people phone one number if they want to vote yes on a topic and a different phone number if they want to vote no. Similarly, the *click-in poll* is a poll on an Internet site that a person happens to visit, such as the CNN Web site's QuickVote poll on its topic of the day. These polls are not considered valid, because only interested people participate and because affected parties can distort the results by participating multiple times, though their advocates argue that these problems are not necessarily more serious than the problems of other survey modes.

Some pseudopolls are not ethically valid. The American Association of Public Opinion Research (AAPOR) has made strong policy statements against companies that claim to be conducting a survey as a means of gaining people's attention when the company has no interest in the person's answers. In particular, AAPOR has attacked the practices of using purported surveys as entrees for selling products (known as "sugging," for "selling under the guise" of a survey) or soliciting money (known as "frugging," for "fund raising under the guise" of a survey). Finally, a *push poll* is propaganda that looks like a poll. Some political candidates have hired firms to call people just before an election, claim they are conducting a poll, and then ask voters if they would vote for the candidate's opponent if they were told that he or she had done some terrible deed or supported some unpopular policy. This may resemble a poll, but the real objective is to change opinions rather than to measure them (Traugott and Kang 2000).

Computer-Assisted Survey Information Collection

It is useful to discuss computer-assisted survey modes in more detail before proceeding. Early survey research techniques were primarily paper-and-pencil (P&P) based, but computer-assisted survey information collection (CASIC)—a term coined by researchers for the federal government—is now common. Particularly important in this shift was the devising of procedures for computer-assisted telephone interviewing (CATI). CATI began in 1971 with Clifton Research Services (Couper and Nicholls 1998) and moved into the academic arena through the efforts of Merrill Shanks at the University of California, Berkeley. The federal government's statistical agencies began to adopt CATI by the early 1980s. The new personal computer technology permitted small companies to run their own CATI operations, leading to rapid expansion of the number of private survey research firms through the 1980s and 1990s.

CATI has several advantages over conventional paper-and-pencil techniques. It permits the interviewer to enter responses directly into a computer, eliminating data entry as a separate stage of the survey process. It controls the order in which questions are asked, facilitating the use of complex branching strategies in which the answer to one question determines the next question to be asked. Two additional innovations have provided further advantages for CATI. One is the development of computer programs to control the assignment to interviewers of phone numbers to call, including keeping track of appointments when respondents ask to be phoned back for the interview at a particular time. Another is the ability to do random experiments, using a computer to randomize question order, the order of answer categories, and/or which versions of questions are asked of particular respondents.

Computerization has the further advantage of moving the data editing process (see chapter 11) to the interview itself. The computer can be programmed to catch implausible combinations of answers and have the interviewer correct one of the entries or ask the respondent to resolve the apparent inconsistency. Additionally, since the data are entered directly into a computer, CATI permits fast analysis of surveys, as epitomized by the reporting in the morning news of CATI surveys taken the previous night. Thus CASIC has increased the speed and efficiency with which surveys are conducted, along with the completeness and consistency of the data that are collected. Computer use does, however, add some complexity to the interviewer's task, requiring him or her to pay attention to the machine and not just the respondent.

Computerized surveys are particularly useful for panel surveys in which researchers want to ask respondents about changes from their previous interview. Relevant answers from previous waves can be preloaded into the computer, so the interviewer can tell when changes have occurred and can prompt the respondent's memory. Rather than asking what jobs the respondent has had since the last interview in September two years earlier, the interviewer can say, "You were working at the XYZ plant when we talked two years ago. Are you still there, or have you had other jobs since?"

The development of notebook computers permitted the development of computer-assisted personal interviewing (CAPI). CAPI was first implemented by government statistical agencies that were committed to face-to-face interviews by local field interviewers (Couper and Nicholls 1998). An extra advantage of CAPI is the ability to send survey data from a notebook through a modem to a centralized facility, as opposed to having to mail completed paper-and-pencil questionnaires back to headquarters.

The computerization of surveys has also affected noninterviewer surveys. Computerized self-administered questionnaires (CSAQ) have respondents go through a questionnaire themselves on a computer. This is done in e-mail surveys and Internet surveys, which require respondents to have access to both computers and the Internet. Knowledge

Networks, an Internet survey organization, even supplies Web TV devices and Internet connections to members of their sample in exchange for participation in their surveys. CSAQ can also be achieved in CAPI: the interviewer can hand a notebook computer to the respondents and have them go through some questions by themselves. This can be particularly useful for sensitive questions (such as about drug use) that the respondent may not want to answer in front of an interviewer. The invention of small personal digital assistants (PDAs) leads to further possibilities, with respondents being asked to click through the survey on a PDA. This technology has been suggested as a way to automate election exit polling, so that responses would be recorded automatically and could be uploaded quickly and accurately to a central computer that would do the tabulation and reporting.

Further innovations include interactive voice response (IVR), which includes both voice recognition entry and touch-tone telephone data entry. In the ideal version, people are first contacted by a human interviewer or through the mail. They are asked to phone a special number (or are transferred to that number if they are contacted by phone), and they enter a PIN so that access to the survey is controlled. Respondents are asked questions by an automated unit, and, depending on the system, they can respond either verbally or by pressing numbers on the phone keypad. Voice recognition software can analyze people's oral responses so that question flow can be regulated interactively according to the answers given to previous questions; however, it cannot yet handle complex answers. As an example of the importance of this system, the Gallup Organization conducted over one million IVR interviews in 1999 (Tourangeau, Steiger, and Wilson 2002). Dillman (2000, 402–12) is positive toward this technology, to the point of his predicting that it will mostly replace CAPI (353).

It should be emphasized that CASIC has disadvantages as well as advantages. In particular, good programming of complicated surveys is expensive and time consuming. It is easy for programmers to make logical errors in computerizing the questionnaire, leading to questions' being skipped or the wrong question's being asked. In addition to the expense of computers, there is the cost of a computer network and the personnel costs for maintaining the computer system and ensuring that it is always available for interviewing. It is essential that interviews be saved, even if the system crashes. All in all, the fixed cost of computerizing a survey is high, even if the variable cost of increasing the number of respondents is relatively low once it is computerized (Groves and Tortora 1998). As a result, especially when few interviews are needed, researchers often do not employ CASIC capabilities, instead having interviewers read questionnaires to respondents.

Interviewer-assisted computerized surveys still differ from computer surveys that are not interviewer assisted. CATI and CAPI were intended originally to improve data quality, and they clearly have been successful for interviewing the mass public. On the other hand, CSAQ and IVR are generally intended to reduce costs and speed data collection, but they have not yet been shown to be as successful in surveys of the mass public (Nicholls, Baker, and Martin 1997).

Internet Polling

Internet polling is still a fairly new mode of computer-assisted data collection. Like many other aspects of the Internet, its potential is exciting, but it is too early to tell how successful it will be. Some approaches are effective, but others have not worked well. Putting a poll up on the Internet can be inexpensive, so many groups put up polls without paying enough attention to quality. However, quality research is never cheap. In the long run, quality standards will be established for Internet surveys. Indeed, there are already several interesting attempts to create high-quality Internet surveys. This section will briefly mention some poor approaches for Internet polls, after which some of the better approaches will be described.

Problematic Internet Surveys

The worst type of Internet polls has already been mentioned: click-in polls. Some popular Web portals conduct daily polls. Respondents to click-in polls are volunteers, and there is no reason to think that they are a representative sample of a broader population. Interest groups can alert their membership when a click-in poll topic is of importance to them, in an attempt to boost the desired numbers. Further, a person can respond repeatedly on these polls.[3] Having said that, even those types of Internet polls have their defenders, who ask whether telephone polls with 50% response rates are really any more scientific than click-in polls. The issues involved in click-in polls will be considered repeatedly throughout this book.

There are several Internet sites that connect people with Web surveys that pay for participation. This leads to volunteer samples of respondents who are more concerned with earning money for participation than with giving honest answers. While it is increasingly common to provide incentives for participants in all types of surveys, the sample's representativeness becomes dubious when many respondents are people who frequent pay sites regularly.

Some survey sites allow free posting of surveys, with the Web sites getting their funding from ads shown to respondents. However, these sites have no incentive to facilitate high-quality polls. Surveys posted on such sites are usually static questionnaires that ask respondents all the questions, even if their previous answers show that some of the questions are not appropriate. The respondent is usually expected to skip those questions, though skip instructions are not always clear. If the respondent answered some questions that should have been skipped, the software may still include those responses in tallies of answers.

An inexpensive form of Internet survey is the e-mail survey, in which the questionnaire is simply embedded in an e-mail message. The respondent has to hit the reply button, put X's inside brackets when choosing answers, and then hit the send button. However, it is very easy to get rid of these surveys by hitting the delete button. Yet e-mail is still one of the simplest ways to get surveys to computer-literate respondents.

There is an unfortunate tendency to ignore the usual ethical guidelines when posting surveys on the Web. It is so easy to put up surveys on the Web that appropriate ethical guidelines are often not noticed.

Better Internet Surveys

As an example of a reasonable procedure for Internet surveys, some universities now allow their faculty and students to fill out internal surveys on the Web. Respondents can be notified by e-mail that they have been randomly chosen to participate, with an indication that they can answer the survey on the Web if they prefer that to a phone interview. They are given the Web address for the survey along with a unique password, so that only designated respondents can participate. Using the Web version of the survey can be a popular alternative because university personnel and students are already very familiar with computer use and since many would prefer doing the survey at off hours rather than when an interviewer happens to call them. There can even be incentives to participate online, such as a chance to win a free campus parking permit. The main disadvantage is that people often decide to take the Web version later but then forget to do so, so reminders must be sent after a week, and phone interviewers then must call people who still forget to do the survey.

There are two important innovative national Internet surveys that will be referred to in several later chapters. One is Harris Interactive, which obtains a pool of respondents by sending out large numbers of invitations offering a chance to win monetary prizes in exchange for participation in polls. Clicking on the ad leads to the Harris Interactive Web site and a short survey that obtains basic demographic data on the person, along with assurances of confidentiality. The person's e-mail address is put into Harris's database of panel members, and those people are then sent frequent notices about particular surveys. The e-mails have subject headings that indicate a chance to win a large prize, such as "Win $6,000: Evaluate Ecommerce sites" and "Win $4,000: Home Purchases Survey." Additionally, participants earn points that can be exchanged for merchandise. A good prize requires about 6,000 points, with each survey participation being worth 100 points and panel members being given the opportunity to participate in one or two surveys each month. The e-mail invitations provide the Web address of the survey along with a unique password that gives access to the survey. The known demographics of participants in each survey can then be compared to known population parameters to weight these Internet surveys appropriately (chapter 8). Reminders are sent out in a few days to panel members who have not yet participated in the survey. There are not yet definitive independent studies of the validity of this approach, but it is a highly visible attempt to produce a quality Internet survey.

Knowledge Networks is also doing innovative Internet surveys (Lewis 2000). It obtains a pool of respondents by conducting a random national telephone survey in which respondents are offered a Web TV receiver and free Internet connectivity in exchange for participating in its Internet surveys. This approach minimizes the coverage problem of many homes' not having computers or Internet access by providing both to respondents. Therefore its sample does not have the middle-class bias that an Internet survey might otherwise have. A free Web TV and Internet access are an attractive incentive for many people, but there are still some potential coverage issues, such as the unwillingness of some older people to learn how to use a computer. Once the Web TV device is installed in the home, respondents answer profile surveys about the demographics of each household member. Each panel member is expected to answer one short survey per week; the Web TV receiver is removed from the home if respondents skip eight surveys in a row. Participants are rewarded with incentives, including cash, gift certificates, movie passes, airline bonus miles, and phone cards. Forty percent of the Knowledge Networks phone sample were contacted for the initial telephone interviews, participated in that survey, agreed to join the panel, and actually had the Web TV receiver installed (Krosnick and Chang 2001, table 1, note a). Again, there are not yet definitive judgments of the validity of this approach, but it is a serious attempt to harness the Internet for quality survey research.

Some of the most innovative uses of Internet surveys involve multimedia. In particular, survey participants who are on broadband networks can be sent streamed video. For example, they could watch a presidential speech as it occurs and continually record their reactions to it, and the television networks can relay the compiled response to viewers by the end of the speech. Yet this approach is controversial, as the participants are essentially volunteers and may not be representative of people who were not at home during the speech; further, the sample is biased to Internet users who can afford broadband access.

Responsible Internet surveys may eventually be a major mode of survey research, with the potential of supplanting telephone surveys. The ability to do inexpensive surveys on the Web, however, will continue to encourage low-quality surveys as well. In many ways, the status of Web surveys in the years between 1998 and 2005 resembles the status of telephone surveys in the mid-1970s, when procedures for quality phone surveys were still being tested and perfected.

Mode Limitations

The myth of the "ideal survey mode" is that there can be an ideal survey mode that is always best. In practice, each survey mode has its limitations. Face-to-face surveys over large geographical areas have complicated personnel and organizational needs that lead to large expenditures and a long time to complete the interviewing. Mail questionnaires are much less expensive but have always been plagued by low response rates, which in turn necessitate keeping the study going for several weeks as attempts are made to increase the response rate. CATI surveys are less expensive than face-to-face interviewing because they save transportation costs and can be faster to complete, but in fact they can sometimes be more expensive because of high programming costs; moreover, response rates are falling rapidly on telephone surveys. Internet surveys can be less expensive (except again for programming costs) and quick, but they have real coverage issues. Unfortunately, the mode with the highest response rate (face-to-face interviewing) is also the most expensive and time consuming, whereas the least expensive modes (mail questionnaires and Internet polls) have the lowest response rates and the least satisfactory coverage.

These trade-offs suggest that a multimode approach can sometimes be useful, and more studies are employing multiple modes. Fortunately, most large survey organizations can conduct surveys using more than one mode. The organizations that do face-to-face interviewing generally also have telephone capabilities too, and phone survey organizations usually are also able to send out mail questionnaires. Similarly, many long-standing survey operations have decided to experiment with Internet surveys. A 1995 survey of a sample of U.S. survey organizations found that all used some mixture of modes, with nearly all using at least one CASIC mode (Groves and Tortora 1998), making true multimode surveys possible.

These various survey modes will be discussed in more detail in later chapters, along with further statements of their advantages and disadvantages. After several types of errors are discussed in the next eight chapters, chapter 12 will explicitly examine the effects of the different modes.

PART II

Response Accuracy Issues

4

Measurement Error Due to Interviewers
THE DEBATE OVER INTERVIEWING STYLE

We begin with measurement error—the error that occurs when the researcher does not obtain accurate measures of the phenomena of interest. Measurement error can be due either to interviewers or to respondents. Interviewer-related error occurs to the extent that responses are different because of the interviewer; the statistical effects of different types of interviewer error will be described in this chapter. Interviewer error is a potential problem in both face-to-face and telephone surveys, but self-administered questionnaires, mail surveys, and Internet surveys are immune to interviewer effects.

Beginning with interviewer error does not mean that interviewers are a large source of error in surveys. Skilled, well-trained interviewers obtain high-quality data. However, interviewers are often not given enough training and supervision, so error due to interviewer effects does occur. Thus, considering interviewer-related error in this chapter provides an opportunity to discuss interviewer management as well as the role of the interviewer more generally. This chapter begins with a discussion of different interviewer styles, followed by treatment of interviewer effects and then interviewer management and supervision.

The Theory of the Interviewer

Understanding interviewer errors requires understanding what an interview is and how to conceive of the role of the interviewer. A lively debate has developed in recent years over the proper interviewing style, with supporters of the conventional standardized interviewing approach being challenged by advocates of more flexible conversational interviewing.

Two Early Approaches

The earliest view was that the interview is intended to obtain honest reporting of facts and attitudes and that this is facilitated when the interviewer develops a good personal relationship with the respondent. The underlying philosophy was based on psychotherapy, more specifically an analogy with a psychological therapist and a patient, with the patient opening up to the therapist better if they first develop a good personal relationship. The term that is usually used to describe this relationship is *rapport*. A warm professional relationship is necessary, with the interviewer being respected and trusted and acting both accepting and nonjudgmental. Indeed, for many respondents, the interviewer is the reason the respondent is doing the interview. The interviewer does not have to become their close friend, but the interviewer has to develop an easy question style so respondents will feel that they can give candid answers.[1] This view of the interviewer was developed in the days when interviewing was exclusively face to face; such rapport is harder to achieve in telephone interviewing. Rapport became controversial when some said it led to overly personalizing the interviews. Still today, however, many of the best interviewers are those who are able to develop rapport with respondents and use that to conduct high-quality interviews.

A *motivational* approach to interviewing was advanced in a series of research studies by Charles Cannell and his associates (e.g., Cannell, Oksenberg, and Converse 1977). Their emphasis was in motivating respondents to answer accurately through building a sense of commitment. The underlying logic is that being a survey respondent is a rare activity, so respondents do not know what is expected of them, and that it is possible to help them understand how they should behave in this setting. They found four specific techniques useful in training respondents to perform as desired.

First is reinforcement of constructive behavior by the respondent. Cannell's previous research (Marquis, Cannell, and Laurent 1972) had found that interviewers were giving positive feedback to respondents whenever they answered questions, even when their answers were not responsive. Indeed, interviewers sometimes gave the most positive feedback in that situation, accidentally rewarding behavior that was not desired. Instead, the interviewer should train the respondent by reinforcing answering behavior with positive comments only when the respondent answers in a thorough and thoughtful manner. A second technique they emphasize is having interviewers stress at the beginning of the interview how important it is that answers be complete and accurate. Third, their research suggests that it is useful to have the respondent promise to try to answer the questions completely and accurately. Fourth, they suggest a technique called *modeling*: giving respondents an example of how they are supposed to behave in answering questions, such as first playing a tape recording of a respondent who is concerned about giving complete and accurate answers. Obviously, these techniques cannot all be used in every type of interview; modeling, for example, is not feasible in a phone interview, but it is possible when a roomful of people is instructed in how to answer a survey.

The Standardized Approach

Cannell's motivational approach is considered insightful, but the dominant view of the role of the interviewer today is that the interviewer should ask questions in a *standardized* manner. The standardized approach to interviewing the mass public emphasizes the importance of the interviewer's establishing a professional relationship with respondents, rather than developing comfortable rapport with them. The underlying principle is that every respondent should be asked the identical questions in the identical manner and that the interviewer's role is to do just that.[2]

The interviewer in standardized interviewing is basically a "relayer of questions" (Houtkoop-Steenstra 2000, 45). Lavrakas (1993, 132) likens the interviewer to an "intelligent automaton," and that description aptly conveys the view of the interviewer's role from the standpoint of standardized interviewing. Indeed, the emphasis often is more on the interviewer as automaton than as intelligent. For example, in standardized interviewing an interviewer is supposed to repeat all answer categories to a respondent whenever it is not clear exactly which category the respondent has chosen. Say approval of something is being measured: the interviewer must repeat, "Do you strongly approve, somewhat approve, somewhat disapprove, or strongly disapprove?" even if it is clear that the respondent is just deciding between "strongly approve" and "somewhat approve."

Rules for Standardized Interviewing
_____

1. The interviewer should read the questions exactly as written.
2. Probing of incomplete answers should be neutral.
3. Answers should be recorded verbatim.
4. Interviewers should be nonjudgmental about answers.

Having interviewers follow a script exactly is particularly important for survey operations that hire hundreds of telephone interviewers but want to minimize training costs. Too often, however, a respondent can recognize when the interviewer is merely reading from a script, especially when the script is read poorly with words mispronounced and the interviewer is seemingly lost whenever the respondent asks about the meaning of a question. Respondents are less likely to take interviews seriously when the interviewer does not seem fully competent. Obviously this is not a problem with standardized interviewing per se, but it emphasizes the challenge that standardized interviewing operations face in making sure that their interviewers sound professional.

Several procedures have been developed for conducting standardized interviewing. The first is to establish the proper context with the respondent. This includes explaining the purpose of the interview, setting the proper tone, and communicating the goal clearly (Fowler and Mangione 1990, chap. 4). The interviewer should be professional, while encouraging a free flow of communication in order to obtain high-quality answers. This approach is based on Kelman's (1953) work on three ways to achieve influence relationships: compliance (through reward or punishment), internalization (linking the desired behavior with the person's own values), and identification (behaving consistently with role expectations). Achieving the interview through a compliance relationship (as when a boss orders a subordinate to answer the questions in a survey) is thought to lead to poor interviews if respondents do not see any reason to give good answers. Internalization can be more effective for producing accurate answers if the interviewer can establish a link between the survey and the respondent's own values and goals. The identification route requires respondents to learn a role in which good answers are provided, as expectations of what good answers should be like are supplied—such as answering closed-ended questions with one of the offered choices.

Maynard and Schaeffer (2002a) refer to survey interviews as "messy interactions," though hopefully good enough to obtain the needed information within available cost constraints. The "messy interactions" phrase usefully points out that interviewing is never perfect; standardization is often stated as an ideal, but it is often not achieved. A study of twelve university survey centers that purported to use the standardized approach found that only one consistently followed standardized interviewing procedures (Viterna and Maynard 2002).

One instance of the problem with standardized interviewing is what Houtkoop-Steenstra (2000, chap. 7) describes as the tendency of interviewers to behave like "ordinary conversationalists" by "normalizing" awkward situations. Her example involves a literacy survey in which respondents were to talk about their reading incompetence. In order to maintain rapport with respondents, the interviewers often rephrased questions, praised indications of progress in reading, made comments about how the respondents' literacy problems were normal, and sometimes indicated that they shared some of the respondents' problems. This may be an extreme case, but it shows how interviewers will often help respondents save face, even when the interview is supposed to be standardized. The general point is that interviewers tend to work through awkward interview situations, finding ways to make them less awkward, even if this departs from standardized interviewing. This situation may not be desirable, but it emphasizes the need for researchers to take such potential problems into account in writing their questions so there is no need for interviewers to do this.

The Conversational Approach

Standardized interviewing is now the dominant approach, but some researchers instead favor *conversational interviewing* (or *flexible interviewing*). Advocates of this approach argue that there will always be ambiguities in question wording, so it should be the job of the interviewer to elicit answers that will be as close as possible to what the researcher is really trying to measure. The goal is that all respondents understand the questions in the same way, but the inevitable differences between respondents suggest that interviewers will have to deal with different respondents differently to achieve that standardized understanding. The result is to give the interviewer a more active role than under standardized interviewing.

The conversational approach harks back to an early description of an interview as a "conversation with a purpose" (Bingham and Moore 1924). Conversational interviewing emphasizes the need to work with the respondent to accomplish that purpose. Suchman and Jordan (1990) particularly argue that standardization prevents interviewers from handling unpredictable situations and that more valid results are obtained when the *meaning* of questions is standardized rather than their *wording*.

The differences between standardized and conversational interviewing styles are most apparent when respondents indicate that they do not know how to interpret a question. In a standardized approach, the interviewer explains that respondents should interpret the questions themselves, whereas in a conversational approach the interviewer is supposed to collaborate with respondents in achieving the intended meaning of the question. Houtkoop-Steenstra (2000, 61) usefully describes this as the interviewer acting as a "spokesperson for the researcher," obtaining the information the researcher really wants, even if that means departing from standardized wording when the respondent misunderstands the question. As Maynard and Schaeffer (2002a) argue, interviewers must employ their "tacit knowledge" in implementing rules for interviewing. Commonsense understandings are achieved between the interviewer and the respondent, understandings that go beyond pure standardization. The notion of tacit knowledge incorporates the view that interviewing is partly art, depending on the capabilities of the interviewer.

Conrad and Schober (2000) report an experiment in a fact-based national telephone survey involving purchasing behavior, where respondents were interviewed twice, with the interviews taken a week apart. Respondents were about their housing and household purchases, using questions from the federal government's Consumer Price Index–Housing Survey and Current Point of Purchase Survey. The first interview was always standardized, whereas the second was standardized for a random half of the sample and conversational for the other half. Conrad and Schober looked to see how answers changed between these interviews and which best fit the official definitions for terms in the study, such as that buying a floor lamp does not constitute buying household furniture. Respondents were twice as likely to change their answers from the first interview when the conversational approach was used for the reinterview (22% of answers changing versus 11% when the second interview was also standardized), and the changed responses were more likely to fit the intended definitions when conversational interviewing was used.[3] For example, 57% of answers about purchases were in accord with official definitions in the standardized interviews, versus 95% in the conversational. Conrad and Schober interpret this evidence as demonstrating that conversational interviewing produces more valid answers, at least on factual material. As should be expected, the added cost of this approach was in terms of longer interviews (from a median of 5 minutes to a median of 9), due both to explaining the conversational procedure to respondents at the beginning of the interview and to the time required to help respondents understand what was being asked in each specific question.

The Choice of Interviewing Approach

The advocates of standardized interviewing would strongly attack the conversational approach as leading to data of questionable validity. They would argue that interviewer effects would be larger with this approach, since interviewers would likely each develop their own styles which would affect all interviews they conduct. The data from conversational interviewing would not be comparable between respondents because they are given different stimuli. As a result, those who favor standardized interviewing would contend that greater question clarity is what is needed, combined with complete standardization; better question wording would make conversational interviewing unnecessary.

In this vein, Houtkoop-Steenstra (2000, chap. 1) contends that "interviewer effects" may really be "questionnaire effects," so that, by extension, interviewer errors may really be questionnaire construction errors. The implication is that researchers should take the nature of interviews as conversations into account when designing their questions. Interviewers depart from standardized interviewing practices when the questionnaire does not make adequate allowance for the reality of interviewing. Thus, flexible interviewing might not be necessary if the questions were formulated so that interviewers and respondents could feel more comfortable with a fully standardized approach.

The standardized and conversational interviewing styles play through differently for fact-based and attitude surveys. In fact-based surveys, it is important that respondents correctly understand what facts they are supposed to provide. Take the question "How many credit cards do you have?" The question seems simple, but respondents may not fully understand what to include in their answer. Should they include debit cards in the count? What about cards that their spouse has, or that they and their spouse have together? Should store cards be included or only bank cards? What about cards they have but never use? Wording the question more carefully could address some of these concerns: "Not counting debit cards that you use only for withdrawals, how many bank credit cards did you yourself use last month?"

Survey organizations differ in the instructions they give to interviewers as to how to handle questions from respondents on the meaning of fact-based questions. Two intermediate approaches can be delineated between the standardized approach of just repeating the question and flexibly working with respondents to develop a proper understanding of the question. A semistandardized approach would be to give interviewers the exact replies they are supposed to give to particular questions that respondents might ask, such as whether to include credit cards used jointly with a spouse. A semiconversational approach would be to allow interviewers to use their best judgment in responding to the respondent's questions but not to initiate clarifications themselves.

It is more awkward for the interviewer to give clarifying information to respondents in an attitude survey than in a fact-based survey. Take the question "In politics today, do you generally consider yourself a liberal, a conservative, or a moderate?" If the respondent asks what those terms mean, it would be hard for an interviewer to respond in a nonbiased manner. A question like this one is usually intended to find out the respondent's ideology, based on how the respondent interprets the terms. There are no standard meanings to the terms, and spontaneous efforts by the interviewer to provide meanings would only throw off the results.

Interviewing approaches also differ in their *probing* styles. Respondents often give answers that are incomplete, unclear, or not fully responsive. Interviewers should be trained to encourage the respondents to improve their answers, without making respondents feel that their answers are inadequate. The best probes are neutral and nondirective. Typical nondirective probes for open-ended questions would be "How do you mean that?" "Tell me more about that," or "Anything else?" Or the interviewer might repeat the original question or simply remain silent. Directive probes that can be answered "yes" or "no" are particularly frowned upon, given that respondents are more likely to answer "yes." With closed-ended questions—items that give respondents a limited set of response options—the proper standardized response to a request for clarification is to repeat all the answer choices, whereas conversational interviewing would permit the interviewer to help the respondent understand the choices. Some advocates of standardized interviewing would argue that probing detracts from standardization (Foddy 1995), since some respondents are being given different stimuli than others.

The two approaches also differ in handling questions that ask for material that has already been provided in the interview. For example, what if the respondent has already indicated his or her birthplace before it was asked? Standardization normally requires that respondents be asked a question even when they have happened to answer it in response to an earlier question. The Census Bureau and some other survey organizations instead permit the interviewer to ask the respondent to confirm their earlier answer, a process known as *verification*. For example, the interviewer might be permitted to ask, "I think you said earlier that you were born in Minnesota; is that correct?" and go on to the next question after a short pause unless the respondent corrects that information. This departs from standardized interviewing, but it is a way of showing the respondent that the interviewer is paying attention to the conversation rather than being an "incompetent conversationalist" (Houtkoop-Steenstra 2000, 85) who does not listen to what has been said. The conversational interviewer might be allowed to skip the question entirely and just record the previously offered information.

The success of the conversational approach depends on high-quality interviewers who have undergone considerable training on the questionnaire so they understand the information that each question is meant to elicit. By contrast, standardized interviewing is the only option for market opinion research firms that hire large numbers of inexperienced interviewers and give them minimal training.

The debate between advocates of standardized and conversational interviewing is presented in an edited volume (Maynard et al. 2002). That volume also nicely illustrates the use of conversational analysis (Houtkoop-Steenstra 2000) and ethnomethodology in analyzing taped interactions between interviewers and respondents. The standardized approach remains dominant today, but conversational interviewing has attracted interest as researchers realize that interviewers often depart from fully standardized interviewing.

Interviewer Effects

The use of interviewers affects answers given by respondents in several ways (Groves 1989, chap. 8). Interviewers can administer the questionnaire in different ways, such as by rewording a question rather than asking it as written. They can emphasize different words or use different intonations in reading the question. They can probe differently with different respondents. Finally, interviewer demographic characteristics can affect the answers that are given, such as if people give more candid answers to interviewers of their own social group.

Interviewer Errors

Interviewers inevitably make some actual errors. Some are mechanical slips, such as the *skip errors* in pencil-and-paper surveys that occur when interviewers accidentally skip questions or do not record answers. One study (Sebestik et al. 1988) found that 90% of interviewer mistakes in pencil-and-paper surveys involved not recording answers. Skip errors are particularly common when the interviewer is supposed to skip some questions because of an answer to a previous question, such as skipping questions about experiences in the military for respondents who have never served. This problem is usually due to the layout of the written questionnaire, with unclear skip instructions making it difficult for interviewers to follow the question order correctly. Analyzing Detroit Area Studies surveys, Sanchez (1992) finds that error in skipping questions is due to the design format of the questionnaire and is not diminished by interviewer experience. Computer-assisted interviewing avoids such errors by making sure that the interviewer records the answer to a question before going on to the next question.

Another mechanical source of interviewer error involves recording the answer incorrectly. *Recording errors* do occur, but their rate is usually thought to be fairly low. Lepkowski, Sadosky, and Weiss (1998) construct a model in which recording accuracy is a function of question characteristics, respondent behavior, interviewer behavior, and the conversational dynamics of the interview. They checked the responses on 200 interviews that they had tape recorded. In simple exchanges (such as closed-ended questions with "yes" and "no" answers), they found that only 0.1% of the answers were recorded inaccurately, with the same rate for CATI and non-CATI interviews. Most of the inaccuracies involved recording "don't know" when the person gave a "yes" or "no" answer to a question, but the researchers deemed these to be random errors. In more complex interviewing exchanges, the error rate was somewhat higher for non-CATI phone interviews than for the CATI interviews, 6.3% versus 4.5%, with CATI interviewers being less likely to not record an answer or to record the wrong answer. They also found that recording errors were mainly associated with the behavior of the respondents, particularly when they gave an answer outside of the actual response choices, with fewer recording errors being associated with the behavior of the interviewers. Recording errors involving entering a different response from the one the person actually gave are usually random errors. However, sometimes interviewers choose to record a different answer that allows them to skip follow-up questions and therefore finish the interview more quickly; this would be considered interviewer falsification and research misconduct, as discussed later in this chapter.

The more serious type of interviewer error has to do with how questions are asked. Interviewers are supposed to ask questions as written, but studies often find that a large proportion of questions are not asked as written. Studies counting the extent to which interviewers change question wording obtain estimates ranging from 20–40% of the time (Fowler and Mangione 1990, 34) to 30–95% (Groves 1989, 383). Interviewer training and experience are not highly predictive of whether questions are read exactly, but interviewers vary considerably. The four best interviewers in one study delivered 11% of questions with minor changes, whereas the four worst interviewers averaged 67% (Oksenberg 1981). These changes are usually random and therefore do not usually bias results. A later study of 79 interviews at the University of Maryland Survey Research Center in 1989 (Presser and Zhao 1992) found that 91% of the questions were asked exactly as worded, and 8% involved only minor alterations, but error was greater for longer questions and for questions asked in series. Interviewer experience was again, surprisingly, not found to make a difference. Whereas most accounts emphasize that rewording questions can affect the results, Groves (1989, 387) maintains that rewording of questions by interviewers does not necessarily increase measurement error, since answers to questions often do not vary with minor changes in question wording.

The main interviewer-related recording errors found by Lepkowski, Sadosky, and Weiss (1998) occurred when incorrect probing techniques were used and, in particular, when the interviewer did not probe. This emphasizes the importance of training interviewers on desired probing techniques.

Statistical Consequences of Interviewer Effects

As described in chapter 3, a key distinction is between uncorrelated and correlated measurement error. Uncorrelated error occurs when an interviewer happens to ask questions somewhat differently from interview to interview in a random fashion, which increases the variance of those items and therefore decreases their reliabilities. Correlated error occurs if each interviewer asks questions in an idiosyncratic manner across all of her interviews, and this has more serious statistical consequences.

The *intraclass correlation*, rho (ρ), is commonly used as a measure of the variance associated with interviewers (Kish 1965).[4] Computing rho is equivalent to computing an analysis of variance in which the treatment is the interviewer, thus obtaining a measure of how much of the variance is due to the interviewers. Rho has a maximum value of 1.0, with lower values indicating that less of the variance is associated with the interviewers. This estimation is accurate so long as the assignment of respondents to interviewers is random (known as an *interpenetrated design*), so there is no reason to expect that respondents assigned to different interviewers should vary.

Rho is used as a part of a multiplier, inflating the variance of a variable. Even a small rho can lead to a large multiplier. The statistic used here is called a *design effect* (or "deff"), and it equals $1 + (\rho(\bar{n} - 1))$, where $\bar{n}$ is the average number of completed interviews per interviewer. The design effect is evaluated in terms of the variance, so the appropriate multiplier for the standard error is its square root, commonly referred to as a "deft": $\sqrt{1 + (\rho(\bar{n} - 1))}$. The deft shows how much the standard error of a variable is increased by correlated measurement error.

As an example of the insidious nature of correlated measurement error, say that the rho is a seemingly trivial 0.01. If the average number of interviews conducted by an interviewer were 34, the design effect would be 1.33, increasing the variance of a variable by a third. The corresponding deft is 1.15, a 15% increase in its standard error, which makes it harder for results to appear statistically significant. The rho will vary between questions in a survey, and obviously this is a problem only on questions for which there are interviewer effects. There is no reason, for example, to expect an increase in the standard error on determining the respondent's gender, but there is likely to be an increase on attitudinal questions.

Consider a situation in which a survey organization is deciding whether to assign an average of 20 or 30 interviews per interviewer. Let's still assume a small intraclass correlation, 0.01, to see what damage even a small rho value can do. The deft with 30 interviews would be 1.136, whereas it would be 1.091 with 20 interviews, which amounts to a third less inflation of the standard error. If 300 interviews were to be taken, having interviewers take only 20 apiece would require a staff of 15 interviewers, rather than just the 10 that would be needed if each interviewer were assigned 30 interviews. Hiring 50% more interviewers (15 rather than 10), though, would cut the inflation of standard errors on variables susceptible to this effect by one-third, though the actual differences in the two defts is small enough to not make a big difference. Thus when survey organizations decide how many interviewers to hire for a project, they are implicitly deciding on the trade-off they can accept between the number of interviewers to train and the amount that some standard errors will be inflated. As Biemer and Lyberg point out, there is a general "lack of information in the research community regarding the damaging effects of [correlated] interviewer error" (2003, 168); interviewer load has direct implications for survey error that should be taken into account.

Varying Interviewer Effects

Interviewer effects can vary across modes of interviewing, across types of questions, between types of respondents, and because of interviewer characteristics. This section will summarize the differences that have been reported along these four different dimensions.

As to *mode differences*, measurement error due to interviewers has been found to be greater with face-to-face interviews than phone surveys. Groves (1989, chap. 8) reports that personal interviews from the University of Michigan's Survey Research Center have an average rho of 0.031. Actually most of the questions have ρ's below 0.02, but the ρ values were higher on sensitive topics. He also reports an average 0.01 value for phone surveys, with the highest ρ being 0.0184, but even the 0.01 rho for phone surveys means an increase in 10% in the variance for every 10 interviews added to an interviewer's workload.

Using a survey mode that does not use interviewers will, by definition, completely eliminate interviewer error. Indeed, the Census Bureau's concern over the effect of census enumerators on the 1950 census was one factor that led to its adopting self-enumeration in subsequent censuses, since the measurement error between households is uncorrelated when people fill out the census forms themselves (Hansen, Hurwitz and Bershad 1961). Doing away with interviewers is a drastic method for avoiding measurement error due to interviewers, but this consideration does explain why some researchers prefer self-administered surveys (as through the Internet) rather than telephone surveys.

The evidence is mixed on whether interviewer effects are greater on some *types of questions*. Fowler's review of this literature concludes that "on average, there is very little evidence that either the content or the form of the question *per se* is a good predictor of where [i.e., the types of questions on which] interviewer-related error is most likely to be found" (1991, 275). Several studies find smaller interviewer effects on factual items than on attitude questions, but in nine University of Michigan Survey Research Center studies the interviewer effects were not different on the two types of questions (Groves 1989, 375). Also in these studies, open-ended questions that respondents answer in their own words were not subject to larger interviewer effects than closed-ended questions (respondents choose among response options that are part of the question). The effects that were found have to do with giving a second mention on open-ended questions, supporting Fowler and Mangione (1990) on failure to probe open-ended questions.[5]

Interviewer effects could also vary between *types of respondents*. Groves (1989, 379–80) expected poorly educated respondents to be more affected by the behavior of the interviewer than highly educated respondents, but the rho he obtained was not larger for less educated respondents. He also expected older respondents to be more influenced by the interviewer. The Michigan Survey Research Center data do show a higher rho for older people. However, that measures variation in nonresponse as well as response errors over interviewers, so the higher rho could just reflect higher nonresponse variation among elderly respondents.

Interviewer characteristics have generally not been found to affect data quality (Fowler and Mangione 1990, chap. 6). Because there are not consistent correlates of interviewer quality, interviewer-related error cannot be minimized by selecting interviewers on the basis of demographic or psychological factors (Hox, de Leeuw and Kreft 1991). The common exception has to do with younger interviewers' producing poorer quality data, but that is probably due simply to insufficient training of student interviewers. Fowler and Mangione (1990) conclude that interviewers do not need specialized knowledge of the topic to get quality data; indeed Fowler (2002, 122) argues that interviewers with such knowledge are not desirable, since they are likely to read more into unclear answers instead of probing.

An extreme example of respondents' tailoring their answers to the apparent sympathies of interviewers comes from the famous "pen experiment" conducted by Howard Schuman (1990) in an attempt to explain the very disparate results of polls conducted before an election in Nicaragua at the time of that country's civil war (Miller 1991). Interviewers randomly used pens showing the leftist Sandinista government logo, the opposition logo, or a neutral design. The results confirmed that clues about the political leaning of the pollster can affect responses, at least in civil war conditions.

Still, some differences related to interviewers have been found, often when interview characteristics relate to questions in the survey. For example, respondents are less likely to express anti-Jewish feelings if the interviewer is identifiable as being Jewish. Interviewer effects are most likely, of course, when respondents know (or think they know) the demographics of the interviewer. Thus, gender effects and race effects are both likely in face-to-face interviews; race may matter less in phone interviews if respondents do not know the interviewer's race.

There is evidence of some gender effects, but the findings are sometimes inconsistent. Female respondents gave more feminist answers to male interviewers than to female interviewers on women's rights and roles, whereas male respondents gave more conservative opinions to male interviewers (Groves 1989, 401).[6] In a consumer attitudes survey, respondents were more optimistic when the interviewer was male, regardless of respondent gender (Groves and Fultz 1985). Female interviewers are more likely to be rated excellent, friendly, and professional (Fowler and Mangione 1990, 104). Huddy et al. (1997) found differences of about 5% in answers on a wide variety of gender-related survey questions administered by female and male interviewers. Kane and Macaulay (1993) report that people are more likely to give egalitarian gender-related attitudes and more criticism of gender inequalities to female interviewers, though interactions between interviewer and respondent gender tend not to be significant in multivariate models. Catania et al. (1996) find that respondents (especially men) indicate higher rates—and presumably more truthful responses—on some sexual behavior questions (such as same-gender sex, extramarital sex, and sexual problems) to interviewers of their gender. Yet the evidence on effects of matching the education and social-economic status of respondents and interviewers is not conclusive, and there is no evidence of age-of-interviewer effects.

The most interesting research on interviewer characteristics has to do with race. Starting with Hyman's (1954) early research, several studies have found more effects of race of interviewer on racial attitudes than on other questions. Respondents have been found to be less likely to express feelings against the interviewer's race. The differences are usually interpreted to show that respondents are more open with interviewers of their own race. Most of the research involves face-to-face interviewing, but some similar effects have been found with perceived race of interviewer in telephone surveys.

Table 4.1 shows differences in responses by whites to racial attitude questions, with a 46% difference in response on attitudes on interracial marriage (Hatchett and Schuman 1975–76). This example is old, but it shows that the race of the interviewer can have a strong effect on face-to-face interviews. There are also social desirability effects due to respondents' trying to save face with the interviewer, especially in face-to-face interviews, with blacks overreporting voting turnout to black interviewers (Anderson, Silver, and Abramson 1988a) and those living in predom-

Table 4.1 Support of Whites for Integration by Race of Interviewer, Adapted from Hatchett and Schuman (1975–76)

	Black Interviewers	White Interviewers	Difference
Interracial marriage	72%	26%	46%
School integration	91%	56%	35%
Residential integration	100%	69%	31%

inantly black areas who were interviewed by blacks in a preelection survey actually being more likely to vote.

Table 4.2 reproduces a classic result from Schuman and Converse (1971), showing that more than 2% of the variance in answers by black respondents in Detroit was accounted for by race of interviewer on 32% of racial attitude questions, whereas that much of the variance was accounted for by race of interviewer on only 3% of nonracial attitude questions. Differences of 13–28% were obtained on questions relating to militant protest and hostility toward whites. Although these differences look large, it turns out that it may partly reflect other differences between the black and white interviewers. The white interviewers in this study were graduate students doing interviews for the first time, whereas the black interviewers were professional interviewers (Groves 1989, 399–400). Schuman and Converse state their belief that the same race-of-interviewer result would have held in the year of the study (1968, after Martin Luther King's assassination) regardless of interviewer training. Still, this example illustrates the importance of looking for alternative explanations of effects that seem to be due to characteristics of interviewers and testing, to make sure that effects are not due to differences in how interviewers are selected.

As another example, Anderson, Silver, and Abramson (1988b) find that African Americans interviewed by whites in face-to-face National Election Study surveys were more likely to give responses indicating warmth toward whites than were African Americans interviewed by African Americans. Also, they report that blacks were more likely to say to a black interviewer that civil rights leaders were pushing too slowly and were more likely to favor increased spending on blacks, welfare, and food stamps; there were not significant differences on several other issues. Caution is appropriate in evaluating this finding, for respondents were probably not assigned randomly to interviewers, leading to the likelihood that black interviewers were more likely to be interviewing in areas in which more blacks lived. Thus some of the differences could be due to neighborhood effects rather than interviewer effects.

Table 4.2 Percentage of Questions with 2% of Their Variance Explained by Race of Interviewer, by Type of Question, in Schuman and Converse (1971) Study (with Number of Questions of Each Type Shown in Parentheses)

Type of Question	Racial Topic	Nonracial Topic
Attitude question	32% (40 Qs)	3% (29 Qs)
Fact question	14% (14 Qs)	8% (47 Qs)

An important study by Darren Davis (1997) documents significant race-of-interviewer effects on over 60% of the attitude questions in the 1984 National Black Election Study survey. He used a two-stage least squares procedure to correct for race of interviewer leading to correlated errors. As an example, other things being equal, African Americans interviewed by African Americans were nearly twice as likely to say they were planning to vote for Jesse Jackson than ones interviewed by whites. Yet it is not clear whether this means they were underreporting their plans to vote for him when interviewed by a white or overreporting when interviewed by a black.

One possible interpretation of these race-of-interviewer effects is that the validity of the data would be greater if black interviewers were assigned black respondents and white interviewers were assigned white respondents—race matching. However, this ignores the real possibility that interviews among same-race actors may yield overreports of extreme positions on racial attitude questions, which would become a bias in the study. Given that there is not necessarily a "correct" answer on racial attitude questions, it is important to be able to assess the extent of interaction effects by having some same-race and some opposite-race interviews. Advocates of standardized interviewing draw yet a different inference. Fowler and Mangione conclude: "The fact that interviewer race can predict answers means that the measurement is not standardized across race of interviewers. When answers are significantly affected by interviewer race, measurement error would be reduced if interviewers were all of the same race" (1990, 101). Although the implication that Fowler and Mangione derive in that last sentence seems to be a way of minimizing measurement error, it also could be read as inadvertently justifying discrimination in hiring interviewers. Furthermore, standardizing on race will not diminish measurement error if people are more likely to give valid answers to interviewers of their own race. Given the several different interpretations of this effect that are possible (Rhodes 1994), it is critical that further research be done before accepting it as true and using it as a basis of interviewer hiring and assignment.

Interviewer matching is rarely used in the United States, except when it is necessary to use interviewers who can speak another language to interview non-English-speaking respondents. Interviewer matching is more necessary in some other countries, as in Arab countries where it would be considered inappropriate for an interviewer of one gender to speak with a respondent of the other gender.

One other interviewer-related problem occurs when panel surveys have the same interviewer repeatedly interview a respondent. Mensch and Kandel (1988) find an *interviewer familiarity effect*, with people interviewed previously by the same interviewer in the National Longitudinal Study of Youth (NLSY) being less likely to admit to having used marijuana and cocaine. Presumably the respondents were less willing to admit to these behaviors with an interviewer whom they already knew. The use of the same interviewers in long-term panel surveys is usually seen as useful for increasing the retention rate of respondents, but this finding shows that it can negatively affect the accuracy of reporting on sensitive questions.

Another interviewer-related issue is *interviewer experience*. Paradoxically, some studies of face-to-face interviewing seem to find that experienced interviewers do not obtain as complete information from respondents as new interviewers (e.g. Hughes et al. 2002). This fits with a study finding that interviewers who completed more interviews obtained fewer reports of known hospitalizations than those with smaller interview loads (Cannell, Marquis, and Laurent 1977). Groves et al. (2004, 273) suggest that the mechanism for these results may be that more experienced interviewers may try to maximize response rates and productivity more than answer quality.

There could also be social psychological and sociological explanations for response error associated with interviewers, such as the role of *interviewer expectations*. Empirical studies find little support for these expectations' affecting answers, but the studies have their own limitations, such as looking at sets of questions rather than individual questions (Groves 1989). Leal and Hess (1999) find that interviewers can be biased in what they think of respondents. Looking at the American Citizen Participation Survey, the Latino National Political survey, and the 1996 National Election Studies survey in which interviewers were asked to evaluate respondents, they find that even with controls for the respondent's actual level of political information, interviewers considered respondents of lower socioeconomic status to be less informed and intelligent. There were also some racial effects, with blacks being seen as less intelligent by interviewers in the participation and NES surveys, again with controls for objective levels of information. However, there is no direct evidence of how these biases translate into survey error.

A possible implication of the research on interviewer effects is that the selection of interviewers matters, but there are too few empirical studies, and many of those studies have methodological problems. There is need for more careful research designs in which comparable sets of respondents are assigned to different types of interviewers, but in the end it may be impossible to obtain a definitive determination of the importance of interviewer effects.

Interviewer Management Issues

Fowler and Mangione (1990, 105) argue that the focus should be on interviewer training and supervision more than on interviewer selection. Interviewer-related error can be minimized by a survey organization through its procedures for managing interviewers, including training procedures, supervision, and incentives for interviewers.

Interviewer Training

A common question is what types of interviewers should be hired. Some survey organizations begin the hiring process by testing the ability of prospective employees to read survey questions exactly as written. Additionally, it is important to hire people who will show up reliably. Beyond these minimal requirements, training may be more important than hiring. This is apparent in the choice some organizations face between hiring professional and student interviewers. There are benefits to both, assuming the students are given adequate training. Having experienced interviewers is important to professional survey organizations. Yet student interviewers are often easier to schedule for part-time work and easier to add (or drop) with the ebb and flow of survey business, but more money is spent on training since student interviewers rarely last longer than a few semesters. Many academic survey organizations use a mixture of the two types of interviewers, often with the hope that the presence of professional interviewers will help give student interviewers the appropriate attitude toward their job. Even a few large commercial organizations such as Gallup supplement their professional staff with student interviewers, in the belief that proper interviewer training is what is really important.

Interviewer training is particularly urgent when interviewing is not carried out on a regular basis. A survey organization that does interviewing every week of the year is better able to maintain a well-trained staff than one that conducts a survey only once every few months. Election exit polls pose the most extreme training problem: they generally occur only once every other year in the United States, which essentially requires hiring and training a new interviewing staff each time. Thus it is not surprising that some of the blame for exit poll errors in the 2000–2004 U.S. elections was ascribed to inadequate interviewer training. Indeed, exit polls should be considered the most difficult type of survey to conduct well, because of the need to start virtually from scratch every few years.

Training Guidelines

Since measurement error due to interviewers is an important component of survey error, it is essential to train interviewers so as to minimize interviewer-related error. Interviewer training includes both training new interviewers in standard procedures and giving them training on new surveys. Every survey organization handles this training somewhat differently, but high-quality survey organizations all emphasize the training they give to interviewers.

The Council of American Survey Research Organizations (CASRO) has developed a set of guidelines, available on its Web site, http://www.casro. org, as to what should be covered in general interviewer training. They include an explanation of the research process, the importance of professionalism of the interviewer, the need for confidentiality, and the importance of maintaining positive feelings by the respondent about survey research.

The general training given interviewers includes how to obtain interviews and how to conduct them. Obtaining interviews is not easy, especially telephone interviews, since people find it easy to hang up when they hear that they are about to be interviewed. As will be seen in chapter 8, survey organizations develop standard responses that interviewers are supposed to give when people ask particular questions such as "How did you get my number?"

The training on conducting interviews emphasizes the interviewing philosophy of the survey organization. Interviewers will be told to read questions exactly as they are written, but differences arise in how interviewers are told to handle questions from the respondent. Organizations that follow a standardized approach will emphasize the need to repeat the question as originally worded without ad-libbing. Organizations that follow a conversational approach will instead explain how to collaborate with the respondent in establishing a meaning for the question that fits the research objectives. Similarly, the interviewer training should explain the type of probing that is used by the survey organization. In any case, the importance of recording the respondent's answers verbatim is stressed. Academic survey organizations that employ the total survey error approach often include an introduction to that approach in their interviewer training, so that the interviewers recognize the importance of their effort to the overall goal of minimizing survey error.

Some survey organizations use videotapes as part of their training. Such tapes can include an introduction to the organization and its history, a statement of the importance of conducting high-quality interviews, and an explanation of the interviewing philosophy that is employed. Training tapes often include examples of how not to interview as well as examples of interviews that are conducted properly.

Interviewer training must also include specific training with any technology that is used in the interviewing. Phone surveys nowadays are generally computer assisted, as are many face-to-face interviews, so the training for such interviewing must include familiarization with the hardware and software to be used.

Interviewer training should include opportunities to take practice interviews. The first practice interviews are typically with fellow trainees in a role-taking mode, with one trainee sometimes taking the role of an uncooperative respondent. Later practice interviews are usually taken with real people, with these interviews not included in the final data.

In addition to general training, it is important to train interviewers on new surveys. Such training includes both going over the purposes of the survey and giving interviewers a chance to practice the survey. Again, most survey organizations have interviewers take a few practice interviews with real people who are not part of the sample so as to learn the interview schedule.

CASRO lists several specific points to address when briefing interviewers on new projects. Interviewers are to be instructed on the sampling method that was employed, how they should handle materials, the anticipated length of interviews, skip patterns for controlling the question order, specific directions on clarifying questions and on probing, and the approach to validating interviews. Additionally, the briefing sessions should include reading the interview instructions and reading the questionnaire itself, followed by practice interviews. Even an experienced interviewer cannot be expected to get a high-quality interview the first time on a new questionnaire, so practice is always important.

Studies of Interviewer Training

Several studies have examined the effectiveness of interviewer training. A Belgian study (Billiet and Loosveldt 1988) showed that interviews conducted by trained interviewers had less missing data and more probing. Fowler and Mangione (1990, chap. 7) conducted an extensive experiment with different amounts of training for face-to-face interviewers. Interviewers with less than one day of training did worst. There was little difference between interviewers with two to ten days of training, except for probing. Probing is one of the hardest parts of interviewing, so training pays particular benefits in improving the ability to probe. Also, interviewers with more training were seen as more task oriented, whereas those with less training were seen as more interpersonally oriented; however, interviewers with the least training were marginally more likely to be rated by researchers as excellent. Still, two and five days of training produced more standardized results than one or ten days. Fowler and Mangione concluded that ten-day training is too long, with interviewers becoming tired of being trained, but two to five days is appropriate. They generally found better quality when interviews were tape recorded. Only the largest survey organizations would even consider having five to ten days of interviewer training, but this study usefully implies that even small survey organizations would benefit from doing more than one day of training. Additionally, interview supervisors are often instructed to give special attention to new interviewers, so that, in effect, the interviewers get additional training in their first few days on the job.

As a means of measuring interviewer compliance with training guidelines, Cannell, Lawson, and Hausser (1975) developed several categories for "interaction coding" to keep track of the interactions between interviewers and respondents. Interviews are monitored or tape recorded, and then tallies are made to count how often the interviewer asks questions exactly as written, questions are repeated correctly, definitions and clarifications are given correctly, the interviewer delivers short and long feedback correctly, and the pacing of the interview is correct. The interviewer's overall clarity is also recorded, in terms of how naturally questions are read (unnatural reading could involve awkward inflection, wrong emphasis, monotone voice, or wooden delivery) and mispronunciations.[7] Interaction coding can be used to identify questions that are problematic, as well as interviewers who do not perform in a standard manner. However, Groves (1989, 389) cautions that there is no direct evidence that behaviors measured in interaction coding are related to measurement error.

Interviewer Supervising

Interviewer supervision provides another opportunity to work on the quality of the interviewing so as to minimize measurement error due to interviewers. The organization of supervision varies between survey organizations, often with first-level supervisors being responsible for shifts and reporting to higher-level supervisors or directly to a field manager.

First-level supervision consists of many tasks, from scheduling and problem solving to performance monitoring and interview validation. The goals of a supervisor include maximizing productivity, increasing the data quality, and answering questions raised by interviewers.

Supervisors for both telephone and face-to-face interviewing are responsible for assigning interviews to interviewers when computer programs do not handle the assignment. In phone interviewing, they get the site ready before the interviewers come in for each shift, they keep track of forms filled out by interviewers when people refuse to be interviewed or ask to be called back later, and they deal with problems that come up during their shift. If the survey is on paper (rather than on a computer), the supervisors collect the survey forms after they are filled out and check that the interviewers are doing a good job. In phone surveys, the supervisors make sure that the interviewers are calling continuously and that interviewers keep appointments (as when a respondent indicates that she can do the interview at a specific time). At the end of a shift, supervisors submit a shift report to the field manager, including the numbers of completed interviews and refusals for each survey, as well as which interviewers did not show up for their scheduled work time.

Interviewer supervision also provides an opportunity for performance monitoring. Fowler and Mangione (1990) suggest that a trained supervisor monitor one in ten interviews. Many telephone survey organizations have developed standardized forms for monitors to fill out so as to provide systematic evaluation of the interviewing. As an example, the interviewer monitoring form (IMF) used by Ohio State University's Center for Survey Research has a supervisor record how the interviewer did in the introduction to the survey, including information given to the respondent, handling the request for the interview, enthusiasm, confidence, sincerity, answering respondent questions, and types of questions asked for scheduling the interview. The interviewer's performance in the actual interview is scored in terms of assertiveness, professionalism, pace, clarity, tone, and overall performance. Additionally, the supervisor tallies errors made by the interviewers in terms of minor or major deviations from question wording, not clarifying ambiguous responses, interpreting questions for the respondent, making extraneous remarks, offering response options improperly, and failing to probe open-ended responses. Ideally the supervisor meets with the interviewer immediately after the interview to go over the form together. Monitoring is often more informal than this, with the supervisor listening in and keeping only a mental note of items to mention to the interviewer afterward. In practice, these are often supportive comments, such as consoling an interviewer who has just survived a difficult respondent.

Some organizations have had good experiences with tape-recording interviews. An experiment by Fowler and Mangione showed that taping the interviews did not have negative effects on the interviewer's feelings and had no effect on the respondent's feelings, but it did result in better probing for one of the groups as well as less bias. In computer-assisted personal interviews using laptop computers, it is now also feasible to tape record interviews with the computer's built-in microphone, thereby improving quality control. Taping and supervision are considered to have positive effects even for the best and most experienced interviewers, partly to correct their tendencies to feel they have developed enough skill to substitute their own judgment for standard procedures.

Interviewer supervision is most effective with phone interviews from a single central location. It is least effective when the interviewing is dispersed, as in national election exit polls where one interviewer is assigned to each sampled precinct, making it impossible to determine how well (or even if) interviewers are doing their job. Some survey organizations now conduct telephone interviews through a "distributed network" with CATI interviewers phoning from their homes (or outsourced to installations in another country) while using a broadband computer connection to the Internet to access the survey. This innovative procedure may make it easier to recruit interviewers and/or lower interviewer costs, but it requires further efforts to maintain effective interviewer supervision.

Supervision should include some interview verification, to make sure that the interviewer is not faking interviews. Interviewers sometimes pretend to be doing interviews, expecting that they will not be caught. Many telephone interviewing operations have the capability of a central supervisor's listening in on the interview, which provides one means of assuring that real interviews are occurring. Modern monitoring programs actually keep track of some aspects of the phone calls, such as how long the person is on the phone and which question is being asked, and this allows supervisors to make sure that appropriate progress is being made. Tape recording interviews is another means of verification. Some centralized facilities use video monitoring of interviewers. Validation through recontacting respondents is common also for face-to-face interviewing, with supervisors phoning a sample of respondents to make sure that they were actually interviewed. CASRO guidelines state that "standard industry practice" involves validating 15% of the interviews. Large survey firms sometimes employ an independent organization to validate interviews in order to assure the client that the validation was real. Obviously interviewers who are found to fabricate interviews are terminated, and interviews that they conducted are closely examined to determine which must be discarded. In an unusual case, several students were disciplined by California State University–Stanislaus after they admitted fabricating interviews in a class survey on the bias of prospective jurors that was used by a judge in deciding to move the 2004 Scott Peterson murder trial out of Modesto, California.

The Office of Research Integrity of the U.S. Department of Health and Human Services has ruled that falsification by survey interviewers is an act of scientific misconduct. That action has led to work on a best practices document on interviewer falsification (American Statistical Association 2003). In addition to fabricating interviews, the document includes as falsification the deliberate misreporting of disposition codes (such as recording a refusal as ineligible so as to make the refusal rate look low), deliberate miscoding of answers (as when the interviewer does not want to ask follow-up questions), and deliberate interviewing of a person who was not sampled (so as to increase the number of interviews obtained). The document emphasizes the importance of promotion of the proper values in interviewer hiring and training, adequate supervision, quality control, avoidance of excessive workloads, and a pay structure that is not based exclusively on number of completed interviews.

Interviewer Incentives

Interviewer pay is one of the largest expenses for surveys. Some survey organizations skimp on pay, but most higher-quality organizations believe that it is better to pay more—well above minimum wage—and hire fewer but better interviewers.

Survey organizations have tried different motivating systems, some emphasizing incentives and others emphasizing demerits. Given how difficult it can be to get respondents to cooperate for phone interviews, giving incentives for completed interviews is a good way to keep interviewers motivated—either extra pay or points that can be accumulated for prizes. Overreliance on incentives for completed incentives, however, gives interviewers a reason to fake interviews. Demerit systems are especially common when using student interviewers, with supervisors keeping track of shifts missed without calling in and interviewers being fired if they get a certain number of demerits a month. Some survey organizations try to motivate the callers by getting them involved in the survey process, but most maintain a clear distinction between the interviewing and the rest of the operation.

Interviewer-Related Error and Survey Constraints

The interviewers' job is difficult because it is multifaceted. They must first convince the respondents to participate, then motivate the respondents to give full answers, and at the same time record their answers accurately. This requires a combination of human skills, quick thinking, and organization, as well as a personality that does not get depressed by refusals.

At the same time, it is not possible to do away with interviewer effects. Some interviewer error is inevitable, as illustrated by studies showing that even standardized interviewers routinely deviate from the exact question wording. Additionally, there can be effects associated with the characteristics of the interviewer, effects that cannot be eliminated since no interviewer characteristics are entirely neutral.

Interviewer-related error is one of the few sources of survey error that are under the direct control of a survey organization. Proper training of interviewers, appropriate supervision, and good pay all can help keep the quality of interviews high. Each of these is, however, costly, which illustrates the continual trade-off between survey errors and costs. When a researcher gives a survey contract to the lowest bidder, the low bid is often due to cutting corners in these respects, which can lead to lower data quality. The trade-off is probably most severe in attitude surveys that include many open-ended questions requiring skillful interviewing.

As an example of the trade-offs that are involved between survey errors and costs, Fowler and Mangione (1990, 144) estimate the effects of different ways to reduce survey errors. Say that the goal is to decrease the standard error on variables by 10%. The standard way to do so is to increase the sample size, which would require taking about 20% more interviews, which increases interviewing costs by 20%. Alternatively, they suggest that the standard error on questions that are most affected by interviewers could be cut by 10% through use of three procedures that have been discussed in this chapter: adding another day to interviewer training, tape recording interviews, and/or reducing the number of interviews assigned to each interviewer. These procedures are not costless, but they increase cost by much less than 20%.

Time is also a relevant constraint. Hiring more interviewers, for example, allows more interviews to be done in a shorter period of time, which will be important to some clients. On the other hand, extensive interviewer training takes time—and some interviewers would not bother returning for lengthy training before conducting actual interviews.

Ethics pose additional constraints. Interviewers can fake interviews; this makes it important to motivate interviewers with information about the importance of the research and through good pay so that they will have incentives to do good work. Additionally, emphasizing the role of the interviewer in helping minimize total survey error is a way of making interviewers understand the importance of their contribution to the research effort. At the same time, validating some of the interviews—with interviewers knowing that this is done—lessens the likelihood of interviews' being faked. Interviewers also face conflicts between increasing response and honoring the respondents' right to voluntary participation; the temptation is to push the person into participating so as to maximize response. Another important ethical requirement is that interviewers are supposed to keep information they obtain from respondents confidential, not telling friends the names of respondents or what they said.

This chapter's consideration of interviewer-related error led to consideration of how to view the interviewer's role: whether the interviewer should act like a therapist, friend, automaton, or partner. Chapter 5 turns to the other side of the coin: the role of the respondent.

5

Measurement Error Due to Respondents, I
QUESTION WORDING PROBLEMS

Measurement error can be due to respondents as well as interviewers. Respondent-related error occurs to the extent that the respondents are not providing the answers they should, given the researcher's intentions. This form of measurement error is most often due to question wording and questionnaire construction. Question wording and questionnaire construction are both major topics, so measurement error due to respondents is discussed here in two chapters, one on each.

How the question is worded will directly affect how much measurement error there is on that question. We minimize measurement error due to respondents by seeking the best way to word questions, and we do that by understanding how respondents process questions.

This is one of many survey-related topics where our way of treating the topic has changed in recent decades. Nonsystematic lore on how to write good survey questions developed in the early years of survey research. Later approaches included standardization in questions (Fowler and Mangione 1990, chap. 5) as well as developing rules for writing questions that would lead to good answers. However, at the same time, several anomalies were discovered, such as apparently similar questions that elicited different distributions of answers, and these anomalies called for a greater understanding of the response process. Fortunately, this occurred around the time that psychologists' study of the processing performed by the human brain led to the "cognitive revolution" in psychology. The increased understanding of that processing has changed many social science fields, including the writing of survey questions.

Chapter 5 begins with a presentation of the various theories of the survey response process. Next, different types of survey questions will be discussed, followed by different question formats. Chapter 6 turns to the questionnaire construction side of measurement error due to respondents.

Theories of the Response Process

The appropriate place to begin is with theories of the response process. Viewed from the perspective of the interviewer in chapter 4, an interview is a form of conversation. But viewed from the perspective of the respondent, an interview is also a cognitive task.[1] That makes it important to understand the cognitive psychology of information processing.

Today there are two complementary sets of approaches to understanding how respondents process survey questions: multicomponent approaches and two-track theories. The most influential of the multicomponent approaches (Tourangeau 1984) divides the survey response process into a series of separate steps. The two-track theories (e.g., Krosnick 1991) suggest that there are two different levels of effort that respondents can make in answering survey questions. One is a "high road": exerting real thinking about the question, following each of the different steps of the multicomponent approach. The other is a "low road": trying to get away with minimal thought, short-circuiting some of those steps, and giving a plausible answer rather than fully thinking through the question (a way of responding known as *satisficing*). The argument becomes that measurement error due to respondents is largely due to respondents who satisfice rather than working through all of the steps. The multicomponent approaches will first be detailed in this section, after which the two-track theories will be discussed.

Multicomponent Approaches

The multicomponent approach breaks the process of replying to a survey question into a series of steps. Tourangeau (1984) gives four components.[2] The first is *comprehension* of the question. The second is *retrieval* of relevant information from the person's memory. Third is using that information to make the *judgment* required by the question. Fourth is *selection* of an answer. These components have a clear order to them, but respondents need not necessarily follow this order: they can carry out multiple components simultaneously, they can backtrack from later to earlier components, and they can skip components altogether. Table 5.1 summarizes the specific processes that constitute each of these components.

Table 5.1 The Response Process

Component	Specific Process
Comprehension	Attend to questions and instructions
	Represent logical form of question
	Identify question focus (information sought)
	Link key terms to relevant concepts
Retrieval	Generate retrieval strategy and cues
	Retrieve specific, generic memories
	Fill in missing details
Judgment	Assess completeness and relevance of memories
	Draw inferences based on accessibility
	Integrate material retrieved
	Make estimate based on partial retrieval
Response	Map judgment onto response category
	Edit response

Source: Tourangeau, Rips, and Rasinski (2001, 8, table 1.1). Copyright © by Cambridge University Press. Reprinted by permission.

This theory combines the cognitive psychology work on memory with social psychology research on accurate reporting. Each step can cause response effects. At the first stage, the respondent may misunderstand the question. At the second stage, the recall may be inaccurate, such as when a respondent is asked to recall an event that occurred in the distant past. As to the third stage of judgment, the respondent may not integrate information accurately, such as how many doctor visits the person has had in the past year, or may not have a well-formed attitude, so that the response is semirandom. At the final stage, the respondent may map the answer onto the response options incorrectly or may "edit" the answer, such as when respondents feel that their answers must be acceptable to the interviewer and therefore filter out answers (such as overtly racist answers) that they think the interviewer would disapprove.

Comprehending the Question

The first component is understanding the question. Belson (1981) hypothesizes that respondents interpret interview rules to require them to answer each question, but choose the possible meaning for a question for which they can more readily form an answer. Ferber (1956) shows that people will answer questions even when they do not know the meaning of terms, so it is important to make sure that the words in the question can be understood. Researchers often overestimate the familiarity of people with the topic they are studying, and as a result they word questions over the heads of respondents. For example, researchers often assume that the publicity given a bill in Congress means that the public will be familiar with it, but many people may not have heard of the bill and so will not understand questions about it.

Obviously questions should be stated simply enough that respondents can understand them, but there are several pressures that lead to writing complicated questions. The need to keep the interview short leads researchers to ask broad questions rather than the specific ones that would be easier to understand. For example, with the exception of economic surveys, most surveys cannot afford to ask a lengthy series of separate questions about a person's income from employment, dividends and interest, and so on. Instead surveys often simply ask about the person's income, letting people understand that term however they choose. A second pressure involves the need to write balanced questions, giving both sides of a public debate. The respondent is expected to keep both sides in mind when answering the question, though some respondents would not view the two as opposites.

Several additional problems intrude on question comprehension. Words in questions are sometimes open to interpretation, as when asking people how often they use butter and people reasonably wonder if that includes margarine use. Questions often presume something that does not fit with the respondent's understanding of a topic, such as questions that ask people to view something as a trade-off (as between national security and civil liberties) that they don't think of as a trade-off. Questions that include terms that people will not understand or that require careful definition also involve comprehension problems. Examples include asking people if they know anyone in a blended family (which presupposes they know that term) or asking the number of rooms in their house (which is a vague concept). Scales that ask people how often they do something (very often, often, not very often, etc.) suffer from severe vagueness problems (Bradburn and Miles 1979), and these terms are applied differently by different races and by different education and age groups (Schaeffer 1991). These problems can be handled in conversational interviewing, where the interviewer can help the respondent work through an understanding of the question, but in standardized interviewing they can lead to error (see chapter 4).

Respondents take the low road, exerting minimal effort at the comprehension stage, when they do not even bother to try to understand the question, just answering "Agree" or giving some other answer that fits the question without paying attention to its content. The solution mainly requires motivating respondents better, but it also helps to make sure that the questions are easy to understand. Survey questions should be tested to look for comprehension problems. A particularly interesting testing approach is measurement of *response latencies*—the time it takes to answer questions (Bassili 1996). Latencies are affected by the salience of the question but also by how difficult the item is to understand. Since people are likely to take longer to answer questions that require more effort to comprehend, questions with longer latencies are likely to be more problematic. Bassili and Scott (1996) provide evidence that people take longer to answer poorly worded questions, particularly questions stated in the negative and double-barreled questions (questions that actually ask two things at once; e.g., "Do you favor reducing illegal immigration to this country by building a border fence?" asks about both reducing immigration in general and a specific technique for doing so). Computer programs that grade the grammatical complexity of sentences can also be used to locate questions that are particularly likely to pose comprehension problems.

Comprehension involves not only understanding the question but distinguishing it from other questions in the survey. That topic will be addressed more directly in chapter 6, dealing with question order effects. As will be seen, sometimes the respondent feels that a question is asking for information similar to that provided in answer to a previous question (termed an "assimilation effect"), and sometimes the respondent believes the question asks for something different from what was asked in the previous question (a "contrast effect").

Retrieval of Information from Memory

The second stage is retrieval of information from memory. Retrieval depends on a number of factors, such as how well the topic has been encoded in the person's memory, how much experience the person has in retrieving it from memory, and whether the question provides sufficient cues to facilitate the retrieval (Alba and Hasher 1983). One might expect that newly acquired information would be the easiest to retrieve, but it may be least well encoded in memory and not previously retrieved, so that would not necessarily be the case.

ENCODING IN MEMORY. How information is retrieved from memory depends on how it was encoded in memory. Encoding is the forming of memories. Social psychologists speak of a variety of encoding processes. One theory is that sets of interrelated memories are stored in an organized cognitive structure called a *schema*. The full memory on a topic is not stored, but rather an abstracted version, a mental map or template, with interpreted meaning being stored rather than full facts. New information can be integrated into a schema to create a single cognitive representation. One implication is that recall can often be improved if the schema used to encode the information is invoked.

Another method for storing frequently experienced events economically in memory is called a *script*. Scripts give the expected sequence for frequently experienced events. For example, people have scripts regarding what to expect when going to a doctor's office or to a lecture class. Events that follow the script are unlikely to be recalled vividly, whereas departures from the usual script are more likely to be easily recalled. Both schema theory and the concept of scripts suggest that people will not be able to retrieve full details of events from memory because what they store is incomplete.

Forgetting is a problem for survey research, since people cannot retrieve information that they have forgotten. Further, it is not possible to distinguish whether someone has forgotten some information or never encoded it in the first place. What is important, though, is to utilize the appropriate cue to evoke the memory if it exists. As will be seen later, this sometimes involves asking the respondent to reconstruct the process leading to the behavior of interest, as a means of helping the person pull the information out of memory.

RECALL. Groves (1989, 422) states that recall of past events is affected by four factors: the length of the recall period, the salience of the event to be recalled, the task difficulty of the recall, and the respondent's attention or motivation. As would be expected, earlier events are usually not recalled as well as more recent ones.[3] This pattern has been verified in surveys using reverse record checks: official records (such as hospital records) are sampled, and then the researcher sees whether the respondent reports those events. More salient events are more likely to be recalled, though the dates of past events are generally not very salient. Recall will be worse for more difficult tasks, such as providing full details or combining several separate pieces of information. Finally, the respondent's attention and motivation affect recall, so respondents tend to recall better when they are asked to be very accurate. One approach to stimulating recall is to ask a long question in order to give respondents more time to search their memories. It has also been suggested that retrieval of an event from memory will be affected by the respondent's mood at the time of the event (Eisenhower, Mathiowetz, and Morganstein 1991).

Many survey questions ask people to retrieve elements about themselves in the past, often termed *autobiographical memory*. Questions about what they ate yesterday or where they vacationed last summer are examples of tasks calling on retrieval from autobiographical memory. Social psychology studies show the limits of such retrieval. For example, people are likely to report a script rather than specifics to some questions, such as answering what they usually eat in a day rather than what they ate yesterday (Smith, Jobe, and Mingay 1991). The occurrence of more events can interfere with memory of earlier events of the same kind, so that asking someone what they ate a week ago becomes very problematic; people may instead infer what they *probably* ate. Additionally, events that had high emotional impact are more likely to be recalled than bland events, and pleasant events are recalled more than unpleasant ones (Wagenaar 1986). A review of the literature on retrieval from autobiographical memory led Tourangeau, Rips, and Rasinski (2000, 97) to conclude that memories of personal experiences often get combined with inferences about what probably happened, particularly after the passage of time. Distinctive events are more likely to be encoded in such a way that they can be recalled better, whereas it is harder to distinguish among routine events.

A key distinction is between recall and recognition, with recognition requiring less cognitive processing. People often can recognize items that they would not be able to recall. For example, the first national political survey on congressional voting found that most people could not *recall* the name of their member of Congress (Stokes and Miller 1962), which led to the view that voters must be voting for Congress on the basis of political partisanship since they could not be voting on the basis of the candidates. Later work, however, showed that 93% of the public could *recognize* the name of their member of Congress (Mann and Wolfinger 1980), and since names are on the ballot and need not be recalled in order to vote, that meant that people could still be voting for Congress on the basis of the candidate factor.

Memory cues can be used in a survey to help people remember past behaviors. An example involves asking people how they voted for Congress. National Election Studies (NES) respondents routinely overreport their vote for House incumbents, even when shown a list of candidates. Box-Steffensmeier, Jacobson, and Grant (2000) demonstrate that this overreport is due to people's being more likely to recognize the name of the incumbent and that the overreport is reduced when the candidates' parties are included in the question or when they are next asked which party they voted for. Political party serves as a memory cue for a behavior that is difficult for many respondents to remember. Photographs and maps can also be useful as memory cues.

Saying that the person must retrieve information from memory does not mean that the person must do an exhaustive search. Krosnick and Alwin's (1987) low road is relevant here: "cognitive misers" will give minimally satisfactory answers rather than do additional cognitive processing. They will retrieve just enough information to make a reasonable decision, without going to the trouble of pulling up all relevant information from memory. An example of this problem is the "retrospective bias" that occurs when respondents asked what they *used* to think about an issue simply give their current attitude instead of bothering to try to recall their past views (Bem and McConnell 1974).

One mechanism that has been suggested for shortcutting a full memory search is keeping an "online tally" of reactions on particular topics. Take, for example, evaluations of political candidates. When voters are asked how much they like a particular politician, do they retrieve everything they know about that candidate from their memory? Or do they instead keep a running tally of their net impressions, which is updated whenever they absorb some new positive or negative information about that politician? The early research assumed that people should be able to retrieve everything they care about in forming their attitudes about a candidate, but later work (Lodge, McGraw, and Stroh 1989) instead shows that online tallies are used.

TEMPORAL EVENTS. The retrieval stage is also relevant to questions about past events, as in asking people how often they went to the doctor in the past year. This type of question will be examined in detail below because it is susceptible to several potential problems.

The term *telescoping* has been adopted to refer to respondents' reporting events as occurring earlier or later than they actually occurred. To give an idea of the potential magnitude of this problem, one survey found that about 40% of household repair jobs were subject to telescoping errors (Neter and Waksberg 1964). As will be discussed later, the most common form is forward telescoping, in which the respondent compresses the time period. One technique for handling this problem is *bounded recall* tasks in panel surveys, in which people are asked about the events that happened since the last interview. Another technique is *landmarking*, in which people are asked about the events that happened since a notable day, such as a holiday (Loftus and Marburger 1983). Calendars can also be useful aids for improving recall of the timing of events. Other techniques are to ask the respondent to retrace occurrences from an initial period to the point of interest and to ask the person to remember other activities that they did at the time of the behavior being asked about.

There are other possible effects associated with memory tasks in multi-wave panel surveys. The term *conditioning* has been adopted (Neter and Waksberg 1964) for the situation in panel surveys where the number of events reported decreases with more interviews, possibly because respondents want to shorten the interviews. If respondents learn that they will be asked several questions about each doctor's visit they report, for example, they may choose to report fewer visits in order to decrease the "respondent burden." Another possible problem in panel surveys is a *reactive effect*: being asked questions in one wave affects subsequent behavior that is reported in the next wave. For example, Clausen (1968) shows that one reason that NES postelection surveys find higher voter turnout rates than official data show is that its preelection surveys motivate some people to go to the polls who would not otherwise vote.

PROXIES. If these theories show the limitations of people's ability to answer questions about themselves, the ability to answer questions about other people is even more limited. The *respondent rule* in a survey consists of the criteria for eligibility to answer the questions. Three types of respondents can be distinguished: a self-respondent, a "proxy respondent" who answers questions for the selected person, and a "household respondent" who answers questions about the household as a unit. People are usually expected to report only their own attitudes, since they would not know what views the question evokes in other people. However, other people may be better able to report some other information, such as health matters; proxies have been found to report more health problems than do the people themselves (Mathiowetz and Groves 1985). Also, some roles, such as grocery shopping and banking, are often differentiated in households, so it is important to ask the person who would know the most about the behavior in question.

Hoch (1987) lists three factors that affect how well people answer questions about other people: (1) their own attitude or behavior, (2) how similar they believe the other person is to themselves, and (3) other information such as statements the other person has made in conversation. Sudman, Bradburn, and Schwarz (1996, chap. 10) interpret the overall evidence about proxy reports as positive, suggesting that proxy reports are not less accurate than self-reports in many cases, even for some attitude measures, depending on the relationship between the proxy and the person. However, there are clear limits to the accuracy of proxies. In the political field, for example, a 1965 survey of high school seniors and their parents (Niemi 1974) shows that each generation tended to assume that the other had the same party identification that they had, inferring from their own attitudes. Additionally, the students tended to assume that their parent's party identification was the same as their vote, presumably because they were more likely to have heard their parents discuss how they were voting than their partisanship. Blair, Menon, and Bickart (1991) suggest that proxy reporters may be more likely to satisfice but less likely to give socially desirable answers. Finally, proxies are less likely to know sensitive facts about the other person, especially if the distance between the two people is large.

Respondents in "establishment surveys" (surveys of businesses) can similarly be seen as proxy respondents in that they are answering for the establishment rather than for themselves. Ideally, those respondents use official records to fill in answers, but in fact respondents are likely to vary in their level of access to the requested information and may instead use their best guesses. Establishment surveys often try to get around this problem by specifying the job title of the person who the researchers expect to be most knowledgeable and/or have the greatest access to the needed records, but it is someone in the establishment itself who decides who will fill out the questionnaire. Respondent motivation is an important concern, particularly when filling out questionnaires takes employees away from money-earning activities. Many establishment surveys are government surveys, but that does not guarantee that care will be taken in filling them out, except possibly in situations where false information can result in fines.

Judgment of Appropriate Answer

The third component in answering questions is using the retrieved information to make a required judgment. The judgment process differs for different types of questions. Tourangeau, Rips, and Rasinski (2000, 10–13) describe the judgment stage for factual questions as involving five processes that respondents use to combine and/or supplement the information they have retrieved from memory. These include judging the completeness and accuracy of the retrieval, making inferences from the retrieval process itself (such as that the event being remembered may not be real if it was very hard to recall), trying to reconstruct memories to fill in gaps in the retrieval, putting the separate retrievals together into a single judgment, and adjusting estimates for omissions (such as remembering recent events and then extrapolating from their frequency to the full period being asked about).

Judgments for attitude questions are seen as similar, with people retrieving considerations from memory and then amplifying those considerations in line with these processes. However, judgments also depend on specifics of the question wording. Intriguingly, Wänke, Schwarz, and Noelle-Neumann (1995) show that the direction of a comparison affects judgments as to the appropriate answer. For example, in a mall intercept survey, when asked, "Would you say that industry contributes more or less to air pollution than traffic?" only 32% said industry contributes more versus 45% saying traffic, whereas when asked, "Would you say that traffic contributes more or less to air pollution than industry?" 57% said industry contributes more versus 24% saying traffic. The authors interpret the results as showing that people bring different things to mind in the two comparisons, taking into account unique features of the first item while neglecting features of the second that the first does not bring to mind.

By contrast, judgments of the frequencies of events are made by such processes as recalling an exact count, recalling specific events and counting them, estimating from occurrence rates, and estimating from general impressions including both guesses and using the middle response category as an anchor and adjusting from it (Tourangeau, Rips, and Rasinski 2000, 146–47). As to event frequency, Blair and Burton (1987) suggest five considerations involved in which estimation strategy people employ: the amount of effort required for recalling and counting each event, the motivation for doing so, how accessible the event-specific information is, the availability of such information as a preexisting tally or rate in memory, and the nature of the task in terms of question wording and instructions.

The low road at this stage would be to accept the first retrieval from memory as the answer to the question, without making an effort to judge its accuracy. This would occur when respondents respond to a question with the first answer they think of, without taking the trouble to think whether that really is their best answer. This occurs when people say their spouse is about fifty years old without thinking through his or her exact age, or blurt out their first thought about a political candidate without weighing their other views about that candidate, or give a rough count of how many times they've seen their doctor in the last three months instead of counting carefully. This is why it is important that the survey and the interviewer try to motivate respondents to give good answers, not just their initial thoughts.

Tversky and Kahneman (1974) argue that people use *heuristics* in their decision making—shortcuts to reduce their cognitive processing. They describe three heuristics that respondents can apply in judging appropriate answers. The availability heuristic is the use of the most accessible memory, which can cause poor judgments if the most accessible memory is atypical. For example, events are judged to be frequent if instances of them can easily be retrieved from memory; people may say, for example, that they go to concerts more often than they really do if one instance was particularly memorable. Tversky and Kahneman's second heuristic is "anchoring and adjustment": people choose a preliminary estimate and then adjust it to the specifics of the question. Their last heuristic is a "representativeness" heuristic, which can lead to overgeneralizing from small samples.

An interesting example of the representativeness heuristic was found when Nadeau, Niemi, and Levine (1993; see also Sigelman and Niemi 2001) looked at respondents' estimates of the size of various minority groups. Blacks were 12% of the U.S. population according to the 1990 census, but three-quarters of the respondents in a 1991 NES study gave answers of 20% or higher when asked, "In the country as a whole, what percent of the U.S. population today would you say is black?" Similarly, Census Bureau figures were that 9% of the population was Hispanic, but three-fifths of the respondents answering the question gave estimates of 15% or higher. Only 2–3% of the U.S. population is Jewish, but three-fifths of respondents estimated it as 10% or higher, and one-fifth thought at least 30% was Jewish. In accord with the representativeness heuristic, people who lived in areas with concentrations of those minorities were particularly likely to overestimate, as were people whose answers to other questions showed that they felt threatened by these groups.

Probabilities and likelihoods are very difficult for many respondents to estimate. Experiments show that people overestimate low risks and underestimate large risks (Lichtenstein et al. 1978). Similarly, people overstate the differences between different events that are all very unlikely to occur and also exaggerate the differences between different events that are all virtually certain to occur (Tversky and Kahneman 1979). Furthermore, probability estimates often do not follow the laws of probability. Probabilities of mutually exclusive and exhaustive events must sum to one, but many people give answers that sum to greater than one when NES surveys during presidential nominating campaigns have respondents estimate the separate probabilities of several different candidates' winning their party's nomination.

Selecting and Reporting the Response

The final component of the answering process is selecting the response. After people make a judgment, they must decide which survey answer comes closest to that judgment. In part, this is a mapping problem, deciding how judgments map into the available answers, such as whether "strongly disagree" or "disagree" better fits the person's position on an issue statement, or whether the respondent's rating of a political personality is 80 or 85 degrees on a 100-degree thermometer scale. The considerations involved in this process vary for different types of questions, so they will be taken up within a later discussion of open- and closed-ended questions.

EDITING RESPONSES. Another part of this final stage is editing the response. There are several psychological reasons that respondents may feel that they cannot give their true answer to a question and instead edit their answer. An example is the *acquiescence bias*, in which people are more likely to agree to a question regardless of its content. Comparing across several studies, Krosnick (1999) estimates the average magnitude of the acquiescence effect to be about 10%, such as by finding that across ten studies 52% agreed with an assertion while only 42% disagreed with the opposite statement. The acquiescence bias has been explained in several different ways: some people's being predisposed to be agreeable, deference of respondents to higher-status interviewers, satisficing when respondents find reasons to agree with a statement and are not motivated enough to think through reasons to disagree with it, and satisficing without any evaluation of the question.

The acquiescence problem can be handled in at least two ways. A common method is to phrase half of the questions that are to be used in an index or scale in a positive way and half in a negative way, so that always agreeing does not lead to an extreme score. Another way is to phrase the question in a balanced manner in a forced-choice format, without giving either side a wording advantage (Converse and Presser 1986, 38–39). An example of the difference that balanced questions can make involves how the phrasing of issue questions was changed in the NES surveys in the mid-1960s. An overwhelming proportion of people answered "yes" in the 1950s NES surveys when asked if they favored federal aid to education. Surveys in the 1960s, taken after the federal government began funding education programs, found a lower proportion of people taking that position when asked, "Some people favor federal aid to education while other people feel that education should be a state or local matter, what about you?" Posing the question with two balanced alternatives eliminated the acquiescence bias that contaminated answers to the unbalanced question in the 1950s. Schuman and Presser (1981, chap. 7) find that simply adding "or do you oppose this" to a "do you favor ___" question does not change the distribution of answers, but adding substantive counterarguments leads to shifts of 4–13% of respondents.

A second type of editing involves respondents' trying to provide consistent answers. If respondents feel that they have already given an answer to a related question, they may choose an answer to the current question that fits with that previous answer. Researchers often try to ask a series of questions on a topic so that they can scale answers to differentiate people who take extreme positions from those who are moderate, but people who are trying to give consistent answers are likely to end up being scaled as extreme.

Another important type of editing has to do with *social desirability effects*, which occur when respondents feel that their true answer would not be acceptable to give to the interviewer and substitute a more socially acceptable answer. An example of this would be in reporting racist attitudes: many respondents with such attitudes may feel that they cannot report them to the interviewer. This effect has been seen in polling for several political races in which a black candidate had a steady lead in the polls but lost to the white candidate in the election, such as when Douglas Wilder lost the Virginia governorship in 1989. A study of the Wilder race found a social desirability effect of about 10% on the vote intentions of white respondents associated with the race of the interviewer (Finkel, Guterbock, and Borg 1991).

The extent of social desirability effects depends on both the respondents and the questions. Some respondents have greater needs for approval from the interviewer than others, and some questions deal with more sensitive topics than others. The result is an underreporting in surveys of activities that society considers undesirable, including drug use, alcohol consumption, smoking behavior, criminal behavior, and so on.[4] Social desirability effects also lead to overreports of such activities as church attendance and voting turnout at elections, with people not being willing to admit to an interviewer that they did not do activities that they should have done. Social desirability effects can be due to respondents' either trying to present themselves to the interviewer in a positive light (impression management) or trying to preserve their own self-esteem (self-deception).

A related problem is *fear of disclosure effects*: respondents may edit their answers for fear of the consequences of providing a true answer. This problem occurs in establishment surveys when companies want to make sure that some information does not become available to their competitors. It also affects personal interviews when people fear that legal authorities might use their answers against them, as in checking for illegal aliens or tax cheaters.

MINIMIZING SOCIAL DESIRABILITY EFFECTS. The simplest method to minimize social desirability effects is to include reasonable excuses (known as "forgiving words") in the question wording, to make it easier for the respondent to give a true response. For example, questions about voting turnout are usually worded to make it easier for people to indicate that they didn't vote, as in the NES question before 2000: "In talking to people about elections, we often find that a lot of people were not able to vote because they weren't registered, they were sick, or they just didn't have time. How about you—did you vote in the elections this November?"[5] Or the excuse could be made into one of the potential answers, such as asking people whether they voted in the election, with choices of "yes," "no," and "I usually vote but did not this time."[6] In order to avoid social desirability effects on sexual behavior questions, Kinsey's original research (Kinsey, Pomeroy, and Martin 1948) "loaded" the question wording to assume that respondents had engaged in the behavior of interest, as by asking people the frequency with which they masturbated rather than whether they masturbated. Longer questions lead to a larger number of reports of socially undesirable behavior, and open-ended questions are better than closed-ended questions when asking about frequencies of those behaviors (see the discussion of both types of questions later in this chapter), so as to avoid people's tendency to avoid the extreme categories on a list of frequencies (Bradburn, Sudman, and Wansink 2004, 103–4).

In face-to-face surveys it is possible to have people check off their answers on a secret ballot, which they are told to seal and/or put into a ballot box. The Gallup Organization found fewer undecided voters when using such a system (Benson 1941). There is also evidence that fewer people give socially desirable answers when a secret ballot procedure is used (Turnbull 1947). This system is often used in election exit polls when the only question of interest is how the person voted, and it produces results that are closer to official election returns (Bishop and Fisher 1995). More generally, self-administered questionnaires give fewer social desirability effects than interviewer-administered surveys (chapter 12).

The usual advice is that social desirability effects are lessened when interviews are conducted without other people being present, since the respondent does not feel pressured to give responses that are acceptable to people who are listening in. In practice, however, it is not always possible for the interviewer to control whether there is someone else present. An early assessment was that "survey data generally are immune to the presence of third parties" (Bradburn and Sudman 1979, 146), but it would be more accurate today to summarize the literature by saying that there are few effects of third parties' presence.

Studies focus particularly on spouse presence during the interview. Smith (1996) found that spouse presence had virtually no effect on questions relating to gender roles, marriage, and sexual attitudes in the General Social Survey. Similarly, Pollner and Adams (1997) found no significant main or interaction effects involving spouse presence in a Los Angeles survey in which emotional support from the spouse and household strain were examined. However, analyzing the National Survey of Families and Households, Aquilino (1993) found spouse presence led to slightly more positive answers on the utility of marriage and husbands' giving lower probabilities of divorce, but greater tendency to report sensitive facts, such as cohabiting before marriage. In a British study, Zipp and Toth (2002) do find that spouse presence increased agreement between the pair on several attitudinal items, including party affiliation, as well as on division of household labor, but did not lead to more socially desirable answers (such as claiming that women do less ironing). Thus the direction of effects due to spouse presence depends on the question, and often there is no spousal effect.

Several techniques have been developed to minimize social desirability effects in surveys on sensitive topics (also known as "threatening questions"). The "bogus pipeline" technique makes respondents think that inaccurate answers can be detected. In laboratory research, subjects are sometimes wired to a physiological device and are told that the device can detect their true feelings. A meta-analysis of studies using this approach found that it does reduce social desirability biases, but that researchers should seek to obtain the subjects' estimate of the output of the device and the extent to which they believe the explanation, in order to control for the subjects' motivations and attentiveness (Roese and Jamieson 1993). That meta-analysis also found that that the use of this procedure in attitude measurement can lead to crystallization of attitudes that were only partially formed, because the procedure seems to be so elaborate as to deserve deep thought. A less extreme version of this procedure involves asking respondents for permission to check an official record to validate their claimed behavior, such as a voting turnout record. Accuracy of responses improves, even when researchers do not actually carry through with the validation. For example, more teenagers report smoking when they believe that false answers can be ascertained (Murray et al. 1987).

Another technique for minimizing social desirability effects is the "item count method" (Miller 1984). Respondents are shown a list of activities and asked how many of the activities they have done. The list includes the sensitive behavior of interest for only a random half of the respondents. The difference in the number of activities that the two halves of the sample perform gives an estimate of what proportion of the sample has done the sensitive activity. For example, the question might ask people how many of the following activities they have done in the past month: gone to a movie, been in a car accident, dieted, and exercised; half of the people are given the added alternative of using cocaine. If the average answer for the control group is 1.81 while the average answer for the group with the drug item included is 1.92, the difference estimates that 11% of the sample (1.92 − 1.81) used cocaine. This technique estimates the proportion of respondents performing the sensitive behavior, but it does not allow the researcher to know which respondents performed that behavior. Additionally, it leads to a larger variance, since only half of the sample is being asked about the sensitive behavior.

Warner's (1965) randomized response technique provides another method for handling sensitive topics by ensuring that the interviewer does not know the meaning of a respondent's "yes" or "no" reply to a question. People might be asked to perform a randomization technique (such as tossing a die), without telling the interviewer the outcome (which face came up), and then to answer one of two questions depending on the outcome. The original version had the person give either a true or a false answer to a sensitive yes/no question, depending on the outcome, such as giving a true answer if the die came up 1 or 2 and a false answer otherwise. An unbiased estimate can be obtained in this situation, so long as the random device has not been set at a 50% chance of giving the true answer. A later version instead has people answer either the sensitive question or an innocuous one (e.g., "Were you born in January?"), depending on the outcome of the randomization.[7] Again, this technique tells only the proportion of people who engage in this behavior, and the variance is much higher than if the question were asked of the whole sample, though less than with Warner's original version. In any case, the randomized response technique is so unusual that respondents do not know what to make of it, so many still suspect that the researcher will somehow be able to tell if they engaged in the behavior. Still, Zdep et al. (1979) find that it gives plausible results in terms of proportions using marijuana, such as higher usage among younger people and by men.

Summary

The four components of the response process provide a useful way to think about how respondents process questions. When one is writing survey questions, it is important to focus on each step, making sure that the questions can be comprehended, make reasonable retrieval demands, will lead to appropriate judgments, and will not lead to severe editing by respondents.

These issues are important for attitude questions but also for fact-based questions. Take, for example, the question of the person's income (Moore, Stinson, and Welniak 1999). First, respondents must comprehend what is being asked in terms of what constitutes income—their own income or the total family income, their main job or that job plus assorted odd jobs, wages or also interest and dividends, and so on. The respondents' memory storage may be challenged at the retrieval stage, since people may store only some parts of the desired information, such as their hourly wage rather than the income total they reported on their last year's income tax return. At the judgment stage, the addition of the several subcomponents of income might be too challenging a task. Finally, at the reporting stage, many people regard income as a sensitive question, so they may refuse to respond or may purposely misreport either by overstating a low income in order to save face or by understating a high income if they distrust the confidentiality of the survey.

Two-Track Theories

Another perspective on the response process is a multitrack theory that suggests there is more than one possible route to developing an answer. There are several two-track theories, with one track being a high road that involves real thinking and the other being a low road that entails a more casual processing of the question.[8] These two-track theories are really about respondent motivation. They assume that respondents could answer questions accurately if they wanted to, so the problem for the survey researcher is to motivate respondents enough to use the high road.

An early two-route theory was Cannell, Miller, and Oksenberg's (1981) process theory. Their high road involves careful processing, based on comprehending the question, cognitive processing, evaluating the accuracy of response, evaluating the response based on the respondent's other goals, and then reporting the answer to the interviewer. By contrast, their low road involves superficial processing, when answers are based on superficial features of the interview situation, such as the interviewer's appearance, which could lead to various biases in response. Another two-track theory is Strack and Martin's (1987) distinction between making a new judgment versus basing answers on existing judgments. Making a new judgment can be considered as following a high road through the steps in the Tourangeau theory, whereas just giving an existing judgment is seen as a low-road response that does not involve much work for the respondent.

The most influential two-track theory was developed by Krosnick and Alwin (1987). Their high road is an *optimizing* approach, in which the respondent works through the best answer to the question, presumably in line with Tourangeau's four steps. Their theory is best known for their low road: a *satisficing* approach in which respondents seek to understand the question just well enough to provide a plausible answer. The satisficing concept was developed by Herbert Simon (1957), who suggested that this is a common economizing strategy in decision making. Decision makers often do not want to expend enough resources to learn fully about all possible alternatives, so they stop getting information when they have obtained enough to make a reasonable decision, as, for example, when consumers get enough information to decide to buy a particular car without getting full information about all cars on the market. Applying the satisficing notion to surveys is an acknowledgment that many respondents are "cognitive misers" (Taylor 1981) who try to minimize the amount of processing they must perform.

The notion of respondents as satisficers is captured in a quote from Tourangeau, Rips, and Rasinski (2000, 315): "We view respondents as pragmatists—at times, as opportunists—in their approach to these mental steps. Respondents have relevant experience in dealing with a range of questions in everyday life, and they use the strategies they've developed in the course of this experience to make sense of the questions in surveys and to satisfy the perceived demands of the interview." Satisficers are opportunists who adopt strategies that get them through the interview relatively painlessly.

Krosnick (1991) differentiates two forms of satisficing in surveys. *Weak satisficing* occurs when respondents go through the four steps described in the previous section but offer the first answer they consider that seems acceptable (such as the last alternative in an oral list of response options). Simply agreeing with the question is also seen as weak satisficing, since respondents are thinking of reasons they could agree with the statement and stop as soon as they come to such reasons without bothering to think about reasons to disagree. *Strong satisficing* instead involves skipping the retrieval and judgment steps and proceeding to interpret the question and select an answer that would appear to be reasonable. Respondents use the question wording as a cue to an answer that could easily be defended and choose such an answer without reference to their own attitude, belief, or behavior. Giving the same answer to all alternatives in a rating scale would be an example of strong satisficing, as would be agreeing with all agree-disagree items, choosing the status quo or middle of rating scales, choosing answers randomly from available response alternatives, and giving "don't know" answers.

Krosnick (1991) hypothesizes that respondents will be more likely to satisfice when the task is more difficult, when their own abilities are lower, and when their motivation to optimize is lower. Task difficulty involves such factors as the complexity of the question, challenges with retrieving information, the interviewer's pace, and the occurrence of distracting events during the interview. Respondents who can perform complex mental operations, have given thought in the past to the question, and have preformulated judgments on the topic stored in memory have higher abilities. Motivation is affected by the respondents' need for cognition, the personal importance of the topic, beliefs as to the importance of the survey for society, whether the interviewer encourages optimizing, and fatigue from answering prior questions in the survey.

Education and age are seen as demographic factors related to satisficing. Education is important because it affects the sophistication of cognitive processing as well as the respondent's knowledge base. For example, a common finding is that less educated respondents give more "don't know" responses. Older people may satisfice more because of memory loss with aging, but age effects could simply be due to lower educational levels among older respondents. Two meta-analyses illuminate these sources of differences by reanalyzing many experiments on question wording effects reported in Schuman and Presser (1981). Narayan and Krosnick (1996) find education effects, with lower education being related to acquiescence and other satisficing behavior. Hypothesizing that the Narayan and Krosnick result was due to education differences across generations, Knäuper (1999) tested for age effects and finds response order effects (choosing the first or last offered choice) were influenced by age, though education may have a separate effect. Both education and age differences can be interpreted in terms of sophistication of the respondent and working memory capacity (Knäuper 1999, 367, reporting on a personal communication with Krosnick), so both education and age can reasonably be seen as affecting satisficing.

Types of Questions

The following sections examine specific memory issues encountered with two important types of questions: attitude questions and temporal questions.

Attitude Questions

Constructing valid attitude questions should be based on an understanding of what attitudes are. Attitude research originally believed that people have "preformed attitudes" that exist independent of being interviewed. The *automatic activation* approach assumed that survey questions automatically activate the relevant attitude, so asking an attitude question would obtain meaningful responses. In rejecting this model, Wilson and Hodges (1992, 38) term it a "file drawer" model since it treats people as if they keep their attitudes on particular topics in mental files that can be opened whenever they are questioned about them.

As shown below, this theory does not fit well with later research by social psychologists. Additionally, answers to survey questions sometimes do not reflect real attitudes, as shown when Wilson and Hodges (1992) found that asking people the reasons for their attitudes can cause people to change those attitudes. Furthermore, many people will answer attitude questions even when they do not have an attitude on the topic. They may feel that the interviewer is expecting an answer and that they will be judged as more intelligent if they answer the question rather than say "don't know." Take the question "Do you approve of U.S. foreign policy toward Madagascar?" It is hard to imagine that many Americans know U.S. foreign policy toward Madagascar, or even know where Madagascar is, so people may just answer on the basis of whether they like the general outlines of U.S. foreign policy.

The Nonattitude Problem

Philip Converse (1964) has termed responses such as those on foreign policy toward Madagascar *nonattitudes*. Converse found anomalies in some policy domains when he looked at the correlations between people's responses in different waves of a panel study. In particular, the correlation between people's positions in 1960 and 1956 was the same as the correlation between their positions in 1960 and 1958, whereas an orderly attitude change process would lead to lower correlations between more distant time points. He reasoned that there could be three basic types of people in such panel studies: those whose attitudes stay the same, those whose attitudes change systematically (such as from opposing a policy in the first study to becoming more and more positive toward it in later waves of the panel study), and people whose positions change randomly. Converse was able to fit a model with only the first and third groups (no change and random change) to data patterns on some policy issues, so change that was found was random rather than real. Because of the prevalence of random changes, he argued that some answers to those questions were nonattitudes. In Tourangeau's model of the response process, this corresponds to not having an attitude to retrieve.

There has been debate over the years about Converse's interpretations. Achen (1975), in particular, claimed that the problem was really measurement error in the questions. Achen argued that a person's views are really a distribution of opinions rather than a single fixed opinion and that variability is exacerbated when survey questions are vague. Still, Converse's basic point serves as a warning for attitude measurement: people may not have attitudes on some topics, and survey research is misleading if it measures nonattitudes.

Schuman and Presser (1981, chap. 10) imply that the nonattitude problem, as seen by low correlations on the same question over time, is partly one of lack of attitude crystallization. Their experiments find higher response consistency for people with stronger attitudes. As a result, they urge that measures of attitude strength and/or centrality be included in surveys.

As an example of the nonattitude problem, Bishop, Oldendick, and Tuchfarber (1980) included a question in a survey about whether the 1975 Public Affairs Act should be repealed. One-third of respondents answered the question even though that act was fictitious. In another experiment on little-known policy questions, such as the Monetary Control Bill of 1979, Schuman and Presser (1981, chap. 5) find that the least educated are more likely to answer these questions whereas the more educated are more willing to admit ignorance when questions are truly impossible.

A real example of this problem occurred in mid-2001, when the topic of human embryonic stem cell research first arose (Clymer 2001). Media polls were showing support in the 54% (Gallup Poll for CNN and *USA Today*) to 69% (NBC News/*Wall Street Journal*) range. However, when first asked a screening question to ascertain if they knew enough to say whether they favored or opposed stem cell research, 57% of the Gallup Poll respondents admitted that they did not. Results of polls sponsored by interest groups varied more widely, from a high of 70% favoring research in a survey sponsored by the Juvenile Diabetes Foundation, as part of the

Coalition for the Advancement of Medical Research, to a low of 24% in a survey conducted for the Conference of Catholic Bishops. Question wording becomes all-important in this instance. The question used in the Catholic bishops' poll indicated that "the live embryos would be destroyed in their first week of development" and spoke of using federal tax dollars for experiments, whereas the NBC/*Wall Street Journal* question spoke only of "potentially viable human embryos." Asking about this highly technical policy area before there had been much publicity on it virtually invited people to respond with nonattitudes, with a high potential effect of loaded question wording.

Nonattitudes can pose a serious problem for survey research, but even tentative attitudes can be worth measuring. Blais et al. (2000) provide evidence that even people who profess to know nothing about a topic may be influenced by it. Looking at feeling thermometer ratings of party leaders in Canada, they show that those ratings affect the vote, even for people who say they know nothing about those leaders. The authors interpret this finding as showing that much of the fluctuation Converse observed was not evidence of nonattitudes but instead is indicative of very tentative attitudes that can have behavioral effects.

The prevalence of nonattitudes leads some people to argue against an overreliance on surveys. People answering survey questions do not have the expert information and understanding of trade-offs that policy makers possess, so the argument goes that polls should not determine public policy. At the same time, of course, it is important that democracies be responsive to the public, so it would be inappropriate to dismiss poll results out of hand. What is necessary is to take into account the level of the public's knowledge on a topic, recognizing that views expressed when knowledge levels are low are primarily nonattitudes.

Attitudes as Temporary Constructions

Many contemporary scholars no longer view nonattitudes as a major problem, though they still view attitudes as "temporary constructions" (Wilson and Hodges 1992) made to satisfy interviewers. They regard answers to survey questions as what happens to come to the "top of the head" when the questions are asked, in line with Taylor and Fiske's (1978) argument that many judgments are based on whatever information is most easily brought to mind. These answers may not be random, but they are not preinterview positions. This can be viewed as a theory of the retrieval process (Tourangeau's second component). This approach to attitudes was developed fairly simultaneously by Roger Tourangeau and by John Zaller. Their work will be considered here together because of the similarity of their positions.

Tourangeau, Rips, and Rasinski (2000, 178–94) call their approach a *belief-sampling* theory of attitudes. They hypothesize that people have several beliefs about a topic and that they take a sample of those beliefs when asked to give their answer to a survey question. While Tourangeau, Rips, and Rasinski write of judgments as averages of the beliefs that are evoked by a question, they argue that the process is actually one of successive adjustments. The first belief sampled forms the basis of any judgment, which is then adjusted when the next belief is brought to mind, and so on in a process that leads to a result equivalent to an average.

In Zaller's (1992) formulation, attitudes are based on *considerations* that come to mind when people consider survey questions. Considerations are anything that might cause a person to choose one position or another when asked an issue question. These considerations combine cognition and affect, such as "The president's proposal to handle domestic terrorism is reasonable." To Zaller, a number of considerations may be relevant when people think about any topic, and which they retrieve is based on which are most accessible at the moment. Zaller feels that people average the considerations they retrieve and base their answers on this average.

This view of attitudes as temporary constructions has implications for attitude change in panel surveys, providing an alternative explanation of the patterns that led to Converse's idea of nonattitudes. People might draw on different considerations at different times or might weight the considerations differently and therefore might report different attitudes even if the same basic considerations were present (Zaller 1992), but are more likely to give the same answer twice if the considerations that are evoked at the two points are highly correlated (Tourangeau, Rips, and Rasinski 2000). However, asking people to think deeply about the question before answering can actually lead to less consistent answers over time, because people with a larger set of considerations may draw inconsistent considerations to mind.

Studies by Zaller and Feldman (1992) and by Tourangeau and Rasinski (reported in Tourangeau, Rips, and Rasinski 2000, 186–94) support the considerations approach and find evidence for its implications. Respondents in panel studies were asked to respond to attitude scales and to report what came to mind as they considered the questions. The correlation of attitudes across panel waves was very high for people with consistent considerations but more moderate for people with balanced considerations. The Tourangeau and Rasinski work included an experiment in which people first read arguments about both sides of a fictitious issue, and the correlations were lower among people who were given more arguments to read.

This view of attitudes agrees with the nonattitude interpretation that survey respondents create answers on the spot. It differs from the nonattitude view, however, in that it is memory based, assuming that there is a lot of information to retrieve from the person's memory. Memory-based processing is thought to be particularly common as regards attitudes on issues. When surveys find differing results on issues, this theory implies that one reason is because different events that occur before the surveys lead to different considerations' becoming salient, even when underlying attitudes are actually quite stable.

Other Attitude Theories

Some additional theories of attitudes are worth mentioning. The first is that attitudes are *impression driven.* Sanbonmatsu and Fazio (1990) argue that attitudes on some topics are "on-line attitudes" (see also Hastie and Park 1986). An example is Lodge, McGraw, and Stroh's (1989) experimental evidence that evaluations of political candidates are overall impressions that are updated as people pick up more information. People do not store in memory the specific pieces of information that led to their evaluations, but they can retrieve and report their overall tallies. According to this position, survey answers are preformed, not created on the spot as in nonattitudes, but the explanations people give of their evaluations are rationalizations since they cannot remember the underlying information. On-line processing is thought to be particularly important in forming judgments about people.

Another position is that attitudes are based on respondents' predispositions, such as their values and group attachments. Response variability can occur because people acquire new information and because they hold multiple predispositions simultaneously. Alvarez and Brehm (2002) differentiate conditions under which holding multiple predispositions leads to ambivalence, uncertainty, or equivocation on the part of survey respondents. They describe ambivalence as occurring when the person's values on a particular issue are in conflict and more information widens response variability, as when views on religion and adults' autonomy over their own choices conflict on such topics as abortion and euthanasia. Uncertainty exists when values are in conflict, but more information narrows response uncertainty, as Alvarez and Brehm argue is the case as regards racial attitudes. Equivocation results when people have multiple predispositions without recognizing that they cannot be achieved simultaneously, as when people favor both cutting taxes and increasing government services without seeing a tension between these objectives.

Putting together these different perspectives on attitudes: many attitudes are meaningful. Preformed attitudes obtained when people have high levels of information are not the only type of meaningful attitudes. Attitudes derived from memory-based processing, as described by Zaller, are meaningful, even if the person's answer to the survey question is created on the spot, and answers based on on-line processing are meaningful, even if the person has little information in active memory. Nonattitudes certainly do exist, but they are not necessarily prevalent, particularly on established issues.

In the end, surveys will certainly continue to ask large numbers of attitude questions, but it is always important to be wary about accepting specific results. The meaning of finding that 58% of the public supports some policy depends on how much people know about the topic and how the question was worded. As a result, it should not be surprising if a later poll reports a different result on the same topic, since public knowledge may have changed and the question may have been worded differently.

Temporal Questions

Time-related questions pose different memory problems involving the retrieval stage of the Tourangeau four-stage model. There are several types of temporal questions. Time-of-occurrence questions ask when something happened, such as "In what year did you first have back pains?" Frequency questions ask how often an activity occurred, such as "How many times have you been to a doctor this past year?" Elapsed-time questions ask how long it has been since an activity, such as "How long has it been since you last went to a dentist?" Duration questions ask how long some activity lasted, such as "How long did your last migraine headache last?" Each of these questions assumes that people organize their memory in time terms, but some of these tasks are very demanding.

Tourangeau, Rips, and Rasinski (2000, 109) list four kinds of information that people retrieve in answering temporal questions: exact information (particularly dates of occurrence), relative order (such as after another event that can be dated more easily than the one being asked about), other details which lead to temporal implications (such as an event's occurring when there was snow on the ground), and impressions due to the retrieval process (such as a weak memory due to the event's having occurred much earlier).

Time Frame Issues

Not only do people forget past activities, but also they have a hard time relating memory tasks to a particular time frame. For example, asking how many times an activity has happened in the past year is awkward since people do not organize their memory on that basis. Bachman and O'Malley (1981) compared self-reports of alcohol and drug use by students in the previous month and in the previous year and found that the average monthly rates implied yearly rates two to three times as high as were reported. This could be due to uneven usage over the course of the year, but a more likely interpretation is that respondents were understating their usage over the year because of memory problems. The clear implication is that it is best to keep the time frame short in such memory questions. Asking how many times the person went to the doctor in the last three years would be an unreasonable task, whereas a three-month time frame would be reasonable.

Asking people when something happened is also susceptible to memory problems. People tend to think that some events occurred more recently than they actually did (which is termed *forward telescoping*) and that other events occurred earlier than they actually did (backward telescoping). For example, an analysis by Johnson, Gerstein, and Rasinski (1998) finds that because of forward telescoping, early use is underreported when people are asked their age when they first used alcohol. Sudman and Bradburn (1973) explain telescoping by a time compression theory in which people consider events over a longer time period than they are being asked about, whereas Rubin and Baddeley (1989) instead explain it with a variance model in which people are more uncertain about the timing of events that are more remote in time, so errors are most likely to occur on earlier events. Similarly, panel studies that ask people to relate events in each month since the previous wave find more change between months asked about in different waves than between months asked about in the same wave, a telescoping phenomenon known as a *seam effect*. Tourangeau, Rips, and Rasinski (2000, chap. 4) provide a broad discussion of temporal questions in surveys, but they are not satisfied with any theoretical explanation of the telescoping phenomenon.

One solution to these problems is to pose temporal questions with reference to "landmark events" that stand out in the person's mind, such as "How often have you been to a doctor since Christmas?" Successful retrieval may depend on using the same calendar boundaries that the respondent does. Students and teachers are likely to organize their memories in terms of the school year and break periods between terms, but business professionals will utilize a different organization of their year. Using landmark events works to the extent that those events are anchored similarly for all groups of interest, though a British study challenges that assumption by showing systematic differences between groups in their dating of significant events, such as the resignation of a prime minister (Gaskell, Wright, and O'Muircheartaigh 2000).

Another solution used in some panel surveys is called *bounded interviewing*. For example, people might be asked in a first interview about which jobs they have held in the past year, and then they are reinterviewed every year asking which jobs they have held in the past year. The answers in the first interview are actually ignored, except as a comparison so that the researcher can tell how many job changes the person has had. As a means of reducing forward telescoping, *bounded recall* questioning reminds respondents of events they reported in earlier interviews before asking them for events since then (Neter and Waksberg 1964). Some ongoing computerized surveys store the person's most recent information from previous waves in the computer-assisted interviewing file, so that the person is actually asked, "In our last interview you were working as a sales clerk at ___. Are you still doing that, or has your job changed since then?" Making use of data from a prior panel wave is called *dependent interviewing*.

A related procedure is to use two time frames in a single survey, first asking about an earlier time period and then about a more recent one as a means of bounding the recall and reducing telescoping (Sudman, Finn, and Lannom 1984). Loftus et al. (1990) validated answers and found that there is less overreporting of medical procedures in the last few months when the respondent is first asked about a previous time period.

Frequency Issues

Surveys often ask people to estimate frequencies of events, such as how often they read medical stories in the newspaper. These questions require judgments, in the third stage of the Tourangeau four-stage theory. Tversky and Kahneman's (1973) work on judgment under uncertainty is relevant here, particularly the availability heuristic. What matters, according to this theory, is how available the information is to the respondent's active memory. If the information being asked about is relatively accessible, the person is more likely to assume that the event is frequent. If the information is not very accessible, the person will assume it is not frequent. The answer is thus based on the information that first reaches the person's working memory, possibly modified by conscious consideration of accuracy if the person bothers to think about whether that first information is typical.

The most important determinant of whether people count events or estimate their frequency turns out to be the number of items they are trying to recall (Burton and Blair 1991). For example, when asked how many times they have written checks in a given period of time, people who have written just one or two checks generally say that they have counted, whereas those who have written more than ten generally say they have estimated. Counting is probably more likely when the period being asked about is short, which suggests it is best to use short time periods if the researcher desires an actual enumeration rather than an estimated rate. Still, it is better to ask for frequencies of events, using categories such as "two or three times a week" when asking people how often they read the newspaper, rather than to use vague response categories, such as "sometimes," that people will interpret differently.

Survey Questions

The theories discussed in the previous sections and the accumulated experience of surveys lead to several principles for designing survey questions. This section will discuss question wording generally as well as problems associated with particular question formats. The multicomponent and two-track theories both have implications that should affect question wording and the choice of question formats. In particular, the satisficing problem should be kept in mind when one is preparing a survey. A variety of question formats should be used, so that the respondent has to pay attention to changes. Additionally, questions should not be all asked in the same direction, so that the respondent has to pay attention to what is being asked rather than merely agree to all the questions. The overall survey should be kept short, since satisficing becomes more likely when a long interview bores the respondent.

Question Wording

Caution is always necessary in question wording, because seemingly equivalent ways of asking questions can obtain different answers. An early example comes from surveys before the United States entered World War II (Rugg and Cantril 1944). By a 41%–33% margin, most people said "yes" to "Do you think that the United States will go into the war before it is over?" whereas by a 44%–30% margin most people said "yes" to "Do you think the United States will succeed in staying out of the war?" Another well-known early example involves the nonequivalence of "forbidding" and "not allowing." When asked, "Do you think the U.S. should forbid public speeches against democracy?" 54% of respondents said "yes"; however, 75% of the public answered "no" when asked, "Do you think the U.S. should allow public speeches against democracy?" (Rugg 1941).

The problem is that question wording is tied directly to measurement error. Bad wording can produce biased results, but even seemingly trivial differences in question wording lead to different results, which serves to emphasize that there is always some random error in survey answers. There was initial optimism that question wording would not affect the correlation of questions with other variables in the survey, even if it affects the distributions of answers. However, hopes for *form-resistant correlations* were dashed when Schuman and Duncan (1974) demonstrated that the relationships between variables can depend on the exact questions asked.

Fowler (1995, 4) cites five characteristics of good questions: that the questions be consistently understood, the questions be consistently administered, the questions consistently communicate what is an adequate answer, all respondents have the information needed to answer the questions, and respondents be willing to provide the answers. Each of these points has clear implications. For example, the need for respondents to have the information required to answer the question suggests that questions about the person's actual experiences are more likely to obtain valid results, whereas hypothetical questions are likely to produce meaningless answers.[9] The importance of consistently understood questions implies the need to be sure that only one question is asked at a time, without double-barreled questions or questions that make assumptions about respondents that may be wrong. The need for consistently understood questions underlines the importance of using words that all respondents will understand or giving definitions if the meanings may not be shared. The importance of respondents' understanding the type of answer to be given implies asking the question so that the desired answers are clear, such as asking people the year in which they got married rather than a vague "When did you get married?" which could get such varied answers as "When I was eighteen" or "After I got out of the army."

These considerations lead to several accepted rules for writing survey questions so as to facilitate comprehension (see such sources as Payne 1951; Converse and Presser 1986; Weisberg, Krosnick, and Bowen 1996;

Dillman 2000, chap. 2; Bradburn, Sudman, and Wansink 2004). The questions should be kept simple and short, without complicated sentence structures. The words used should be ones that people would be expected to understand. Also, questions should not be confusing; this excludes double-barreled questions that involve two questions at once and questions stated in the negative (such as "Do you favor not raising taxes?"). Double negatives in questions lead to results that cannot be interpreted, as when the Roper Organization asked, "Does it seem possible or does it seem impossible to you that the Nazi extermination of the Jews never happened?" (Ladd 1994). Another common caution is against using prestige names in questions (such as referring to "President Bush's policy") because of the potential biasing effect, but experiments by Smith and Squire (1990) suggest prestige names are useful in that they eliminate the effect of education on "don't know" responses while providing a political basis for answers.

It is also important to avoid terms that would be unfamiliar to people, that are ambiguous, or that are vague. Fowler (1992) finds changed results when unclear terms in survey questions are clarified, such as a decreased use of butter when margarine is specifically excluded from a question about butter use and more reported exercise when walking is explicitly included in an exercise question. Nonspecific referents allow people to interpret a question in very different ways. For example, when asked, "Do you think children suffer any ill effects from watching television programs with violence in them?" people interpret "children" as being of any age group from babies to older teenagers, often focusing on their own children or grandchildren (Belson 1986). The problem, however, is that full clarification of what is to be included in answering a question can lead to unwieldy questions that are hard to understand. A laboratory experiment in Britain by O'Muircheartaigh, Gaskell, and Wright (1993) found no effect of some intensifiers, such as asking if people were "very" annoyed rather than just annoyed by television advertisements. Thus while small changes in question wording can have large effects, supposedly important differences sometimes do not have effects. Especially problematic are questions that ask respondents to classify something beyond their abilities to respond to, such as asking people what types of museums they go to, rather than providing a set of categories such as science museums, art museums, and so on (Houtkoop-Steenstra 2000, chap. 6).

Response categories, of course, have to be mutually exclusive, unless multiple answers are desired. Houtkoop-Steenstra (2000, chap. 5) emphasizes the importance of including response options as part of closed-ended questions, so that the respondent does not answer prematurely. For example, asking, "What political party are you closest to: the Republicans or Democrats?" would allow people to answer before the interviewer gets a chance to list the parties, whereas that problem is avoided when the question is phrased this way: "Generally speaking, do you usually think of yourself as a Republican, a Democrat, an Independent, or what?" Response categories for attitude questions should be balanced, as in asking people if they agree or disagree with a statement rather than just if they agree with it. However, using only "agree" and "disagree" options allows respondents to give easy answers that do not require real thought. Petty, Rennier, and Cacioppo (1987) show that respondents think more deeply when questions are phrased as interrogations (such as asking whether the respondent feels a particular policy proposal will work) rather than as assertions (simply asking if the respondent agrees or disagrees with a statement).

Further, question wording should be objective. The importance of finding objective wording was made clear when Tom Smith (1987) showed that the public is more negative when asked questions about people on welfare than when asked the same questions about the poor, apparently because the former involves connotations of government waste and bureaucracy. Similarly, Rasinski (1989) found that a slight alteration of wording in questions can result in large changes in support for proposals. For example, changing "assistance to big cities" to "solving the problems of big cities" increases support by 30%. Abramson and Ostrom (1994) show that by focusing on a short time frame, the Gallup Poll's wording of the party identification question ("In politics, as of today, do you consider yourself a Republican, a Democrat, or an Independent?") obtained more Republicans and fewer independents in the early 1990s than the National Election Studies wording ("Generally speaking, do you usually think of yourself as a Republican, a Democrat, an Independent, or what?").

In developing attitude questions, it is important also to pay attention to attitude strength. The discussion so far has emphasized the direction of attitudes (such as being for or against a policy proposal), but the intensity with which attitudes are held should also be measured. Strong attitudes are ones that are stable over time, are resistant to change, and affect a person's behavior (Petty and Krosnick 1995). A common example of the importance of attitude strength involves gun control in the United States. Polls routinely show that most Americans favor gun control, but Congress rarely passes gun control bills because the minority who oppose such legislation feel more strongly about the topic and are willing to vote on the basis of that one issue when they feel their rights are being threatened.[10] Miller and Peterson (2004) provide a useful discussion of how to measure four aspects of attitude strength: attitude importance, accessibility, ambivalence, and certainty.

Question Formats

There are several different possible structures for survey questions. The basic distinction is between open-ended and closed-ended questions. The *open-ended question* is like an essay exam, in which people can give any appropriate reply that comes to mind, such as "What do you like about the Republican Party?" The *closed-ended question* is instead like a multiple-choice exam, in which people are given a fixed set of possible answers to choose from, such as "In the past year, did your family's financial situation improve, get worse, or stay the same?"

Open-Ended Questions

Open-ended questions have respondents search their memory for possible answers rather than providing them with acceptable alternatives. Researchers who use open-ended questions like them as a way of seeing what is on respondents' minds and also observing their thought processes. These questions show how people think and what is important to them. For example, Converse (1964) coded responses to questions asking people what they liked and disliked about the political parties and presidential candidates in terms of how they conceptualized politics—particularly whether they thought of politics in ideological terms. An open-ended format is also sometimes used for knowledge questions, as in asking who the vice president of the country is.

Differences in how verbose people are can confound interpretation of answers to open-ended questions. Some people give long, elaborate answers when asked what they like and dislike about political parties and candidates, whereas others give short, crisp answers in the hope that the interviewer will quickly move to the next question. Differences in the number of answers people give can thus be due to how verbose people are rather than being due to differences on the variables being asked about. However, Geer's (1988) analysis of the NES open-ended likes-and-dislikes questions finds that differences in the number of comments offered are due to education and interest in politics, with most of the people who give no responses being apolitical. In a subsequent experiment to test the validity of open-ended questions, Geer (1991) showed that people are more likely to repeat information on salient issues when asked open-ended questions, suggesting that responses to open-ended questions do not involve merely reciting recently acquired information when that information is not important.

There also seems to be a tendency for people to choose easy answers on open-ended questions rather than harder ones. This tendency is particularly obvious on questions asking for numerical answers. When asked to estimate numbers, for example, people tend to give round numbers that end in 5 or 0. Thus when asked how many times they went to a doctor in the past year, people who went many times are more likely to give a multiple of 5 (5 or 10 or 15) than to say 7 or 13.

Scale Questions

Scale questions are closed-ended questions that have ordered categories, which can be either numerical or verbal. There are several common scale formats. Agree/disagree questions ask people to indicate whether they agree or disagree with a statement, often along a 5-point scale: strongly agree, agree but not strongly, neutral, disagree but not strongly, and strongly disagree. Semantic differential questions ask people to rate objects on a 1 to 7 scale, where the two end points are polar opposites, such as 1 for good and 7 for bad. Performance evaluation questions (such as student ratings of teachers) often employ 5-point rating scales, where 1 represents the lowest level of performance and 5 the highest. Seven-point issue scales ask people to indicate their position on a 1 to 7 scale, where the two end points are given verbal labels (e.g., 1 for favoring government health insurance and 7 for favoring private health insurance) and the steps from one number to the next are treated as equal intervals. The feeling thermometer asks people to rate objects on a 0 to 100 thermometer scale, with low scores meaning cool feelings and high scores meaning warm feelings. In any case, there should be as many positive as negative categories in a scale, such as asking if people "strongly approve," "approve," "disapprove," or "strongly disapprove" of a policy proposal.

METHODOLOGICAL PROBLEMS. Several psychological mechanisms affect answers to scale questions. For example, there is a positivity bias to many rating scales, as when asking students to rate teachers in course evaluations. There is also a tendency for people to avoid extreme categories, which is sometimes called a response-contraction bias. Wilcox, Sigelman, and Cook (1989) show that NES respondents differ in how positive they are in general when they rate social groups on the feeling thermometer. This suggests a need to use relative ratings rather than absolute ratings when deciding which scores are highly positive toward a group. For example, instead of defining positive ratings as ones over 75 degrees, positive ratings might be defined as ratings 25 degrees above the person's mean rating. Or the ratings given by a respondent may be standardized before they are analyzed further, so correction is made for the mean and variance of that person's ratings.

Krosnick and Fabrigar (1997) interpret the literature as suggesting that 5 to 7 points is the optimal length for rating scales, leading to more reliable and valid scales than when fewer or more points are used. Furthermore they suggest that scales with each category labeled are more reliable and valid than ones that have only endpoints labeled, though this requires selecting labels that correspond to equal intervals and that have precise meanings. However, the evidence is not all on one side. Alwin (1997) compared feeling thermometers and 7-point scales and found that questions with more categories are more reliable and valid, though people tend to use just a few thermometer ratings—mainly multiples of 5. Also, negative numbers are evaluated differently from positive ones. Schwarz et al. (1991) show evidence in a German study that people are less likely to choose responses with negative values, so there are fewer responses of −3 to −1 on a −3 to +3 scale than of 1 to 3 on a 1 to 7 scale, even if the two should be fairly equivalent.

A common variant of these scales is the *branching format*. For example, people can first be asked if the government should spend more money or less money on education. People who want more money spent would then be asked if they favor spending a great deal more money or somewhat more money, with a similar question asked of people who want less money spent. Answers to these branching questions are then made into a 5-point scale: a great deal more, somewhat more, no change, somewhat less, and a great deal less. Krosnick and Berent (1993) argue that using a branching format increases the reliability of answers over using a single scale, as well as being faster to administer.

Anchoring and adjustment are evident in scale questions when respondents are asked to rate several items in a row. The answers that respondents give to the first item serve as an anchor against which later items are rated. For example, if a person rates the first political candidate as a 70 on a thermometer scale, the score given to the next candidate will reflect how much more or less the person likes that second candidate than the first.

Another issue has to do with using rankings instead of rating each item separately. Rankings have people order a set of choices explicitly into their first choice, second choice, and so on. An early study (Alwin and Krosnick 1985) found that ranking yields similar results as to the aggregate relative ordering of the choices as rating, but with different underlying factor analysis results—a bipolar factor underlying ranks versus multiple factors for ratings. Krosnick and Alwin (1988) show that correlations between ratings and some other variables are weakened because of people who do not differentiate much among the stimuli being rated, particularly leading to *end-piling*, in which ratings of several items are piled up at the positive end of the scale. Rating items is easier and faster for respondents, but studies show that rankings provide more reliable and more valid data. Krosnick (1999) argues that ratings are susceptible to satisficing in the form of rating many alternatives equally ("nondifferentiation"). Asking respondents to choose their first and last choices before having them rate the others (a "most-least-rank-then-rate" procedure) improves the quality of rating results (McCarty and Shrum 2000).

NEUTRAL CATEGORIES. There is a vigorous debate as to whether to include middle categories for scale questions. On a strongly agree, agree, disagree, strongly disagree scale, for example, is it better to include a neutral middle category or go with just four categories? This is a matter of the selection component in Tourangeau's four-stage model of attitudes, focusing on how people whose views are mildly in one direction map their judgments onto the response categories—if they can use a middle category or are forced to choose.

Offering a middle alternative is usually found to affect survey results. Experiments in face-to-face surveys reported by Presser and Schuman (1980) found that the proportion of people in the middle category increases 10–20%, and sometimes more, when the middle category is offered. Unfortunately, as they note, the middle alternative was usually the last choice offered the respondent, such as when asking whether laws on marijuana use should be more strict, less strict, or about the same as now, so the increase could reflect its coming last (a *recency effect*, as explained below). Schuman and Presser (1981) found that a middle category did not alter the balance of responses between the two opposite ends of a question, but Bishop's (1987) experiments showed that use of a middle category did make a difference, with larger effects among respondents who are less involved with an issue (Bishop 1990).

As Payne (1951) originally concluded, the middle alternative is useful for sorting out people with convictions but not for finding the direction that people are leaning. Zaller's (1992) work suggests that the middle category is a measure of ambivalence, given when people retrieve both positive and negative considerations on the topic. Bradburn, Sudman, and Wansink (2004, 142) recommend using the middle category so as to learn about the intensity of attitudes, but an alternative to using a middle category would be following up a question with a measure of the intensity of the attitude (Converse and Presser 1986, 36).

Some researchers feel that it is best to offer a middle category so as to allow an easy out for people without opinions. Other researchers counter that it is best to try to force people to give their attitude rather than let them off the hook easily by offering the neutral alternative. Specifically, Krosnick (1991) argues that offering a middle alternative is an invitation for respondents to satisfice rather than to go through the harder work of optimizing their responses. Krosnick and Fabrigar (1997, 148) admit, however, that "it seems sensible to include midpoints when they are conceptually demanded," as when asking whether spending on a particular government program should be increased, decreased, or kept the same. There is currently no uniform practice on this matter, though the recent focus on minimizing satisficing behavior is leading many researchers to avoid middle categories.

Closed-Ended Questions

The final basic question format is the closed-ended question with unordered categories, as when people are asked which of several alternatives are their reasons for some attitude or behavior. For example, people might be asked if the reason that they do (or do not) shop at a particular store is price, location, product quality, product availability, or the brands carried. One problem with closed-ended questions is that people usually restrict themselves to the choices they are offered, even when they are told they can volunteer a response that is not listed, so that people are unlikely to answer the above question in terms of the quality of sales personnel at the store.

The closed-ended format encounters problems because people have a limited ability to think through all the alternatives at once, particularly for long questions. In terms of Tourangeau's four-component theory, judgment between many alternatives is a difficult task. Krosnick and Alwin (1987) argue that people will satisfice in this situation, choosing a reasonable answer rather than going through the work of carefully weighing all the alternatives. There are two possible satisficing strategies in this case, and which is used most seems to depend on the interviewing mode. Studies show that people answering such questions in written questionnaires tend to choose answers early in the list; this tendency is known in the psychology literature as a *primacy effect*. Presumably weak satisficiers are fatigued by considering the list, so they choose the first one that seems plausible. Similar effects are likely in Internet surveys when respondents see all the alternatives at once. By contrast, in phone interviews there is a tendency to select one of the last response categories, because it is hard to remember back to the first few answers that the interviewer listed; this is known as a *recency effect*.[11] Face-to-face surveys that do not show respondents the response categories would be expected to have recency effects, since respondents have to remember alternatives read to them. Both primacy and recency effects show that response order can matter.

One solution to the response order problem is randomizing the order of response alternatives for different respondents. That is particularly possible in computer-assisted interviews. It takes more time and effort to program such randomization, but it would effectively control for such effects in item response distributions and allow one to measure these effects. However, these effects would still affect individual responses, which suggests the importance of keeping response tasks as simple as possible and of emphasizing the need for respondents to concentrate in answering questions.

The response categories in a scale also convey information to respondents as to what typical values are. For example, in asking people how many hours of television they watch per day, Schwarz et al. (1985) compared responses from two studies that used different response categories. When nearly all of the categories were below 2.5 hours (e.g. less than a half-hour, a half-hour to one hour, etc.), 84% of the sample gave answers below 2.5, compared to 63% when nearly all the categories were above 2.5 hours. There is no way of telling which is the correct result, but this example nicely illustrates that answers depend on the alternative responses that are provided. Question wording gives information to respondents that they may use in deciding on their response, in this case providing a frame of reference that they use in estimating the frequency of their own behavior. It would be better in this instance to ask people to give a numerical estimate of their hours of television viewing than to bias their answers by giving them a set of categories.

A variant of the closed-ended question is the "mark all that apply" format used in some paper-and-pencil questionnaires. This format takes minimal space and looks quick to complete, but studies show that it is not optimal. Rasinski, Mingay, and Bradburn (1994) show that high school seniors in the National Education Longitudinal Study chose fewer response options with this format (as when asked reasons they were not planning to continue their education past high school) than when they were asked individual yes/no questions about each alternative. Similarly, the Voter News Service election exit polls were modified after 1994, when it was found that fewer people indicated affiliations (married, Hispanic, etc.) when they were to mark all that applied to them than when they were asked about each affiliation separately (Mitofsky and Edelman 1995; Merkle and Edelman 2000).

Open versus Closed Questions

The wording of questions can be seen as a constraint on the respondent (Schuman and Presser 1981, 299–301). People have to fit their answers into the categories provided in closed-ended questions, which can be awkward when no available category corresponds to their judgment. Even open-ended questions often constrain answers by implying a frame of reference.

Open- and closed-ended questions on the same topic can produce different results. The response options in closed-ended questions can serve as memory cues, helping people retrieve thoughts from memory that they would not consider in answering open-ended questions. This was demonstrated effectively by Schuman, Ludwig, and Krosnick (1986) in looking at answers regarding what is the most important problem facing the country. In four separate telephone surveys, random halves of respondents were asked either the usual open-ended question or a closed-ended version regarding which of four specified problems (the high cost of living, unemployment, threat of nuclear war, and government budget cuts) was the most important. The gross ranking of issues was generally the same in each format, but more people mentioned each of the four specified problems when it was made salient to them in the closed-ended format. Indeed, the problems that were mentioned least often in the open-ended version experienced the greatest increases in the closed-ended version. For example, less than 1% of respondents mentioned budget cuts in the open-ended version in December 1983, while 10% gave that answer to the closed-ended version that month. The closed-ended question did allow people to name a different problem, and an average of 15% did so, but it still looks as if many people who would have had other problems in mind in the open-ended version felt constrained to give one of the listed problems in the closed-ended question. Answer categories to closed-ended questions constrain results.

Additionally, there is a classical debate as to whether open or closed-ended survey questions are better. Open-ended questions allow people to express their answers in their own words, permitting researchers to examine how people think as well as what they think, as when Campbell et al.

(1960) coded answers to what people like and dislike about political parties and candidates in terms of how they conceptualize politics. However, coding open-ended questions takes a large amount of time, so many survey operations prefer using closed-ended questions. A further problem is that people may not know the reasons behind their actions, as assumed by open-ended questions (Nisbett and Wilson 1977; cf. Fiske and Taylor 1991, 270, 399–402). Closed-ended questions permit asking about topics that might have effects on the respondents' behavior that respondents do not recognize. For example, analysis of the closed-ended questions in the 1996 NES show that empathy toward the presidential candidates significantly increased the vote for Bill Clinton (Weisberg and Mockabee 1999), whereas that factor did not get mentioned in responses to open-ended questions (Smith, Radcliffe, and Kessel 1999).

After giving separate analysis of open- and closed-ended questions in the 1996 NES survey, Kessel and Weisberg (1999) provide a useful discussion of the limits of each approach. Both types of questions have limitations. Closed-ended questions make it too easy for people who have not considered a topic to randomly choose an answer so that they go to the next question, and people who have complicated attitudes on the topic cannot offer qualified answers. The answers obtained by open-ended questions are ones that are accessible to the respondent, which could mean merely that they were primed by some event prior to the interview rather than representing a longer-held attitude with their answer. Answer choices to closed-ended questions are limited to ones that the question writer thought of. Questions themselves often have to be written months in advance for national face-to-face surveys, which can be problematic in some settings, such as when having to guess before an actual campaign begins what issues to use for closed-ended questions in an election survey. This is less of a problem when asking broad open-ended questions, such as "What do you like/dislike about the Republican/Democratic Party?" but answers to such questions depend on the person's answering style, so differences between the responses of two people measure their garrulousness as well as differences in their attitudes.

Open-ended questions tend to be used more often in three types of situations: academic surveys, exploratory work, and elite interviewing. Academic survey organizations can afford to use open-ended verbal questions because they generally do not need to get results back quickly to clients and so have time to code answers rather than needing to use "precoded" questions. Open-ended questions are particularly well suited for exploratory work, in order to find the response categories that people would want to use for replying to closed-ended questions. Survey organizations that need to get results quickly to clients, however, rarely have time to explore response alternatives first with open-ended questions. Elite respondents, such as government officials, often feel confined by the categories of closed-ended questions, and open-ended questions permit the investigator to obtain a wider range of information from elites, though at the cost of having to code that information. Open-ended questions are also the norm in nonsurvey interviews, including qualitative, in-depth, and life-history interviews (Gubrium and Holstein 2002, chaps. 4–6).

Question Wording and Survey Constraints

It is important to realize that measurement error is inevitable since question wording is never perfect. People will inevitably interpret questions differently because of their different perspectives. Effort should be taken to word questions well, but perfection should not be expected. The criteria for good survey questions are that they be reliable, valid, and useful. A question is reliable if similar answers would be given if it were asked repeatedly of the same person in a limited time frame. A question is valid if it measures the concept of interest to the researcher.[12] Sometimes a researcher realizes that a question has not been worded perfectly but it is still useful, such as when it is decided that it is better to maintain old question wording so that time trends can be analyzed, even if some alterations could lead to a better question.

Additionally, when attitude questions are asked, it is important to measure attitude strength. Make sure that attitudes measures are "real" by asking only about topics that people would have information about and on which attitudes would be crystallized. Recognize that people might be ambivalent because they have conflicting considerations, so exact results of attitude questions should not be taken overly literally.

The number and types of questions directly affect survey costs. The more questions in an interviewer-administered survey, the longer the interview takes and the more costly is the interviewing. More reliable measurement can be obtained by combining several questions on the same topic into an index or scale, but that also increases the length and therefore the cost of the interview. Closed-ended questions are faster, and therefore more can be asked in a given period of time; Groves (1989, 491) estimates that a closed-ended question takes about 15 seconds while an open-ended question takes 30–45 seconds. Some repetitive questions are even faster, as when people are asked to rate more items on a thermometer scale. Open-ended questions lead to further costs in terms of needing to code answers into categories (see chapter 11), but again, some researchers feel open-ended questions can give higher-quality data than closed-ended questions.

Ethics constitute an additional constraint on survey questions. Some questions may be inappropriate to ask, possibly because they are too prying or too sensitive. As will be seen in the discussion of survey ethics in chapter 14, questions that involve deception of the respondent are seen as inappropriate, unless the respondent is debriefed at the end of the interview. Additionally, questions whose answers could be of interest to legal authorities are sometimes better not asked unless they are essential to the research. Ethics also extends to respecting the person's right to privacy, in terms of keeping responses confidential (chapter 14).

The discussion of question quality will be continued in chapter 6, as part of the treatment of pretesting surveys.

6

Measurement Error Due to Respondents, II
QUESTIONNAIRE ISSUES

Measurement error due to respondents can also be related to question-naire construction problems. As will be immediately evident, these considerations are not as well developed in the literature as are the issues related to question wording that were raised in chapter 5. Much of the focus is on how question order can directly affect survey results. This chapter begins with a discussion of the theory of question order effects and of the nature of question context effects, after which methods for questionnaire development will be treated.

Theories of Question Order Effects

Understanding the writing of questionnaires requires recognition that an interview is a form of conversation and that there are usual conventions for conversation. These conventions have implications for question order, question positioning in surveys, and survey length.

Grice (1975) has suggested that a general "cooperative principle" for conversation is to "make your contribution such as is required, at the stage at which it occurs, by the accepted purpose or direction of the talk exchange in which you are engaged." This leads to four more specific maxims for ordinary conversation. Grice's "maxim of quantity" specifies, "Make your contribution as informative, and no more informative, than is required." His "maxim of quality" is "Do not say what you believe is false, or for which you lack adequate evidence." Grice's "maxim of relation" demands that talk be relevant, staying on the current topic of conversation, unless one person makes a segue to a new topic. Finally, the "maxim of manner" directs speakers to "avoid obscure expressions and ambiguity. Be brief and orderly."

Surveys are not "ordinary conversation," but Grice's maxims still suggest that respondents are likely to engage in certain forms of behavior in surveys because surveys resemble conversations. If they interpret the sequence of questions as a question-answer-response conversation sequence, their answers will be dependent on the order in which the questions are asked, leading to *question-order effects*.

Additionally, the position of a question in a survey can affect how people will answer it. For example, asking a battery of questions on a topic can increase the salience of that topic when another question is asked later. *Salience effects* can be understood as "priming"—making something salient to respondents in the retrieval process that might not otherwise be salient. For example, asking questions about terrorism would prime that theme in the mind of respondents when they are later asked about a war that is portrayed as part of the "war against terror." The problem, of course, is that there might be question order effects regardless of which question appears first; it is very difficult to avoid such order effects completely. It is possible to randomize the order of some questions, but other questions asked previously in the survey could still affect answers to those questions.

A very general possibility is that the respondent will not pay attention to changes in questions and will simply give the same answer to a series of closed-ended questions. This is known as the *response set* problem. After answering "agree" to one question, the respondent may satisfice by answering "agree" to the following set of questions, without thinking through each question separately. This possibility poses a problem for index construction, in that the person might be scored as extreme on an index of trust in government, for example, if he agrees with all the items on that index. The common way to avoid response set is to make sure that some questions are asked in the opposite direction, so that "agree" on some questions has the same meaning as "disagree" on others. That way, a person who agrees to everything would be located in the middle of an index rather than at an extreme.

This leads to a more general consideration of *context effects*, effects due to the context in which a survey question is asked. As Sudman, Bradburn, and Schwarz (1996, 81) point out, this is referred to as an effect rather than as an error since context-free judgments do not exist in the real world either. According to this perspective, attitudes are always context-dependent construals (Wilson and Hodges 1992).

The positioning of questions in a survey can affect answers in two further ways. First, *rapport effects* can occur on sensitive questions. Asking
sensitive questions at the very beginning of an interview, when a trusting
relationship with the interviewer has not yet been created, is likely to lead
to some suspicion, with many respondents not admitting to the behavior.
People are presumably more willing to answer such questions candidly
when they appear later in a survey than when they appear early. Second,
questions near the end of a long survey may suffer from a *fatigue effect.*
Respondents may satisfice more in their answers when they are tired of
the survey and want it to end. This problem limits the possibility of putting
sensitive questions at the end of long interviews.

A final issue has to do with measurement error in later waves of panel
studies. As respondents become familiar with participating in a survey,
there are *panel-conditioning effects* (Kalton, Kasprzyk, and McMillen 1989),
such that their answers can begin to differ from those given by first-time
participants in that survey. This is also referred to as "time-in-sample"
bias, and it is exemplified by people's reporting less crime victimization the
longer they are in the sample of the National Crime Survey (Cantor
1989). There are several possible reasons for these effects. Respondents
might be giving more accurate answers in later waves since they know
what is expected of them, or the opposite may occur: people may avoid
some responses since they realize those responses would lead to several
new questions. Bartels (2000) finds negligible panel effects on eight analyses of NES panel data, while finding panel biases in the 40% range for four
analyses involving campaign interest and voter turnout, variables that are
likely to be related to survey participation. Several Internet surveys rely on
panels of participants, so this can be an added source of error for such surveys unless they use a rotating panel design to measure and adjust for
panel conditioning effects (Couper 2000).

Context Effects

Survey researchers have long realized that the order in which questions
are asked can affect the answers that are given. The classic case in point is
a Cold War example involving whether Russia and the United States
should allow each other's reporters into their country. In an experiment,
a random half of the sample was first asked whether the United States
should allow in Communist reporters and then asked whether Communist
countries should let American reporters in, while the other half of the
sample was asked the same questions in the opposite order. People were

more likely to feel that Communist reporters should be allowed into the United States when that question was asked second rather than first, as shown in table 6.1, which summarizes results obtained by Hyman and Sheatsley (1950) and Schuman and Presser (1981). One explanation that has been offered for this question order effect has to do with which attitudes are invoked. The first question primes the respondent to think about a particular attitude. Asking first about whether the United States should allow Communist reporters invokes an attitude toward Communism, while asking first about Communist countries allowing U.S. reporters instead invokes an *evenhandedness norm*.[1]

Another example of question-order effects involves abortion. In 1979, when asked the general question "Do you think it should be possible for a pregnant woman to obtain a legal abortion if she is married and does not want any more children?" 58% of the public answered yes, but that figure was only 40% in a 1978 survey when people had first been asked, "Do you think it should be possible for a pregnant woman to obtain a legal abortion if there is a strong chance of serious defect in the baby?" There is less support on the general abortion question, probably because people feel they have already shown their reactions when there are potential birth defects, so the general question must be asking about situations where the health of the baby is not an issue. This result was replicated in several studies taken during the 1980s, with an average difference of 11%, showing that results of polls on important topics can be affected by question order (Schuman 1992, 7–9).

Early studies of the prevalence of context effects found such effects to be minimal. Schuman and Presser (1981) examined 113 attitude questions in the 1971 Detroit Area Study, finding significant effects at little more than the rate expected by chance and concluding that only three probably involved real effects. Smith (1991) looked at over 500 variables

Table 6.1 Example of Question Order Effect

Question Asked First	Hyman and Sheatsley (1950)	Schuman and Presser (1981)
A first	36% agree to A	55% agree to A
B first	73% agree to A	75% agree to A

A: "Do you think the United States should let Communist reporters come in here and send back to their papers the news as they see it?"
B: "Do you think a Communist country like Russia should let American newspaper reporters come in and send back to their papers the news as they see it?"
The cell entries show the proportion of each sample who agreed with question A, depending on whether A or B is asked first.

in the 1988 and 1989 General Social Surveys, obtaining a level of occurrence also near chance levels and deciding that only 12 instances of context effects were probably real, with an average difference in response of 7.5%. The implication of these studies was that context effects are rare, but as seen below, later studies instead emphasize situations that lead to context effects.

There is some evidence that context effects and other question-order effects are less serious in self-administered questionnaires since respondents can look through all the questions rather than processing them consecutively. In particular, Bishop et al. (1988) find that question-order and response-order effects are eliminated in mail surveys, though question-form (e.g., "no opinion" filters, offering a middle alternative, or agree-disagree questions) effects remain, whereas Ayidiya and McClendon (1990) find fewer response effects in mail surveys, though a primacy effect was still found as regards response order.

Types of Context Effects

A basic distinction is between assimilation and contrast effects. *Assimilation effects* occur when an earlier question gives meaning to a later question, as in the reporter example, pushing answers toward consistency with answers to the earlier question. Another classical example is the Tourangeau and Rasinski (1988) experiment asking about an obscure Monetary Control bill being considered by Congress. Support for the bill increased when questions about it immediately followed questions about inflation, presumably because respondents inferred from context that the Monetary Control bill was an anti-inflation measure. People give more consistent answers to questions on the same topic that are next to each other in a survey (McGuire 1960). This is an application of Grice's maxim of relation in that respondents interpret the current question as staying on the same subject as the previous one. Additionally, some studies have shown more extreme attitudes on a later question after earlier questions on similar topics, presumably because of repeated activation for objects linked in long-term memory (Judd et al. 1991).

Tourangeau et al. (1989) find similar carryover effects when respondents were asked questions about several policy issues. For example, respondents were 10% less likely to say that the Supreme Court has gone too far in protecting the rights of people accused of crimes when that question was preceded by questions having to do with civil liberties, such as whether "the authorities should stick to the rules if they want other people to respect the law." They find support for an "accessibility hypothesis" that predicts carryover effects are greatest for respondents with well-developed but conflicted views. Additionally, their meta-analysis of six studies obtains a significant buffering effect—context effects are reduced when the questions are further apart in the survey—though the level of agreement with the issue did not affect the extent of context effects.

Contrast effects occur if respondents interpret the second question as invoking a different attitude from the first, so answers are pushed in the opposite direction from the earlier answers. If one asked how many times people have been to the movies in the past month and then asked how many times they've been to entertainment events in the past month, respondents could naturally assume that the second question was meant to exclude movies since that was already asked. The answers in the abortion example above exemplify such a *subtraction effect*, since people believe that the second question is asking for something beyond what they answered on the first. This fits with Grice's (1975) maxims of conversation discussed at the beginning of this chapter: contrast effects occur when respondents avoid redundancy by not repeating what they have already told the interviewer.

Context can affect each of the four steps in the Tourangeau model of the survey response (chapter 5). Comprehension of the question is affected by contrast effects—for example, when people assume that a later general question must be asking something different from what was asked by a preceding specific question. Context affects the retrieval process when an earlier question makes the respondent think about something that is relevant to answering a later question, as when one question makes people recall the political party of a politician before being asked to evaluate that politician. The example about the Communist and American reporters shows how context can affect judgments. Finally, context affects the reporting stage, as when people tend to give consistent answers to attitude questions that are placed next to each other.

Moore (2002) has suggested a typology of context effects. Table 6.2 summarizes his results, with examples from various Gallup polls of the late 1990s. A consistency effect occurs when the rating of the second object is moved in the direction of the rating of the first object, as when President Clinton was viewed as more honest after a question about Vice President Gore's honesty while Gore was viewed as less honest after a question about Clinton's honesty. A contrast effect occurs when the rating of the second object is moved in the opposite direction of the rating of the first object, as when Speaker of the House Newt Gingrich was viewed as less honest after a question about Senate

Table 6.2 Types of Context Effects, Based on Moore (2002)

Type of Effect	Stimulus A	Stimulus B	Year	Question	Change if B Is Second Rather Than First	Change if A Is Second Rather Than First
Consistency	Gore	Clinton	1997	Generally honest?	+7*	−8*
Contrast	Dole	Gingrich	1995	Generally honest?	−8*	+4
Additive	Black people	White people	1996	Thermom- eter	+12*	+10*
Subtractive	Shoeless Joe Jackson	Pete Rose	1999	Make eligible for Baseball Hall of Fame	−12*	−12*

*$p < 0.05$

Majority Leader Bob Dole's honesty, whereas Dole was seen as more honest after a question about Gingrich's honesty. An additive effect occurs if ratings of an object are always more positive if it is asked about second, as when ratings of both white and black people are higher if the respondent is first asked to give a thermometer rating of the other group. A subtractive effect occurs if ratings of an object are always more negative if it is asked about second, as when willingness to make both Pete Rose and Shoeless Joe Jackson eligible for the Baseball Hall of Fame decreased when the respondent was first asked about making the other eligible. This is a useful typology, but more theory is needed as to when to expect each of these different effects.[2]

The Direction of Context Effects

Several views have been advanced as to differences between circumstances when assimilation and contrast effects will occur. Schwarz, Strack, and Mai (1991) hypothesize that contrast effects occur when respondents interpret a general question as asking for new information beyond what was given in a preceding specific question, whereas assimilation effects may occur when specific questions bring things to mind that are useful in answering a later summary question. They also suggest that the assimilation is more likely to occur when several specific questions are asked, whereas the contrast effect is more likely when only a single specific question is asked before the general question.

Tourangeau's (1999) review of previous studies finds three factors relevant to the direction of context effects when such effects exist. First, contrast effects predominate when the task involves deeper processing, because they involve respondents' realizing that a distinction is required, but assimilation effects occur when processing is more superficial, as when respondents are distracted. Second, contrast effects are more likely when conditions foster comparisons, particularly among familiar items. Additionally, assimilation is more likely when the items are of different levels of generality, whereas contrast is more likely when the items being compared are at the same level.

Two different psychological models can be invoked to understand these effects. Tourangeau's belief-sampling model (Tourangeau, Rips, and Rasinski 2000), explained in chapter 5, assumes that survey answers are based on sampling a few beliefs and then averaging them. Assimilation effects occur because of changes in the underlying set of beliefs from which the sample is drawn, but contrast effects occur when some beliefs are excluded from that sample. Schwarz and Bless's (1992) inclusion/exclusion model instead assumes that people form representations of the stimulus being judged and of a standard with which that stimulus is being compared. Context can lead to some information's being incorporated into the representation of the stimulus, leading to assimilation, or it can lead to information's being excluded from the representation of the stimulus, resulting in subtraction-type contrast. Additionally, the context can lead to information's being carried into the representation of the standard, creating comparison-type contrast.

In any case, the focus on context effects is certainly important for understanding the implications of questionnaire construction and question order for survey responses. However, it also emphasizes the difficulty of ascertaining people's real attitudes. Context exists in the real world as well, though in less observable ways than in surveys. People's views about topics will depend on those real world contexts. Minimally, this points to the impracticality of trying to remove context effects from attitude surveys. It instead suggests trying to mimic better the context considerations that occur in the real world. One approach of this type involves framing experiments in surveys (Nelson and Kinder 1996). Random halves of samples are asked questions about the same public policy, but following lead-ins that frame the issue in different ways. For example, people can be asked about government spending on AIDS research after the topic has been framed as either a problem due to the actions of people with AIDS or simply medical research. People give different answers depending on how the question is framed, showing again how "true" answers depend on context, but also suggesting that the range of opinion is between the values obtained by the two frames. Politicians and the media frame policy issues in particular ways, so it may be appropriate to ask survey questions in the same way that they are framed in public discourse.

Questionnaire Development

Questionnaire construction is more than just writing questions, because the order of questions in a survey can affect the answers and the willingness of respondents to complete the interview. The usual strategy is to start with a simple, interesting question that is directly relevant to the survey's stated purpose so that the respondent accepts the legitimacy of the survey. Following Grice's maxim of relation, questions should be grouped by topic, but starting with ones that will be most salient to respondents. Sensitive questions are best placed near the end. Demographic questions are usually left to the end, with the income question often being the last question because of how sensitive it is. Additionally, attention must be made to the format of the questionnaire, and the survey should be pretested as part of its development phase, as will be explained in the remainder of this chapter.

Questionnaire Format

Questionnaire layout is an important matter in self-administered questionnaires, including mail and Internet surveys. Formatting affects the initial decision of people to respond to the survey, missing data on specific questions, and answers to the questions that people answer.

Mail surveys and other self-administered questionnaires should be formatted so as to maximize the chances that people will take the survey. As part of his tailored design method (see chapter 8), Dillman (2000, 82–84) strongly advocates using large-size (either $8\frac{1}{2} \times 14''$ or $11 \times 17''$) paper folded into booklet format. To attract attention, the first page can be a questionnaire cover, with a graphical design, a title, identification of the sponsor, and the address of the survey organization. So as not to distract from the survey, the back cover is best kept simple, with a thank you for participating and a space for additional comments.

It is also important to format surveys so that people are unlikely to skip questions accidentally. That problem leads to item-missing data, which will be considered in chapter 7 along with suggestions for minimizing navigational errors.

Question wording is not the only thing that affects answers; answers will also vary according to layout of the survey, inclusion of pictures in the survey instrument, and the spacing between questions. For example, it is important to leave enough space for answers. People are more likely to give a long answer when a large space is left blank after the question, and less likely when questions are tightly packed on the form or screen.

Internet surveys are particular susceptible to formatting problems. For one thing, questionnaires and response alternatives may look different depending on the respondent's browser and screen resolution, making the survey less standardized than the investigator may realize. Also, care is necessary in choosing question structures and answer formats for Internet surveys, making sure that all answers appear on the screen at the same time, and giving clear instructions on all questions. For example, when pull-down lists of possible answers are used, the top default category should not be a substantive alternative; otherwise a respondent who does not notice a question may be inadvertently recorded as giving that default answer. The appropriate way to use such a format is for the default on top of the list to be "Pull down this menu to choose an answer," so that substantive answers are recorded only for respondents who actively choose them.

Some issues regarding formatting for Internet surveys have not yet been decided definitively. In particular, should the entire questionnaire be visible at once, or should each question be on a separate screen? Dillman (2000, 395) prefers having the entire questionnaire be visible at once, so the user scrolls from beginning to end, which corresponds to how people have become used to dealing with content on the Web. This is faster than having people answer separate related questions in a screen-by-screen approach (Couper, Traugott, and Lamias 2001), and it is easy for a respondent to back up to see or change a previous answer. My own preference is to have each question on a separate screen. This enables the computer programmer to control the flow of questions, instead of requiring respondents to follow complicated skip patterns themselves. The screen-by-screen approach also facilitates sophisticated survey experiments in which different half-samples ("split-ballot experiments") are given different question wordings, question frames, and/or question orders. Best and Krueger (2004, chap. 4) provide a useful discussion of formatting issues for Internet surveys.

Tourangeau, Couper, and Conrad (2004) report on a series of experiments on the effects of the visual display of response options in Web surveys. They find that respondents interpret the midpoint on a scale as a middle response, that people react faster when options are arrayed in a logical progression from the leftmost or topmost item, and that putting questions on the same screen increases their intercorrelations.

Pretesting Surveys

It has long been the practice to "pretest" surveys, trying them out to make sure that they work as planned. Simply reading the questions out loud in an interview situation can find some problems, because questions often sound very different in an interview from the way they "sound" when they are just looked at on a piece of paper. More formal pretests with people like the intended respondents allow researchers to spot questions that are not worded well and to discover how much time the survey will take. Pretesting is also important for making sure that the flow of the questionnaire works. It can be very difficult to get all skip patterns working perfectly in a survey, and pretests are one way to find errors of this type before the full round of interviews. This is very important in computer-assisted interviews, which require extensive debugging before going into the field. This section will describe both conventional pretest procedures and newer "cognitive" techniques.

Conventional Procedures

Conventional pretesting involves trying the survey out on a set of people who are not part of the regular sample. Pretest interviewers are debriefed as to what problems occurred. The researcher is sometimes able to monitor phone pretest interviews or to listen to tape recordings of them and hear how respondents deal with the survey. Additionally, records are kept of the questions on which respondents ask for clarification and the questions that interviewers most often rephrase or misread (Fowler 1991, 276). Answers are inspected to look for questions to which respondents give inadequate answers. These problems should signal to the researcher the need to rethink the question wording before going into the field for the actual interviewing. Pretests can also give a warning when a question is phrased so that there is no variation in answers, which usually is a signal that the question has not been worded so as to pick up a real controversy.

One way of improving the quality of the feedback from interviewers during a pretest is to have them fill out forms answering specific questions about how the interview went. Some large-scale surveys have interviewers fill out such a form after each pretest interview, while the interview is fresh in their mind. The researchers can examine this information to see which questions elicit negative reactions.

Behavior coding (Cannell, Miller, and Oksenberg 1981) has an observer code the interaction between the interviewer and the respondent in the pretest. For example, instances when the interviewer misreads the question are recorded, as are instances when the respondent asks for more information. Fowler and Cannell (1996) give examples of behavior codes for how questions were read: (1) exactly, (2) with a slight change that does not alter the question wording, or (3) with a major change. Examples of respondent behavior codes are (1) interruptions, (2) requests for clarification, (3) adequate answers, (4) answers that are qualified to indicate uncertainty, (5) inadequate answers that do not meet the question objective, (6) "don't know" responses, and (7) refusals. Questions that lead to frequent deviations from the interview script require more attention before the full interviewing takes place, as do questions that lead to inadequate answers. Presser and Blair's (1994) study finds behavior coding is reliable for problems that are found.

Expert reviews and focus groups are also used sometimes as part of preparing the survey questions. Expert reviews have substantive experts review the questionnaire content to make sure that it meets the survey objectives. Additionally, questionnaire design experts look for wording problems, such as vague terms. These reviews should look for likely problems involving reading the questions, instructions, item clarity, hidden assumptions, knowledge, sensitivity, and response categories (Biemer and Lyberg 2003, 263). Presser and Blair (1994) found that expert groups located the most problems with questions at the least cost. Instead of using experts, focus groups assemble a group of people who are like the eventual respondents to discuss the survey topic. This procedure is particularly useful in an exploratory stage in which researchers want to determine how people think about the survey topic, how they define key terms, and what kinds of questions work best. For example, focus group participants can be asked what types of events they would include if they were asked which cultural events they attend, and they can be asked if they can count how many they have attended or if a different answer form would work better.

Cognitive Procedures

Dissatisfaction with conventional questionnaire development procedures led to the Cognitive Aspects of Survey Methodology (CASM) movement (Jabine et al. 1984). CASM is an interdisciplinary approach that uses insights from the cognitive revolution in psychology to reduce measurement error in survey questions. The contributions of this approach are reviewed in Sudman, Bradburn, and Schwarz (1996, chap. 2) and in conference volumes edited by Schwarz and Sudman (1992, 1994, 1996), Sirken et al. (1999), and Presser et al. (2004). Forsyth and Lessler (1991) provide a useful discussion of the different cognitive approaches to surveys.

One of the most important CASM procedures is *cognitive interviewing*, in which researchers study the cognitive processes used by survey respondents as they answer questions. In *think-aloud* protocols, respondents tell what they are thinking as they formulate answers (termed the *concurrent* mode), though this assumes that people are able and willing to give candid readouts of their thought process. This is used mainly in testing question wording while developing questionnaires. This approach is usually contrasted with traditional pretests of surveys in which a small sample of people who would not otherwise be in the sample are asked the questions and interviewers report which questions were problematic. The think-aloud approach is designed to shift the focus to respondents' actual statements, rather than the interviewers' interpretations of them. A related approach is *retrospective interviewing* (also known as a *respondent debriefing*), in which people are asked questions about the interview after completing the questionnaire, such as whether they had problems with specific questions or how they interpreted particular words or questions. Retrospective interviewing is virtually a case of interviewing people about the original interview. Alternatively, after answering a question, respondents can be asked to paraphrase it as a means of seeing how they understood it, with additional questions asking them to define key terms and to explain how they worked through answers. Cognitive interviews are often conducted by the researchers themselves or even by psychologists who are trained in how to probe reactions effectively, with respondents being paid to participate.

Another interesting technique is the vignette, in which researchers pose hypothetical situations to the respondent to learn how respondents would answer questions about them. An example involves testing in federal surveys to make sure that respondents were interpreting employment questions correctly. A 1991 Current Population Survey described some quasi-work situations (such as "Last week, Susan put in 20 hours of volunteer service at a local hospital") to see if people categorize them as work. A third of respondents classified volunteer service as work (Martin and Polivka 1995), leading CPS to reformulate the employment question to ask about "working for pay" (Norwood and Tanur 1994).

Other cognitive interviewing techniques include having respondents rate their confidence in their answers, giving respondents memory cues to aid their recall, and measuring response times (latencies). Latencies are of interest because taking a long time to answer a question is often a sign that the question wording is unclear. Response latencies in CATI surveys can be measured in three ways. Active timers have the interviewer start and stop a timer manually by hitting a key on the computer keyboard; this requires interviewer training and quick reflexes on his or her part. Voice-activated timers are automatically triggered by the respondent's voice, but they require expensive hardware and are thrown off by extraneous noises, which lead to a high proportion of invalid responses, though the results correlate very highly with active timers (Bassili and Fletcher 1991; Bassili 1996). Latent timers record the time required to complete each question from when it appears on the screen to when the interviewer starts typing the response. Mulligan et al. (2003) show that good results can be obtained using latent timers with controls for the interviewers, to take into account differences in their speeds of asking questions.

CASM researchers have also developed several response aids. Using landmarking events (chapter 5) and giving respondents more cues to help retrieve items from memory are both considered response aids. Other aids for temporal tasks include asking people to start with the most recent event and go backwards in time ("backward search"; Loftus and Fathi 1985) and giving people calendars for recording personal life events (jobs, schooling, marriage, birth of children, and so on) to improve recall and dating of events ("life event calendar"; Means and Loftus 1991). A study by Belli, Shay, and Stafford (2001) shows that the life event calendar approach with flexible interviewing led to more accurate answers on employment and health questions without increases in costs.

Asking respondents about associated events can often facilitate recall. For example, before asking a voter turnout question, Belli et al. (1999) asked a random half-sample to think about what would have happened if they had actually voted in the last election—"things like whether you walked, drove, or were driven by another person to your polling place, what the weather was like on the way, the time of day that was, and people you went with, saw, or met while there." Asking people to think in terms of related events reduced the extent of overreports of the vote, particularly for people interviewed longer times after the election.

CASM researchers have devoted considerable attention to techniques for minimizing unintended mistakes by the respondents (known as *slips*), particularly those related to the time process. Breaking up questions into small steps helps minimize slips. For example, say that the respondent is asked to list the items he or she ate the previous day. The person may be trying to answer accurately but may simply forget some items. Asking the person instead to list what was eaten at each meal separately may yield a more accurate list, because the person is focusing on smaller, more discrete aspects.

CASM procedures are somewhat controversial. Willis, DeMaio, and Harris-Kojetin's (1999) review argues that there has been little actual evaluation of these procedures, particularly because of the scarcity of independent measures of the quality of survey questions. In particular, the number of cases in cognitive interviews is usually very small, their coding can never be fully reliable, their laboratory setting may not generalize to the field, and much of the evidence is only anecdotal. Similarly, Tourangeau, Rips, and Rasinski (2000, 21) report considerable skepticism as to what the CASM movement has achieved, even though it has wide acceptance among statistical researchers for the U.S. government. Fowler (2004) does give examples of questions that were improved by cognitive interviewing, but Beatty (2004) shows there is wide variation among interviewers in cognitive testing. DeMaio and Landreth (2004) find differences among teams in assessing results of cognitive interviews, though they find that different cognitive approaches yield similar results.

Yet the real comparison should be between conventional pretesting and cognitive methods. As Krosnick (1999) argues, conventional pretesting is simplistic and not reliable since interviewers' impressions are subjective, so behavior coding (Presser and Blair 1994) and cognitive pretesting seem preferable to standard pretests in finding problems with the questionnaire.

Questionnaire Effects and Survey Constraints

Respondent-related error is partially under the control of the researchers who are conducting the survey when they decide how to word questions and in what order to put them. One of the best ways of handling this source of error is to pretest the questionnaire. The questions may look very good to the researchers, but trying them out on real people often brings problems to light. Questions may not be interpreted as the researchers intend or may elicit queries from respondents as to what the wording means.

Pretests, think-aloud interviews, and behavior coding directly add to the cost of the survey. Yet they can alert researchers early to problems, so that errors can be minimized before the full data collection with its larger costs. Even a pretest of 25 people can be very useful, without adding much to the cost of a study. Surveys that plan to do tens of thousands of interviews can easily afford to take a few hundred interviews at the pretest stage in order to minimize subsequent measurement error; very large studies, such as the U.S. Census, typically do full pilot studies. Cognitive pretesting and behavior coding can be more useful than conventional pretests, though they are also expensive.

Length poses another important constraint in surveys. Researchers invariably want to ask more questions than it is feasible to ask in an interview, so staff of survey organizations regularly explain to researchers that they have to cut back to a reasonable number of questions. After all, survey costs depend on the length of the interview in addition to the number of respondents. Pretests are useful for determining the approximate length of the interview. If the pretests run long, which is usually the case, the survey organization tells the researcher that the number of questions and/or the number of respondents has to be cut in order to stay within the budget. Furthermore, it is important that researchers understand that respondents are more likely to engage in satisficing behavior in a long questionnaire, so that a shorter questionnaire will get higher-quality data. However, there may be problems in the eventual data analysis if the researcher cannot ask all the questions of potential interest. Sometimes researchers handle the length problem by asking some questions of only a random half of the sample. While this decreases the length of the survey, it increases programming costs and prevents the researcher from using answers in one half of the survey as explanatory variables for questions in the other half.[3] In the end, the researcher must set realistic priorities so that the survey length fits within available cost and time constraints.

Finally, context effects raise an ethical issue in reporting surveys. As shown in this chapter, context does affect answers given to questions, but survey reports rarely indicate the question order. Minimally, the survey report should candidly admit when question order effects are likely. Making the survey questionnaire available would allow readers of the survey report to gauge for themselves if there are possible context effects.

Measurement error—whether due to interviewers (chapter 4), question wording (chapter 5), or the questionnaire (chapter 6)—is always an important limitation to surveys. Another important limitation is when survey questions are not answered, which brings us to the topic of nonresponse error.

7

Nonresponse Error at the Item Level
THE SCOURGE OF MISSING DATA

The next type of error to be considered is nonresponse error—the error that occurs when data are missing. The type of nonresponse discussed in this chapter involves respondents' failure to answer every appropriate question, known as item nonresponse. Chapter 8 treats the other form of nonresponse, unit nonresponse, which occurs when it is not possible to interview an intended respondent.

Nonresponse is a problem to the extent that it biases results, so that sample statistics are not on-target estimates of the corresponding population parameters. For example, the true mean of a variable equals its sample mean plus a bias equal to the missing data rate multiplied by the difference between the means for respondents and nonrespondents. The bias will be small if the missing data rate is low or if respondents and nonrespondents are similar. In practice, however, the bias cannot be estimated since the nonrespondent mean is not known. The bias problem is most serious when people who would belong to one answer category are most likely to not answer that question, as when people opposing a candidate from a racial minority do not indicate their voting intentions in a preelection survey.

The first part of this chapter discusses the different ways in which item nonresponse occurs, along with ways to minimize item nonresponse. The second section considers some of the effects of item-missing data. The third section presents procedures for handling missing data. Because item nonresponse can distort data analysis results, this is a statistical topic, though many survey practitioners do not handle missing data in the ways that statisticians usually advise.

Theories of Nonresponse

There are three basic types of item nonresponse in surveys. The most important type for the purposes of this chapter is the "don't know" answer, because it is, unfortunately, a fairly common response. Item non-response also occurs when respondents refuse to answer some questions, which is a much rarer situation. Finally, responses are missing on a particular question whenever interviewers accidentally skip items or respondents skip items on a self-administered questionnaire or Internet survey. That situation is usually referred to as "not ascertained" in order to distinguish it from the situation in which respondents explicitly say they do not know the answer to the question. In order to minimize missing data, survey researchers must be concerned about the cognitive basis of each of these forms of item nonresponse.

"Don't Know"

The "don't know" response is deceptively simple. It seems to be a substantive answer to the question, meaning that the person lacks the information needed to answer the question, but that is not always the case. For example, some people, even in their normal everyday speech, begin many of their sentences, "I don't know, maybe . . ." as in this exchange: "Where should we go to dinner tonight?" "I don't know, maybe we should go to a Chinese restaurant." The answerer here is trying to make a suggestion but happens to use the "don't know" phrase as a way to condition the suggestion. It would be a social mistake if the person's dinner companion cut him off after the "don't know" part of the response by saying, "Well, in that case, let's just go to a fast food place." It is equally important in surveys to be able to interpret whether a "don't know" response is really substantive.

Meanings of "Don't Know"

"Don't know" answers are often a form of satisficing. Respondents view it as an easy answer to give, hoping that the interviewer will just go on to the next question. While it can be a form of satisficing, it is important to recognize that it may have several different meanings. A useful way to conceptualize the psychological sources of "don't know" responses is to return to Tourangeau's four-stage model of the survey response process discussed in chapter 5. Just as Krosnick (2002) points out that "no opinion" responses can be due to each of those stages, so too can "don't know" responses be the result of aspects of each stage.

First, at the question comprehension stage, some people may not understand the question enough to try to answer it, so they simply say "don't know" instead. Respondents who satisfice may not even bother to try to understand the question but will answer "don't know" in order to move on to the next question.

Second, at the retrieval from memory stage, some people will not locate any information in their memory about the topic, which is when a "don't know" response is a real, substantive reply. This is particularly the case for event-based and factual questions, such as "Who is the president of India?" This stage is also involved when the respondent has difficulty recalling the information needed to answer the question or when it is the first time that the respondent has thought about the question. Because of those possibilities, it can be useful to encourage the respondent to take time in answering, to think more before answering, or to try hard.

Third, at the stage of judging the correct answer, some people will retrieve information but will decide that it is not sufficient to formulate an answer. They may feel ambivalent, thinking that "it depends," and a "don't know" answer saves them from deciding which position predominates in their mind. They may have latent attitudes that they are not immediately aware of, and a "don't know" answer permits them to satisfice rather than go through the hard work of optimizing their answers. Meaningful information can be gained in each of these situations if the respondents in oral interviews are prodded to report their views.

Finally, at the stage of selection of an answer, some people will not be able to translate their opinion into one of the available response alternatives, such as when they have a neutral opinion but no neutral alternative is offered. It can also occur as part of the editing process, particularly when respondents are unsure that their answers will satisfy the interviewer, in which case the interviewer can usefully reassure the respondent by saying something like "There is no correct answer to these questions; we are just interested in your opinions." Difficult questions can be preceded with a reassurance that makes respondents more comfortable, so they will not feel intimidated if they do not think they have enough knowledge to justify an answer. Also some respondents may say "don't know" as a form of refusal because their actual opinion would not be socially desirable or would otherwise reflect poorly on their self-image. Berinsky (2004, 30) terms this "opinion withholding." As described in chapter 5, alternative question forms have been developed for some sensitive questions, including the randomized response technique (Warner 1965).

The emphasis in this section has been on the cognitive basis of "don't know" answers, but they can also be seen as measurement error related to either respondents or interviewers. On the respondent side, "don't knows" can reflect question wording, particularly poor question wording which confuses respondents. Proper question wording is particularly important in self-administered questionnaires, since interviewers cannot probe don't knows. Finally, there is some research suggesting that don't knows can be an interviewer effect (Lessler and Kalsbeek 1992, 208–9) in that they occur more often when the interviewer is personable, when the interviewer feels the questionnaire is difficult, and when the interviewer considers the question sensitive.

Minimizing "Don't Know" Responses

The usual view in the survey research literature is that "don't know" responses should be discouraged. If respondents are saying "don't know" because it seems like an easy answer to give, then these answers can be minimized by changing the respondent's response calculus.

The standard strategy for altering the respondent's response calculus is to have interviewers probe don't knows. Respondents thereby learn that the "don't know" reply does not get them off the hook. Because of the many possible sources of "don't know," survey organizations should give interviewers extra training in how to deal with them. Interviewers should be sensitized as to the different meanings of "don't know" and how to sense what that answer means. If, for example, the person seems simply to be saying "don't know" as she thinks through a question, the interviewer can wait a couple of extra seconds to see if she says more. Most survey organizations tell interviewers to employ nondirective probes when people give a "don't know" answer, such as by repeating the question. Requiring interviewers to probe "don't know" answers also removes any subconscious incentive for interviewers to encourage don't knows. While it is not possible to probe "don't know" responses on Internet surveys, some Internet surveys give people a pop-up window encouraging them to reconsider such answers.

Another strategy to minimize "don't knows" is to offer incentives for answering all the questions in the survey. This strategy does not stop respondents from satisficing, but it does get more respondents to answer all the questions. However, Davern et al. (2003) report that $10–$20 incentives did not improve the data quality on the factual Survey of Income and Program Participation (SIPP) conducted by the Census Bureau, as measured by incomplete records and the need to edit responses or impute missing data. This is just a single study, but it suggests that more research is required to ascertain whether incentives would really be useful in cutting down the number of "don't know" responses.

FILTERS. A common way to handle "don't know" situations is to ask *filter questions*, preliminary questions to find out whether the person has an opinion on the topic. For example, "There has been argument as to the conditions under which abortion should be permitted. Do you have an opinion on that?" After this is answered, only people who say they have an opinion are asked the real question. A "quasi filter" includes "no opinion" as one of the response alternatives. For example, "Do you favor more controls on gun sales, fewer controls on gun sales, or don't you have an opinion on that?"

Both the filter and the quasi filter result in fewer people's answering the attitude question that would otherwise. Schuman and Presser (1981, 116–25) find that filtering raised the proportion of "no opinion" answers by an average of 22% in 19 experiments they conducted. In a classic experiment, Bishop, Oldendick, and Tuchfarber (1986) show that the proportion of people giving an opinion on a fictitious "Public Affairs Act" fell 38% when the quasi filter "or haven't you thought much about this issue?" was used.

Filtering can also affect the correlations between separate questions. Schuman and Presser (1981, chap. 4) generally find that filters do not affect the relationships between survey items, particularly between attitude items and background variables. They report only two instances in which the correlation between attitude items was affected—one in which the correlation is increased whereas it was decreased on the other, presumably because of how the no-opinion people felt about the underlying orientations even though they had no opinions on the specific items. Thus, filtering items may not always affect correlations between items, but it could either increase or decrease them when it does affect them.

There is considerable controversy in the research literature as to whether filtering procedures of either type should be used. The usual argument in favor of using a filter strategy is that it makes sure that only real opinions are measured (Converse and Presser 1986, 35–36). However, many survey researchers argue that it is best to try to force people to give their opinions (Krosnick 1991) rather than to let them off easily.

One argument against using filters on attitude questions is that they filter out some people with meaningful opinions. In a classic Swedish study on the effect of filters, Gilljam and Granberg (1993) asked questions on building nuclear power plants. The first version of the question gave as one possible answer that the person did not have an opinion (a quasi filter), while two later versions in the same survey did not offer that choice. Whereas 15% accepted the "no opinion" option on the first question, only 3%–4% gave such answers on the later questions. The correlation between the second and third answers was 0.82 among people who answered the first question, compared to only 0.41 among those who opted out originally. The lower correlation shows that "no opinion" choices do filter out some people whose opinions are less meaningful, but even 0.41 is large enough to show that the filter also removed some people with meaningful views. The authors conclude that the filter removed many people with relatively inaccessible attitudes rather than only people without attitudes; this suggests that filtering should be avoided.

Another argument against using filters is that they encourage satisficing behavior. People who select "no opinion" answers have characteristics, such as less exposure to the news media, that suggest that they are less likely to have real opinions. However, studies show that offering that response option does not increase the question's reliability, and "no opinion" answers tend to be at the same level across a questionnaire, as if question content is irrelevant (see the review of this literature in Krosnick 1999). Krosnick (1991) argues that offering the "no opinion" option encourages satisficing by implying to respondents that they do not need to do the cognitive work required to produce an optimal answer. Krosnick et al. (2002) report results of a series of experiments with and without "no opinion" options that support that conclusion. An experiment with a 1989 NES pilot study with reinterviews found that offering the "no opinion" option did not increase consistency of answers to the same question over time, suggesting that people responding "no opinion" would not have answered randomly had that option not been offered. In all of Krosnick's experiments, the proportion of "no opinion" answers was much higher when that option was offered, with an increase of 23% in a study by the Ohio State University's Center for Survey Research. As satisficing theory would predict, this increase was largest among people with less education, with 35% in the CSR study among that group versus only 16% among the most educated respondents. The education effect manifested itself for questions late in the survey, confirming the expectation that satisficing would be increased by fatigue. Krosnick et al. conclude that that the "no opinion" option may prevent measuring meaningful opinions rather than enhancing data quality by eliminating meaningless answers. Krosnick's work is likely to lead to less use of filter questions in surveys, though it will not be surprising if some later research challenges his conclusions.

Filters are also used sometimes on factual questions, and they can again affect reports of facts. In an experiment with a student sample, Knäuper (1998) shows lower reports of such behavior as witnessing a crime if a filter is employed ("Did you witness a crime in the last 10 years?") than when no filter was used ("In the past 10 years, how many times did you witness a crime?"). The direct unfiltered version implies that the event happens with considerable frequency, whereas the filtered version may imply that the event is a rare occurrence. Furthermore, a filter question can make respondents think that they do not know much about the general topic being asked about and thus discourage them. For example, failing to answer questions about politics lowered respondents' reports of their interest in politics (Bishop, Oldendick, and Tuchfarber 1984).

KNOWLEDGE QUESTIONS. A related issue is whether "don't know" responses should be encouraged or discouraged on knowledge questions, such as "Who is the president of Russia?" "Don't know" is actually a meaningful response on such questions, but there is controversy as to whether such responses should be encouraged. Delli Carpini and Keeter (1996, 305) recommend encouraging such responses on political knowledge questions, as by having the interviewer say, "Many people don't know the answers to these questions, so if there are some you don't know just tell me and we'll go on." With this lead-in, the average rate of "don't know" responses was 31% on their political knowledge items. Delli Carpini and Keeter argue that their political knowledge index is more reliable with the exclusion of don't knows that would otherwise be random responses.

Mondak (2001), however, contends that the propensity to guess is a response set. Respondents who guess tend to have higher knowledge scores than those who do not because by chance they will get some questions right that people who are not guessing will not answer, so knowledge scores would vary with the respondent's tendency to guess. To avoid that bias, he recommends discouraging don't knows, as by having the interviewer say "Many people don't know the answers to these questions, but even if you're not sure I'd like you to tell me your best guess." With this lead-in, the "don't know" rate was under 6%. Because people no longer differ as much in their tendencies to guess, Mondak argues that this approach yields greater validity, even if the index's reliability is diminished.

Item Refusal

Item refusals receive much less attention in the survey research literature than "don't know" replies because refusals are relatively rare. Indeed they are rare even on questions that most people would consider sensitive or personal. There is some feeling that a person's vote is private, but few respondents in postelection surveys refuse to say how they voted. Even surveys on sexual behavior obtain few item refusals. An unpublished study by Tourangeau, Rasinski, Jobe, Smith, and Pratt (reported in Tourangeau, Rips, and Rasinski 2000, 263–64) shows 97%–98% of a sample answered questions on sexual behavior, sexually transmitted diseases, and illicit drug use.

The question that is most notorious for refusals is the income question. People in the United States are reluctant to tell their incomes to strangers. The refusal rate on income is often a quarter to a third of respondents. Several techniques have been developed to ask the income question in a manner that people are more likely to answer. Thus, it is common to give people a set of income categories, asking them to specify the appropriate category rather than give their actual income (Juster and Smith 1997), but there is still considerable missing data on income even when broad categories are used. Indeed, the amount of missing data on income is high enough to show that income is considered so sensitive that some actual answers to that question are probably false, with some people not wanting to admit low incomes and others trying to conceal high incomes.

The general topic of sensitive questions is discussed both in chapter 5 on question wording and in chapter 12 on mode differences. Both of those discussions are relevant to understanding the psychological bases of item refusals as well as other strategies for minimizing them.

Not Ascertained

The "not ascertained" situation can occur whenever interviewers or respondents can skip over questions. Navigational errors (accidentally skipping questions) occur with some frequency in surveys where the interviewers are reading the questions from written questionnaires (as opposed to computer screens in which a single question appears on the screen at a time). That makes it particularly important to be sure that the question sequences in written questionnaires are easy to follow. Short questions should not be put at the bottom of a page where they might not be noticed. Clear arrows can be used when there are complex branching situations in which the next question to be asked depends on the answer to the current question.

Dillman (2000, chap. 3) also provides several design principles for self-administered questionnaires so that respondents understand what is being requested of them. Instructions should be placed with the question, perhaps in italics. Only one question should be asked at a time. Visual navigation guides should be provided, but they should be kept simple. For example, a consistent font should be used, with questions possibly in boldface print. The beginning of questions should be standard, and the questions should be numbered consecutively, with more space between questions than between questions and answers. Answer choices should usually be in a single vertical list, with answer boxes consistently on the right or left side of the page. Arrows can be used to help people move to a question depending on their answer to a previous question. Important words in a question can be underlined, so long as that is done sparingly. The box on this page shows an example of a page in a self-administered questionnaire that follows these principles for keeping the question sequence clear, in an effort to minimize item nonresponse.

Questions are also sometimes skipped intentionally. Interviewers can be embarrassed by questions and decide not to ask them. Interviewers sometimes skip questions when they feel that the answers are clear from previous replies, but this can still be coded as "not ascertained." Respondents in self-administered and mail questionnaires often skip questions that they do not want to answer. By the time the survey data are being analyzed, it is not evident whether skipping questions was acciden-

---

Sample Mail Questionnaire

Judicial Survey

1. How long have you served in your judgeship? _____

2. Why did you initially decide to seek this office?

_____

_____

3. In total, how many years did you practice law before joining the bench? _____

4. How satisfied were you in practicing law?
   ❐ Very Satisfied
   ❐ Somewhat Satisfied
   ❐ Neither Satisfied Nor Unsatisfied
   ❐ Somewhat Unsatisfied
   ❐ Very Unsatisfied

5. In what field of law did you spend the largest proportion of your time before joining the bench?

_____

Source: Adapted from Williams (2004)

tal or purposeful. All that is clear is that the respondent's answer to the question was "not ascertained."

Computer-assisted interviewing (including CATI and Internet surveys) can eliminate the possibility of accidental skips by requiring each question to be answered before the next question can be addressed. However, this does not fully cure the problems that create intentional skips. If CATI interviewers are so embarrassed by a question as to want to skip it, they may read the question in a nonstandard manner, possibly prefacing it with an apology. If respondents do not really want to answer the question, they may refuse to answer it. In voluntary Internet surveys that require answers to each question, respondents sometimes abort in the middle if they find a question that they are unwilling to answer. To minimize that possibility, some Internet surveys are programmed to permit respondents to skip questions. Sometimes the survey goes back and asks the skipped questions again at the end; questions that are skipped a second time are treated as refusals.

Effects of Missing Data

Missing data in the form of "don't know" responses, item refusals, and skipped questions affect data analysis. These effects are simplest to observe in looking at the distributions of variables, but they are most insidious in complex data analysis.

Reporting and Interpreting Missing Data

Missing data should be reported when the distribution of a variable is shown. Instead of showing only the proportions of people who give each legitimate answer, the report can also give the proportions in the "don't know" and "refused" categories. ("Not ascertained" responses would usually be ignored, particularly when they are likely due to interviewers' having skipped questions.) In practice, however, these categories are often not reported, because it is thought they would confuse readers of survey results, who want to see the proportion giving each legitimate answer.

In particular, it can be useful to separate out don't knows from refusals, especially when "don't know" responses are substantively meaningful. For example, when a question involves political knowledge, such as the name of the governor of one's state, "don't know" is a fully meaningful answer, equivalent to giving the wrong answer. It is essential to report the "don't know" rate for such knowledge questions.

Lore has developed in some fields as to how to interpret missing data.

For example, people who say they have not decided shortly before an election in which an incumbent is running are thought to be likely to break in favor of the challenger or to not vote, since they have had ample time to decide whether they like the incumbent enough to support reelection. Also, in elections in which a minority candidate is running, people who identify with that candidate's party but say they are undecided are thought to be leaning toward the other candidate but unwilling to admit that to the interviewer because of social desirability bias. Such lore is not always correct, however, and can even be contradictory, as the two generalizations above are in relation to opposition party identifiers who say they are undecided when the incumbent is running against a minority candidate. In the 2004 presidential election, many preelection pollsters handled undecided voters incorrectly, assuming that they would break strongly against the reelection of President George W. Bush, when in fact they ended up voting only narrowly for John Kerry.

The problems presented by missing data are compounded in multivariate analysis. The variances of the variables can be incorrect, which throws off regression coefficients, significance tests, and other forms of analyses.

These effects can be substantive as well as statistical. As an example, political surveys find that as many as a third of the U.S. population cannot respond to questions asking if they are conservative, liberal, or moderate in politics. Elite discussion of politics is often phrased in ideological terms, but research shows that many people do not conceive of politics in this way (Converse 1964; Knight and Erikson 1997). As a result, multiple regression equations for voting are problematic when ideology is used as one of the predictors. Most standard statistical computer programs simply drop from the analysis any case with missing data on any of the variables (known as "listwise deletion of missing data"), so using ideology as a predictor can result in the equation's being estimated for only two-thirds of the sample. Even worse, that two-thirds is not representative of the general population, for it is the most politically sophisticated part of the sample that can answer the ideology question. Yet regressions including ideology as a predictor of voting are usually reported without cautions about how the missing data limit generalizations from the analysis.

Missing Data Situations

Statisticians have devoted considerable attention to proper handling of missing data. They categorize item-missing data into three situations. The ideal situation is when there are no systematic differences between people who answer a question and those who do not. In this instance, when some data are *missing completely at random* (MCAR), the missing data are considered to be "ignorable." With MCAR, the cases with complete data are a random sample of the original sample. This is a strong condition that occurs when the probability of missing data on a variable cannot be predicted from any of the available data for that person. That is, the probability of missing data on a variable is unrelated to values on that variable and on all other variables in the data set. This condition would clearly be violated if, for example, people with less education do not answer the ideology question in a political survey as often as those with more education. As will be seen below, statistical inferences are safest when the missing data are MCAR.

By contrast, the most serious situation is when the missing data are *nonignorable* (NI). This condition occurs when the probability that a value is missing depends on the (unobserved) missing value, as when higher-income people are more likely not to respond to the income question in a survey. That is, missing data are nonignorable when the probability of missing data on a variable is related to the parameters being estimated. When missing data are nonignorable, the data set is not representative of the population of interest. Statistical tests generally assume that missing data are ignorable, though that assumption is not always true. Sherman (2000) provides tests of ignorable nonresponse in surveys.

An intermediate situation is when the data are *missing at random* (MAR). This condition occurs when the probability of missing data on a variable is related to the observed data but not to the person's actual value on the missing variable. That is, the probability of missing data on a variable is unrelated to the value of that variable when controlling on other variables in the data analysis. An example would be if the probability of having missing data on income were related to the person's education and gender but not to the person's actual income. Some ways of dealing with missing data lead to reasonable statistical inferences under MAR, but these conditions cannot be tested since that would require knowledge of the person's values on the missing data.[1]

Missing-Data Handling

Missing data are handled differently in different disciplines and even by different researchers in the same field. Missing-data handling is, unfortunately, often not evident in research reports and articles or is relegated to footnotes or technical appendices, without admitting that the results of data analysis can depend directly on how the missing data are handled. Three basic approaches to handling missing data will be reviewed in this section: analyzing only the available (nonmissing) data, trying to fill in the missing values, and modeling the process that leads to data's being missing.

Analyzing the Available Data

Some fields analyze survey data statistically without regard to the presence of missing data. There are three simple and common ways of doing so, but they can lead to serious statistical errors unless the missing data are MCAR. The danger is that the estimates will be biased, meaning that their expected value does not equal the value of the population parameter being estimated.

Listwise deletion (also known as *complete case analysis*) is the process of skipping in the analysis any case with missing data on any of the variables being analyzed, such as on any independent variable in a regression analysis. Analyzing only the data from respondents who answered all the questions being analyzed does lead to unbiased inferences when the data are MCAR, though standard errors will be larger so significance tests will be conservative. Biased estimates are obtained under MAR with listwise deletion, but the estimates of regression analysis are still unbiased if the probability of missing data on the independent variables does not depend on the dependent variable (Allison 2002, 6–7). Listwise deletion provides misleading results if the missing data are NI, such as using a survey to estimate average income in a community if higher-income people are more likely to refuse to give their income.

As an illustration of how serious this problem is, King et al. (2001) find that authors of articles in major political science journals in 1993–97 lost, on average, one-third of their observations when they used listwise deletion. King et al. estimate that the resultant estimates of parameter values are, on average, one standard error further from their true values because listwise deletion was used. They argue that these effects of listwise deletion are more serious than the specification bias that would occur if variables with large amounts of missing data were excluded from the regression analysis.

Another simple way of handling missing data is *pairwise deletion* (or *available case analysis*), in which correlations between each pair of variables are calculated on the basis of people with valid data on both. If the missing data are MCAR, the estimates are consistent, so that they will be approximately on target for large samples (Allison 2002, 8). However, biased parameters are obtained under MAR with pairwise deletion, and pairwise deletion provides misleading results if the missing data are NI. Furthermore, pairwise deletion can lead to a set of correlations that could never occur with a full data set, in which case several data analysis techniques, such as regression and factor analysis, cannot be employed.[2]

A third approach that is sometimes used for dealing with missing data in regression analysis is a *dummy variable* approach. Instead of an independent variable's being represented with one predictor, it is represented by two: one equal to the variable when data are not missing and equal to a constant (usually the mean of the variable) when the data are missing, and the other a dummy variable equal to 1 when the data are missing and 0 otherwise. The regression coefficients appear to give the effect of the independent variable as well as the effect of its being missing, but Allison (2002, 10–11) warns that these regression estimates are biased even under MCAR.[3]

In summary, the simple ways of dealing with missing data have limitations, unless the data are MCAR and are therefore ignorable. There is little problem if the cases with available data can be seen as a random sample of the original sample, but other procedures for handling missing data are necessary otherwise.

Imputing the Values for Missing Data

Because of the bias that is introduced in analysis of the nonmissing data when the missing data are not MCAR, statisticians have developed methods for *imputation* of values for the missing data—inserting values for them. Ideally, imputation allows standard data analysis methods to be used as they would be for complete data sets, gives estimates that appropriately handle differences between respondents and nonrespondents, yields standard errors based on the reduced sample size, and shows how sensitive statistical inferences are to different nonresponse models (Rubin 1987, 11).

However, there are several statistical concerns with imputation of missing data. First, the distribution of the variable can be distorted by the imputation of missing data, unless the distribution given to the missing data is the same as that for the actual data on that variable. Second, some imputation procedures lead to underestimating the variance of the variable, which directly affects significance testing. Third, the relationship between the variable with missing data and other variables can be underestimated when missing data values are imputed. Indeed, imputing missing data values for a single variable by itself can lead to implausible combinations in relationships with other variables. All in all, there is a trade-off between bias and variance when one is imputing values for missing data, with some of the methods for eliminating bias in mean values leading to faulty estimates of variance. Brick and Kalton (1996) provide an excellent discussion of the advantages and disadvantages of imputation, along with critiques of alternative imputation strategies.

There are many different imputation strategies. Some of these were developed when computing power was limited, but they are still used sometimes today. The *multiple imputation* strategy that is most approved in modern statistics is the most complicated, in that it actually involves imputing several different values for missing data. Imputation has been called *explicit* when the new value actually goes into the data set that is distributed to other potential users, as is common in some public-use data sets from federal government surveys. Even when imputations have not been explicitly been inserted into the data set, data analysts can use an imputation strategy. The remainder of this section will go through different imputation approaches. Some methods have multiple names in the literature; the names given here correspond to common usage.

Deterministic Imputation

The simplest forms of imputation substitute a deterministic value for the missing data. Since this assignment is uniquely determined, the exact same value would be imputed if the imputation were done a second time. The deterministic procedures described in this section generally lead to underestimates of the variance of variables, which can result in overly narrow confidence intervals and falsely rejected null hypotheses. These procedures vary considerably in terms of how much of the survey data they employ, from using none to using other people's answers to the question or using other answers that the person has given.

ALTERNATIVE SOURCE PROCEDURES. One set of imputation procedures uses information from other sources for the same person. Examples include imputing a person's race from the race of a brother (known as *deductive imputation*), filling in information from the medical records of the same person (*exact match imputation*), and using data from a previous interview with the same respondent (a *cold-deck* procedure). These procedures are highly accurate but obviously not perfect. However, such information is not always available, collecting the extra information may take extra staff time and money, and the quality of the imputation depends on the quality of the other information.

Using information collected in an earlier wave of a panel study is common but poses potential problems. It inevitably underestimates change, so it is reasonable only to the extent that the variable is stable for the population being studied. Also, it assumes that the earlier value was measured without error; this leads to a problem of correlated errors when the variable's values in the two waves are analyzed together. Panel studies have found that people may claim fewer years of education in a second wave than they claimed in the first wave (Asher 1974), showing the limits to treating the original values of even seemingly objective information as error free.

HOT-DECK PROCEDURES. By contrast, *hot-deck* procedures replace missing data with the value obtained for a similar participant (a "donor") in the same survey (Lessler and Kalsbeek 1992, 213–17), where similarity is based on one or more *assignment variables*. This strategy works best to the extent to which a set of criteria for similarity of participants is based on cross-classification of assignment variables into homogeneous categories.

The simplest sequential hot-deck approach would be to use the data for the previous similar case. Say, for example, that we wanted to know the monthly stipend given to graduate student respondents by their departments, but some did not answer that question. Say too that we know there are real differences in stipends paid by different departments but differences within each department are much smaller. We could use the student's department as the assignment variable in this instance. If the second psychology graduate student interviewed did not answer the stipend question, the hot-deck procedure would use the answer given by the first psychology graduate student. We could further refine this procedure if we had more information. For example, if beginning graduate students are normally given a lower stipend than advanced graduate students and if the stipend data were missing for a beginning sociology graduate student, we would impute the answer given by the previous beginning sociology graduate student. This example nicely shows how imputations can be based on combinations of multiple assignment variables.[4]

A related approach is *distance-function mapping*, where the imputation is from the nearest neighbor on the assignment variable (Lessler and Kalsbeek 1992, 218–19). This approach is useful when the assignment variables are numeric rather than categories as above. One could determine the nearest neighbor on the basis of summing the absolute values of differences on the various assignment variables or using more complicated multivariate algebraic functions (such as the Mahalanobis distance measure). Distance-function matching has the advantage of maintaining the statistical association of the missing variable with assignment variables. It is often viewed as another hot-deck procedure, a generalization to numeric assignment variables.

The hot-deck procedure as explained so far works well to the extent that the combinations of multiple assignment variables lead to homogeneous missing-value cases' being grouped together. The above example would not be effective, for example, if there were wide stipend disparities among beginning graduate students in a department. Hot-deck imputation assumes that people with missing data are like the donors, and the imputation is only as good as that assumption. Also, donors are sometimes used to impute missing data for more than one respondent, which creates correlated error problems.

Hot-deck imputation is one of the few statistical techniques that have been accepted by the U.S. Supreme Court. Since 1960, the Census Bureau has been using the hot-deck approach to estimate household size when it is impossible to interview people in a household directly. For example, if one family in an apartment building refuses to cooperate, its size is estimated to be the same as that of the previous household in that building that did cooperate. In the 2000 census, the state of Utah fell just 857 people below the population cutoff needed to get an extra member of Congress, so Utah initiated a legal challenge to the hot-deck imputation procedure as an unconstitutional estimation of population size in violation of the Constitution's requirement for a full enumeration. The Supreme Court decided that this usage of hot-deck imputation does not violate the Constitution (*Utah v. Evans* [2002]).

Hot-deck imputation does not distort the distribution of the variable as much as some other imputation procedures do, and it leads to more realistic estimates of the variable's variance, but this is again a deterministic value that does not include a component for uncertainty in the imputed value. Hot-deck imputation maintains the statistical association of the missing variable with assignment variables but can lead to understatement of its relationship with other variables. This has become known as the "pregnant male problem," since imputing gender on the basis of a set of assignment variables that do not include pregnancy status has been known to lead to declaring some missing respondents to be male when, in fact, they are pregnant!

OPTIMIZATION PROCEDURES. Several deterministic imputation procedures seek to maximize or minimize error. For example, a conservative approach would be to assign in place of missing data the incorrect answer in a knowledge question or the socially most undesirable answer on an attitude question. This strategy avoids errors in one direction but does so by going overboard in the opposite direction. A related approach is assigning a neutral score for missing data, such as 4 on an 7-point ideology scale where 1 means liberal and 7 means conservative. It seems less substantively distorting to assign a neutral value than any nonneutral value, but in fact this assignment method distorts the variable's distribution by building up cases in the center, which distorts its variance and its correlations with other variables.

The most common imputation procedure is *mean imputation*, replacing missing values with the mean of the variable for people who respond. This approach seems reasonable and optimizes in a least variance sense but has several statistical disadvantages. Imagine, for example, a bimodal distribution on a variable, with no cases having the mean value. Assigning the mean value to missing data would distort the variable's distribution, especially if so many cases are missing that the mean becomes the mode. That would occur, for example, if the mean income were used to replace missing data on income in a city in which nearly everyone has either a very high or very low income. As this example also illustrates, adding a concentration of cases at the mean decreases the variable's variance, which would lead to incorrect correlations with other variables. Another major problem with this method is that it assumes that respondents and nonrespondents are on average alike, which is not necessarily true.

A variant of this procedure is *mean imputation within cells* (or *conditional mean imputation*), assigning the mean value in adjustment cells. As with some of the hot-deck procedures, cases are first grouped into adjustment cells so that cases within cells are similar. The variable's mean for complete cases with each combination of assignment variables replaces missing values in that cell. This procedure is usually better than simple mean imputation, though it still distorts the variable's distribution and nonlinear estimators such as variances (which are underestimated) and correlations.[5]

When the assignment variables are numeric, *regression imputation* is possible. A regression equation is generated to predict the variable, when it is observed, on the basis of other variables, and that equation is then used to estimate values for nonrespondents. The better the predictors are, the better this approach works. This is again a deterministic procedure, giving a single fixed value for every instance of missing data, optimizing in a least-squares sense. As a result, it can distort the variable's distribution and variance. Also, it will inflate the missing variable's association with the assignment variables. If multiple regression imputation is used with all imputations based on the other independent variables, then regression coefficients are consistent if the data are MCAR, meaning that they will be expected to approximate the population parameter more closely as the sample size becomes larger.

An example of regression imputation comes from The Ohio Political Survey (TOPS) conducted in 1998. There was little missing data in the TOPS study except on household income. One in four respondents did not answer the income question, either refusing or stating they were uncertain. The staff of the Ohio State University Center for Survey Research imputed the income for those who did not answer, based on replies to several other questions, including race, gender, age, and education. The average imputed income for those who did not answer the question was much lower ($30,371) than for those who did answer ($51,832). Contrary to the usual assumption that high-income people are most likely to refuse to answer the question, this suggests that low-income people were most likely to refuse. However, the linearity assumption of regression may be yielding faulty results here if people with both very low and very high incomes are most likely to refuse to answer the question (Lillard et al. 1986).

Stochastic Imputation

The deterministic imputation procedures described so far treat missing-data substitution as fixed, when in fact there are several possible values that could reasonably be imputed for the missing data. As a result, deterministic methods generally provide biased estimates when the data are not MCAR, though, as already pointed out, the estimates sometimes are at least consistent, which helps for large samples. However, the standard errors are usually underestimated, so test statistics are overestimated and statistical significance can seemingly be obtained when it should not be. A simple way to handle this is to introduce a probabilistic element into the imputation procedure, which increases the variance so standard errors are estimated better and significance tests are more often on target.

A random element can be introduced with many of the imputation procedures that have already been described. In hot-deck imputation, a person can be given a randomly chosen assignment from the values of all the people in the assignment cell (known as *hot deck random imputation within classes*). Similarly, instead of imputing the mean value on a variable, the researcher can add a random error term to (or subtract it from) the mean, so that the imputation distorts the variable's distribution less.

Stochastic imputation can also be used when one is using the regression imputation procedure above, by adding a random error term to the predicted value.[6] This *stochastic regression imputation* procedure does not distort the variable's distribution, though the variance still underestimates the uncertainty. Unlike deterministic imputation, it maintains the statistical association of the missing variable with the assignment variables (Lessler and Kalsbeek 1992, 223). Thus adding a random component to the predicted values of the variables removes the bias, and the estimates are consistent under MCAR and MAR (Little and Rubin 2002, 65), making this one of the best single imputation methods. Still, the standard error estimates are too small, so test statistics are too large, which can lead to false rejection of null hypotheses.

The imputation methods reviewed so far all involve imputing a single value for each missing value. Unfortunately, single imputation methods do not adequately represent the uncertainty about the missing value. Even when the estimates are unbiased and consistent, single imputation generally leads to underestimating standard errors and overestimating test statistics, leading to the finding of statistical significance when it should not be found.

Multiple Imputation

A final method for imputing missing data, which is more complicated but has fewer statistical disadvantages, is *multiple imputation*. This increasingly popular method involves doing multiple nondeterministic imputations, such as regression with error terms. This procedure is repeated several times, and then a standard analysis is performed separately on *each* complete data set. The final estimate is the average of those obtained for each data set, with a variance based on the variances within each dataset and the variance in estimates across the datasets (Rubin 1987; Schafer 1997). Multiple imputation is costly in that it requires saving multiple versions of the data sets. Five to ten imputations are usually considered enough, but Rubin and Schenker (1986) find that as few as two or three imputations may suffice.

Multiple imputation has an important advantage over all the above methods: it permits assessment of the uncertainty due to imputation. For example, if the variable with the missing data is to be used in a regression analysis, the analyses would be run for each version of the data set, and the variance of the regression coefficients can be examined. Ideally the regression coefficients would be relatively invariant across the different versions of the data. If the separate estimates do vary widely, many researchers would decide against using the multiple imputation results, since replication of the analysis could produce very different results.

Further, widely varying separate estimates suggest that the missing data are probably nonignorable, in which case any conclusions based on the data would be fallacious.

Multiple imputation produces estimates that are consistent, asymptotically efficient, and asymptotically normal when the data are MAR. Doing multiple random imputations and using the variability across imputations to adjust the standard error serves as a corrective to the attenuated standard error estimates under single imputation. Allison (2002) also regards multiple imputation as the preferred method for estimating nonlinear models when there is too much missing data for listwise deletion. Multiple imputation is explained in more detail in Little and Rubin (2002) and Rubin (1987), among other sources.

Several public use data sets (data files that have been made available for wide distribution) now employ multiple imputation, but using these data sets requires running the analysis on each version separately and not pooling them. SAS does include a multiple imputation procedure, but most other common statistical analysis packages do not currently include such procedures. One publicly available multiple imputation program is King et al.'s (2001) Amelia program, at http://GKing.Harvard.edu. Allison (2002) discusses other imputation software.

Imputation Procedures

There are countless variations on imputation procedures. For example, there are some special methods for imputing missing data in panel surveys (Lepkowski 1989). Also values can be imputed for "don't know" responses based on people who responded to the question only after probing. Mathiowetz (1998) reports an interesting study that suggests the utility of this approach. She analyzed the University of Michigan Survey Research Center's Health Field Survey with answers validated with data from respondents' actual medical records. Interviewers were instructed to record initial "don't know" responses, even if probing led to a legitimate answer. The mean number of actual outpatient visits for respondents who answered without uncertainty was 4.1, compared to 6.4 for respondents who expressed some initial uncertainty, 7.1 when the interviewer had to use a scripted probe which was successful, and 8.9 when the probe was not successful. The number of visits to be reported seems to be a measure of task difficulty, since it is more difficult to remember a large number of visits than one or two, which fits with the results' showing greater uncertainty for more difficult tasks. Mathiowetz imputed missing data values, using age, gender, education, and self-reported health status as predictors. That imputation resulted in accurate imputing (defined as within one visit accuracy) for only 20% of the people who did not know their total number of visits. By contrast, limiting the donor pool for imputation to people who initially gave missing data but responded with probing led to more accurate imputation (35% within one visit accuracy). Thus, imputing missing data on the basis of "don't know" respondents who eventually answer may improve data quality on some types of variables, since people who give "don't know" replies are most like those who initially answered "don't know."

The imputation procedures described in this chapter can also be employed when unlikely values are found in the editing of data at the post-survey stage (see chapter 11). Data editing is the process of examining whether the values of variables seem right and making corrections when appropriate. Inspection sometimes finds that a few outlier values or out-of-range values are recorded, as when a value of 83 is listed for a person's years of education. When unlikely values are present, they can be purged and then imputed using one of the techniques in this chapter, such as a regression analysis using variables that usually are related to education, so as to obtain a more plausible value for the person's years of education.

Imputation is appropriate for dealing with ignorable missing data by adjusting for observable differences between cases that do and do not have missing data and assuming that the remaining differences are nonsystematic. However, missing data are often not missing at random and are therefore nonignorable, as when people who are drug users are less likely to answer questions about drug use. Since this situation is in fact quite likely, we have to be concerned about how imputation methods perform under NI missingness. Unfortunately, even basic multiple imputation leads to biased estimates under NI missingness. Since the MAR condition cannot be tested empirically, the analyst must decide on theoretical grounds whether to use imputation techniques appropriate for MAR missing data or to model the missing data as NI.

Modeling the Missing Data

When there is an understanding of the causes of missing data, *model-based* analysis can be employed by developing a model for complete data and adding a component for the nonresponse mechanism. These procedures can be used whether or not the missing data are ignorable. This approach is also used for dealing with unit nonresponse (see chapter 8).

One model-based procedure is a Heckman *selection model* (Heckman 1979), in which two regression equations are developed and solved simultaneously using maximum likelihood estimation. One of the regression equations is a substantive equation modeling the variable of interest; the second is a selection equation modeling whether the respondent answered the question. For example, if there were considerable missing data on a question involving drug use, one equation would predict drug use for people who answered the question and a separate probit equation would predict whether the person answered that question (coded as a 1/0 dummy variable). This approach requires good predictors of the response equation and requires some variables that predict answering the question but not the variable of interest. The Heckman selection model is more complex than the above methods for dealing with missing data, but it usefully forces the researcher to think through the causes of nonresponse.

As an example, Berinsky (1999; 2002) has applied a selection model approach to test the difference in effects of "don't know" answers on race-related attitudes. Using one probit equation to estimate actual opinions on race-related topics and a simultaneous probit equation to estimate "don't know" responses, Berinsky (1999) finds that people who oppose integration policies are more likely to respond "don't know." Thus, opposition to these policies seems to be greater than is found when one simply looks at the distribution of stated opinions. As an example of this approach, Berinsky's (1999) equation for planned vote in a preelection poll in the Dinkins-Giuliani 1989 mayoral race in New York City is based on partisanship and a host of demographic variables, while his equation for answering that question is based on those variables plus expressed certainty of voting as well as whether the person voted the previous year. Solving these equations simultaneously leads to a significant 0.66 correlation, meaning that people who answered are more likely to support Dinkins (an African American Democrat), whereas those who did not answer are more likely to fit the profile of Giuliani supporters. The correction for *selection bias* also lowers the predicted support for Dinkins from 55.3% to 51.8%, much closer to the 51.2% of the two-party vote that he actually received. Missing data do not appear to be random in this instance, so it is best to model the missing data.

The selection model procedure yields different types of results from those yielded by imputation. Instead of estimating values for individual cases, it shows how the coefficients for predictors change when the missing data process is taken into account. Also, it estimates the correlation between errors in the outcome and selection equations, and it yields a set of coefficients for an equation estimating whether respondents answer the question. However, this procedure is not feasible if the researcher does not know how to model the missing-data process or cannot find an available variable that predicts whether the person answered the question but does not predict the response to the question itself. Still, this a very useful approach, when it is feasible.[7]

Item Nonresponse and Survey Constraints

Missing data are inevitable in surveys. Experienced interviewers should be able to minimize missing-data problems by probing what people mean when they say "don't know" and by encouraging people to answer questions that they are reluctant to answer. Hiring good interviewers is, of course, costly, so this becomes another place where survey organizations differ. The best survey research organizations train their interviewers in procedures for minimizing missing data, but this adds to the cost of the survey. It is equally important for researchers to lay out written questionnaires carefully, so that questions will not be skipped accidentally. Also, researchers should word sensitive questions carefully, so as to minimize the extent to which respondents refuse to answer them.

There is considerable tension between how item-missing data are usually handled and statistical advice on how it should be handled. Most often only complete data records are analyzed, while cases that involve any missing data on the variables being analyzed are discarded, but this approach leads to biased results. Table 7.1 summarizes several statistical solutions for handling missing data. These solutions are complicated, but a few simple points should be remembered. The better imputation methods lead to less biased results, with better estimates of the variable's distribution, variance, and correlations with other variables. Whether mean imputation or regression imputation is used, adding a stochastic error term leads to estimates with better statistical properties than using a deterministic approach. Furthermore, creating and analyzing multiple versions of the data set with different random imputations permits analysis of the sensitivity of results to the imputation. Model-based imputation is the most sophisticated imputation approach, but it makes the greatest demands in terms of either prior knowledge of relationships with other variables (Bayesian imputation) or of the causes of item-missing data (Heckman selection models).

Statistical solutions to missing data take more time and care than just ignoring missing data. There are computer programs to impute missing data, but purchasing these programs adds to the cost of a survey. One also needs to understand what assumptions the missing-data software is

Table 7.1 Methods for Handling Missing Data

Analysis with missing data	Omitting missing data from calculations:
Listwise deletion	drop cases with missing data on any variable
Pairwise deletion	calculate correlations between each pair of variables using only cases with valid data on both
Dummy variable	include extra variable showing whether data are missing on the variable
Deterministic imputation	Impute single fixed value for missing data:
Deductive imputation	determined from values of other variables
Exact match imputation	using data from another source for same unit
Hot-deck imputation	using value for similar respondent
Distance-function mapping	using value for most similar respondent
Incorrect answer substitution	using wrong answer on the variable
Neutral imputation	assigning middle value on variable
Mean imputation	assigning mean value on variable
Mean imputation within cells	assigning mean value of cases that are similar on chosen variables
Corrected item mean substitution	adjusting item mean for person's "ability"
Regression imputation	estimating value from regression equation
Stochastic imputation	Add random component to imputation:
Hot-deck random imputation within classes	randomly choosing assignment from values of other cases similar on chosen variables
Stochastic regression imputation	adding random term to regression-based values
Combination imputation	adding random term to value from other imputation systems, such as to mean imputation
Multiple imputation	Performing several stochastic imputations:
Multiple hot-deck imputation within classes	randomly choosing values of several other cases that are similar on assigning variables
Multiple stochastic regression imputation	adding several random terms to regression-based values
Model-based methods	Model the missing data as well as the substantive process:
Selection model	develop a separate equation for the missing data process
Bayesian approach	combine previous information with data from survey
Maximum likelihood	solve for values that most likely produced data that are present

making; blindly using missing-data software can lead to false confidence in analysis. The other cost of imputation methods is that they are hard to explain to clients. Imputation methods are commonly used in large federal government surveys, but many commercial clients would not want to fabricate data. Minimally, ethics in reporting survey results requires admitting the extent of missing data as well as being open as to how they have been handled in the data analysis stage.

PART III

Respondent Selection Issues

8

Nonresponse Error at the Unit Level
THE RESPONSE RATE CHALLENGE

Nonresponse error can be at the unit level as well as at the item level. Unit nonresponse occurs when some people in the designated sample are not interviewed. The most intriguing controversy on nonresponse error is the extent to which efforts should be made to maximize response rates. A high response rate has traditionally been considered essential to guarantee that the sample is representative enough to be credible, but studies are beginning to find that lower response rates do not necessarily lead to greater response bias.

Statistically, as with nonresponse error at the item level, nonresponse error at the unit level can bias results if the response rate is low and/or if respondents and nonrespondents are dissimilar.[1] As stated in chapter 7 for item nonresponse, the true mean of a variable equals its sample mean plus a bias equal to the missing data rate multiplied by the difference between the means for respondents and nonrespondents. Thus, the amount of bias depends on the nonresponse rate and on how different respondents and nonrespondents are. Nonresponse error causes bias if nonresponse and survey variables have a common cause (Groves and Couper 1998, chap. 1). Survey results would be biased, for example, if fear of crime caused people to refuse to answer a survey measuring fear of crime, or if Democrats were less likely to participate in a voting survey. Similarly, postelection survey estimates of voter turnout are invariably biased upward because the people who are most likely to vote are also most likely to participate in surveys.

This chapter begins with a discussion of the sources of survey nonresponse. The second section discusses theories of survey participation, and the third section summarizes empirical studies of cooperation along with methods for increasing cooperation. The response rate controversy and statistical solutions to unit nonresponse will be reviewed in the fourth section.

Survey Nonresponse

There are three types of unit nonresponse. *Noncontact* refers to situations in which the designated respondent cannot be located, such as when no one answers the phone or when a mail questionnaire cannot be delivered. Noncontact is basically a logistics problem; it is "ignorable" nonresponse (see chapter 7) if being at home is essentially random with respect to the variables being measured, which is usually the case.

Incapacity is when the designated respondent is incapable of being interviewed, due to such reasons as physical or mental health, speaking a different language from the interviewer's, illiteracy in self-administered surveys, or technological inability to deal with an Internet survey. Incapacity is largely beyond the researcher's control; it rarely causes nonresponse bias, except when it is related to the subject of a survey, as when health surveys cannot interview people too ill to participate.

Noncooperation occurs when the designated respondent refuses to participate in the study. Noncooperation is likely to be nonignorable if people are not willing to participate in a survey because of its topic, such as if people with conservative social values were less willing to answer a survey on sexual behavior. The cooperation issue involves important psychological considerations that will be detailed in this chapter.

Survey reports often include separate contact and cooperation rates. The contact rate is the proportion of cases in which some member of the household unit was reached, whereas the cooperation rate is the proportion of all eligible units ever contacted that were interviewed. These rates can be combined into an overall response rate, which is the proportion of eligible reporting units that led to interviews. Some surveys also report a refusal rate—the proportion of all potentially eligible cases in which the respondent refuses to be interviewed. These statistics are often thrown around loosely, without clarity as to how they were calculated, so the American Association for Public Opinion Research (AAPOR 2004) has prepared a standard set of definitions, which are detailed in the appendix to this chapter.

Contactability

Although the main focus of this chapter is on cooperation, it is appropriate to turn briefly to the other main aspect of unit nonresponse: the problem of contacting the respondent. People are often not at home when interviewers attempt to contact them. Groves and Couper (1998, chap. 4) construct a model of contactability in which the relevant factors are physical impediments (such as locked gates or caller ID machines that impede access), at-home patterns, and when and how often the interviewer tries to contact the household. These factors are influenced by social-environmental factors and social-demographic attributes.

It is sometimes possible to obtain information about contact patterns. When contact is made with the appropriate household but not with the designated respondent, the interviewer may be able to obtain supplemental demographic information about the intended respondent. For example, using information ascertained from neighbors or other family members, Brehm (1993, 44–47) reports that the elderly, people with children, women, and those with higher incomes are more likely to be accessible in face-to-face surveys.

There are also time-of-day and day-of-week patterns to survey participation. Telephone interviewing is more successful during the evening than during the day, except on Saturdays and Fridays. Also, different types of people are home at different times of day, with older people more likely to be at home during the daytime. Lavrakas (1993) gives information on the best times for telephone interviewing. Fortunately, a study of telephone interviewing does not find significant differences between months of the year (Losch et al. 2002).

Tracking polls that report daily statistics are affected when particular types of people are less likely to be home some days than others; for example, younger people are less likely to be at home on weekend evenings. One statistical solution to this problem is to report only three- or four-day moving averages. That solution averages out the effects of sampling any particular day so as to avoid a Friday-night or Saturday-night effect.

Panel Studies

Panel surveys inevitably lose some respondents at subsequent waves (known as *mortality, attrition,* or *wave nonresponse*). Panel mortality is partly a recontact problem if the person has moved. The cumulative effect can be considerable: for example, not being able to follow 10% of the participants in the previous wave of a panel study would cumulate to being able to recontact less than half of the original sample by wave eight.[2] Panel mortality can be a serious problem, biasing the sample toward residentially stable households (Winkels and Withers 2000). Some panel studies try to follow movers, but others do not. Young people are especially mobile, so panel studies become unrepresentative in terms of age if they do not try to track young people.

Following respondents who move is itself a complicated operation (see Burgess 1989). One approach, called *forward tracing*, involves asking people at the initial interview for the name and address of someone who will know their address if they move. An alternative is "tracking" respondents, such as by sending them periodic reports or greeting cards between waves, so they can be tracked by the change-of-address notifications that the post office provides.

Panel studies can examine what types of people drop out by examining the data for their first interview. Attrition is not random: O'Muircheartaigh (1989) points out that the attrition in SIPP was around 20% in five waves for people who rent their room for cash, respondents under age twenty-four, nonwhites, and married people whose spouse was absent. Lepkowski and Couper (2002, 265–68) found that ability to locate the respondent again was related to sociodemographic and geographic variables (with African Americans less likely to participate again), while on an election study panel survey (Lepkowski and Couper 2002, 268–71) the ability to locate the person again was related to accessibility (people who provided phone numbers being easier to locate again) and survey experience (people with less missing data being easier to locate again).

Callbacks

A simple response to survey nonparticipation would be *substitution* for the people who were not interviewed. When someone cannot be contacted, the interviewer would just substitute a neighbor or a respondent who seems similar. Statisticians, however, strongly argue against this approach. Not only is there no way of telling how well the replacements substitute for those who were not interviewed, but there is every reason to believe they would be different. At the extreme, consider a one-evening telephone poll on a breaking news story. Substituting new respondents for people who could not be contacted would further increase the poll's dependence on the type of people who are at home and easy to contact in the evening, making the sample even less representative of the population. An example of an exception when substitution makes sense is when needed information can be obtained by talking to a knowledgeable informant, as when one person monitors the health of his or her entire household. Substitution is most common in marketing research, where the conventional wisdom is that purchase patterns would not be affected by the variables that cause survey nonparticipation.

The alternative to substitutions is callbacks—trying again to contact the desired respondent. This approach is common in most survey modes, though cost often constrains the extent of callbacks. Repeated callbacks can be expensive, which creates a trade-off between boosting the response rate and keeping survey expenses within budget. Survey organizations differ considerably in how many callbacks they do, with five being a common number for phone interviews. Market opinion surveys often do not bother with callbacks at all, whereas some academic survey organizations will call phone numbers back more than a dozen times. Telephone surveys routinely find that the percentage of people successfully contacted goes down with each callback, suggesting a rule of diminishing returns (Groves and Couper 1998, chap. 4). In any case, the number of callbacks is often set to vary depending on the disposition on early calls, for example with more callbacks when a respondent is selected than when no one was at home.

Callbacks are made at different times of the day and days of the week to increase the chances of contacting the household. In face-to-face interviewing, the interviewer will seek to return to the area later and try again to locate the respondent. In computer-assisted telephone interviewing, call-scheduling software can keep track of codes for call outcomes, schedule calls, set priorities on calls, assign calls to interviewers, keep track of callbacks, keep track of refusal conversions, close out cases, refer calls to supervisors, and generate reports for monitoring interviews. This leads to a more systematic history of the calls than is likely when the records are kept by hand.

To give an idea of the success of callbacks in telephone surveys, Groves et al. (2004, 171) show a summary of five separate surveys in which only 40%–60% of those eventually interviewed were interviewed on the first call. In other words, callbacks can double the contact rate. An additional 10%–20% were interviewed on the second call, another 10% on the third call, and about 5% on the fourth, with fewer on each of the following calls. It is difficult to generalize from these results without more knowledge of the specifics of when calls were being made in these individual surveys, but this study has important implications for surveys with short polling periods. Overnight phone surveys will clearly have much lower contact rates than surveys that have time for callbacks. The news media often conduct overnight polls to gauge public reaction to an important news event. The inability to attempt callbacks in such polls is a problem because the type of people who happen to be at home one evening are not likely to be fully representative of the population at large, particularly underrepresenting young men. As a result, overnight polls should not be viewed as careful surveys. Even three-day surveys that the media often conduct are likely to have low response rates, because they are not able to conduct many callbacks and refusal conversions.

Research has also been done on callbacks. Groves and Couper examined how households that were contacted in the first attempt differ from those that were not contacted until later attempts. For example, they found higher contact rates at the first attempt for households with more adults, young children, and elderly people in the household. As to social-environmental factors, Groves and Couper report lower contact rates in urban areas, though that is explained in some of their models by multi-unit structures and the crime rate. Several telephone surveys have compared the demographics of participants according to the number of callbacks required in order to contact them. Typical findings are that women, unemployed, married, older and less educated people are more likely to be reached with only a few callbacks (Merkle, Bauman, and Lavrakas 1991), and more callbacks help mitigate these biases. That finding implies that the extra effort to reduce noncontacts would not introduce bias.

An important caution is that nonparticipants in a survey can differ from people who were more difficult to contact. Studies on different survey topics support this conclusion. Analysis of six British surveys (Lynn et al. 2002) found that the households added by efforts to reduce the noncontact rate were different from those that were not contacted originally, especially as regards age and education. Similarly, when Lin and Schaeffer (1995) used official records on child support payments, they determined that people who could not be contacted differed from those who were difficult to contact. Those who could not be located owed or paid the least amount of support payments.

Unfortunately, working to increase the response rate can also increase the refusal rate. The case in point is a 1986 phone survey by the Census Bureau that found that doubling its usual survey period for the National Crime Survey from two weeks to four increased both the response rate and the refusal rate by 3% (Sebold 1988). Thus, an additional 6% of phone numbers were eventually answered after repeated calls when the survey period was extended, but this increased the cost of the study with only a marginal increase in response rate and a corresponding increase in the refusal rate. Researchers must decide whether the increased effort is cost effective under these circumstances.

Theories of Survey Participation

While contactability is important in surveys, cooperation is a more serious problem. The theories of survey participation focus on what leads people to cooperate by participating in surveys and therefore how to increase the chances that people will cooperate. There are economic, sociological, and psychological theories of survey participation.

The simplest approach is a *cost-benefits* economic perspective, viewing participation as a matter of costs versus benefits. People are more likely to participate if the costs are low (such as a short survey) and if the benefits are high (such as being paid to participate). The cost-benefits logic is certainly important, but by itself it does not provide a sufficient handle on survey cooperation.

Social exchange theory (Blau 1964) is the basis for one sociological approach to survey participation. This theory assumes that people's actions are motivated by the return that people expect to obtain. The exchange concept emphasizes that the survey must have value to the respondents in order to interest them. Social exchange theory is based on establishing trust that the rewards for participation will outweigh the costs. From this perspective, giving the respondent a monetary incentive to participate in the survey can be seen as a kindness that evokes a norm of reciprocity. The designated respondent feels that he or she must participate to reciprocate the kindness of the incentive. This interpretation implies that the nature of the incentive is less important than the act of giving an incentive, so that a token incentive should work as well as a large one.

Several psychological approaches to survey participation have been advanced. The most basic is the notion of *altruism*, viewing survey participation as helping behavior for a stranger. Interviewers often include in their request for participation "It would really help us out" as a way to get people to think altruistically. Another psychological approach is based on *opinion change*, trying to convince the person that the survey topic is salient, relevant, and of interest to the respondent. This approach also implies that simple messages requesting participation are likely to be more successful than ones that are hard to comprehend. However, in line with Petty and Cacioppo's (1986) distinction between central and peripheral routes for persuasion, a direct approach to obtain participation may not work as well as a peripheral approach would.

The request for a person to participate in a survey can be seen as a request for compliance. Looking at the factors that affect compliance with requests more generally, Cialdini (1988) has differentiated six influences on compliance, each of which suggests a peripheral persuasion approach to get a designed survey respondent to participate in the survey. The first is reciprocation; incentives for respondents are a way to invoke this mechanism by making them feel they owe something to the interviewer. The second is consistency; interviewers use this when they first get their foot in the door with an initial question and use that as a lever to get people to continue with the interview. Third is social validation. Interviewers sometimes attempt to invoke this mechanism by mentioning to respondents that many of their neighbors have participated or saying that most people who have participated have enjoyed doing the interview. Fourth is liking, which is invoked if the respondents like the interviewer. Fifth is authority, getting people to comply with people in power, as when a survey emphasizes that it is sponsored by the federal government. Sixth is scarcity, making people value a rare opportunity, as when emphasizing that survey participation is important because the researcher needs to hear the experiences or views of people like the participant. Thus, there are several different possible ways of getting people to comply with the request of being a survey participant. Groves, Cialdini, and Couper (1992) review these compliance approaches in more detail.

Brehm (1993, chap. 4) includes several psychological factors relating to the individual in his model of survey compliance. Brehm's model consists of four predictive factors: the respondent's relationship to strangers (including suspicion and fears regarding confidentiality), relationship to the interviewer (rapport and empathy), relationship to the interview (time and interest in the topic), and relationship to self-image (if the respondent views himself or herself as a helper, is healthy, and is free from family stress).

Subsequent models of survey participation have combined economic, sociological, and psychological considerations. Groves and Couper (1998, chap. 2) develop a conceptual framework that includes four factors that affect survey participation. The social environment is relevant in terms of economic conditions and neighborhood conditions for face-to-face interviews. The survey design, particularly the topic of the survey and how it is administered, affects participation. The household being approached is a third factor, including its sociodemographic characteristics. Fourth, the interviewer is an important aspect of obtaining cooperation, with the interviewer's characteristics, experience, and expectations all affecting the contact between the interviewer and the household. Groves, Cialdini, and Couper (1992) call attention to a fifth factor as well: the interaction between the respondent and the interviewer. The initial conversation between the interviewer and the respondent plays a crucial role in the decision to participate. The decision of a person to participate in a survey thus depends on the interaction between the social environment, the survey design, the household, and the interviewer.

This model is extended in Groves, Singer, and Corning (2000) in their "leverage-saliency theory." This theory recognizes that different individuals will be affected by different matters in considering whether to participate (leverage). Some people give considerable weight to incentives or to the topic of the survey or the authority of the sponsor, while others give more weight to the how big a burden participating will be. Additionally, the request to participate makes particular points salient to people when asking them to participate. The authors model the decision to participate as a function of the products of those factors. For example, in a survey about community matters, the survey has high leverage for a person with high community involvement, so that can positively affect the chances of participation if the community aspect is made salient in the request to participate, whereas making that salient would not increase the likelihood of participation of a person with low community involvement. Note that this logic works to keep response rates low on mail questionnaires: respondents see the topic at a glance, so a mail survey may lose people not interested in the topic unless other factors such as incentives or authority prevail in their minds. By contrast, respondents on interviewer-administered surveys cannot gauge the topic of the survey as easily, so the topic has less negative potential. This leverage-saliency model is likely to provide a comprehensive framework for future work on this topic.

The leverage-saliency theory has an important implication for getting people to cooperate: the request for participation should ideally be tailored to the individual respondent, using an appeal that fits the person's concern about the survey topic, the identity of the sponsor, and so on. The tailoring strategy will be discussed at length later in this chapter.

The Noncooperation Problem

The noncooperation problem has become very serious. Refusals were not a serious problem in the 1950s and 1960s but have increased since then.

For example, a 2001 survey by the Council for Marketing and Opinion Research (Sheppard 2001) found that just under half of the sample said that they had refused to participate in a survey in the previous year, compared to 36% in 1990 and 19% in 1980. The National Election Studies (NES) surveys had refusal rates in the low teens through the early 1970s but has had 20% and above since then (Luevano 1994). The 1990 NES survey had 20% refusal, compared to only 2.5% noncontact and 6% "other noninterview" (Couper 1997). The telephone interview response rate for the Survey of Consumer Attitudes stayed about 70% from 1979 to 1996, but maintaining that rate by 1996 required twice as many calls to complete an interview and required refusal conversion on twice as many cases (Curtin, Presser, and Singer 2000). A Pew Research Center study (2004) found that the response rate on a standard telephone survey that was in the field five days fell from 36% in 1997 to 27% in 2003, with the difference being due to 20% less cooperation of households that were reached. Thus the refusal rate has increased as surveys have become more common. Yet high response rates are demanded in some situations, most notably the 75%–80% rate expected by the Office of Management and Budget (OMB), which must approve most surveys by the U.S. federal government.

Internet surveys were originally seen as a possible panacea, with the ability to contact large numbers of potential respondents at minimal cost. However, as they have become ubiquitous, unit nonresponse has become a problem with Internet surveys too (Vehovar et al. 2002). Krosnick and Chang (2001) provide rare detail on response rates on two Internet pre-election surveys taken in June–July 2000. Only 18% of the 12,523 Harris Interactive panel members who were contacted participated in the survey. A Knowledge Networks survey had a response rate of 28%, calculated as the proportion of people who were originally phoned and asked to join an ongoing panel who agreed to join the panel, had the WebTV receiver installed, and participated in this particular survey. Calculated as a proportion of 7,054 ongoing Knowledge Networks panel respondents who were asked to participate in this survey, the rate of participation was 70%. For comparison, a parallel telephone survey by the OSU Center for Survey Research obtained interviews from 43% of the 3,500 eligible phone numbers.[3]

These figures show that the phone survey had a much higher response and the Harris Interactive survey had the lowest. However, the lower participation rates for the Internet surveys are partially balanced by the larger size of their ongoing panels, usually around 10,000 for Harris Interactive, so that there were more participants in the Internet surveys than in the telephone survey at a lower cost. Survey researchers have to decide how they view this trade-off between response rate and sample size.

Studies of Cooperation

Several studies have sought to assess the importance of some of the factors that have been hypothesized to affect survey cooperation. Some studies obtain a record of the participation process by having interviewers document their initial interactions with people, both those who agree to participate and those who do not. This permits the researcher to be able to estimate a response propensity function and develop models about nonparticipants. Although it is usually not possible to find out information about people who do not participate in a study, some estimates can sometimes be obtained. For example, people who would not cooperate may be like the actual respondents who were more reluctant to be interviewed.

Some research looks at single factors that affect cooperation. For example, studies show interviewer effects on unit nonresponse. Groves and Couper (1998, 211–13) examined face-to-face surveys and found that interviewers with more experience and more confidence had more success in obtaining respondent cooperation. They also found (chap. 8) more negative comments about survey participation when people were contacted on the phone versus more questions about the survey in face-to-face contacts, with fewer of both types of behavior being obtained by experienced interviewers, probably because of their skills, including their ability to anticipate concerns.

In a rare study of survey administration effects on cooperation, Marton (2004) analyzed a series of 53 monthly Buckeye State Poll surveys taken by the Ohio State University's Center for Survey Research from late 1997 through early 2002. After detrending the time series, she found that cooperation rates were lower in months when the survey was longer and in months when there were more new interviewers.

Brehm (1993, chap. 4) was able to do a multivariate test of parts of his survey compliance model through secondary analysis of past surveys, though few surveys have all the needed measures of the respondent's relationship to strangers, to the interviewer, to the interview, and to self-image. As an example of his findings, the relationship to the interview was much more important than relationship to strangers for the 1989 Detroit Area Study. Relationship to the interview was again the most important factor for the 1988 NES, relationships to the interviewer and to self-image were of moderate importance, and relationship to strangers was the least important factor. Although generalizing from such few studies is hazardous, Brehm's analysis suggests that interest in the topic, time for the interview, and having had a bad past experience with interviews are among the most important factors in affecting survey compliance.

An experiment by Groves, Presser, and Dipko (2004) finds some support for the leverage-saliency theory. The researchers sampled from lists of four groups of people (e.g., teachers), developed tailored introductory survey requests that were appropriate for each group, and then randomly assigned the tailored requests to respondents so that the effects of tailored requests could be measured experimentally. They found that the odds of cooperation were about 40% higher on the topics of interest and that monetary incentives did not have significant added effects. There is a potential biasing effect if tailoring leads people who are more interested in a topic to be more likely to agree to be surveyed, but the authors estimate that effect as being only in the 1%–2% range. Thus this study implies that tailoring the request to participate would not lead to serious response bias.

Some types of surveys encounter specific problems. Nonresponse in the election exit polls conducted by the Voter News Service (VNS) was primarily due to refusals. In 1996, for example, the average refusal rate was 33%, plus 10% "misses" when the interviewer was too busy to intercept the voter or when the voter did not pass by where the interviewer was stationed (Merkle and Edelman 2000). Laws in some states regulate how close exit interviewers can be to polling places; in Texas, for example, barring interviewers from being within 100 feet means that people who park close to the polling place never have a chance to be interviewed. By 2000, the VNS response rate fell to 51% (Konner 2003). Merkle and Edelman (2002) find that the response rate was affected by such factors as the number of questions on the front of the questionnaire and how far from the polling place the interviewer must stand. Fortunately, response rate by precinct had little power in explaining deviations of the precinct's exit poll result from its actual vote. Still, nonresponse is considered to be one of the main factors that contributed to exit polls' overestimating the Kerry vote by about 5.5% nationally in 2004.

Human interviewers increase response rates. A recent development in phone surveys is interactive voice response (IVR), the system in which people are phoned and expected to answer recorded questions. That approach is not likely to be very effective, since respondents often hang up as soon as they realize they are listening to an impersonal recording. Beginning with a human contact to request their cooperation for a survey is likely to be more effective, even if the person is switched to the IVR system for the actual survey, though people may still not stay with a very long IVR survey. Tourangeau, Steiger, and Wilson (2002) evaluated four IVR studies by the Gallup Organization that began with a human interviewer. They found that IVR breakoffs were more common than in parallel CATI interviews; breakoffs varied with the length of the survey, with just over 1% on a 6-item survey and 31% on a survey averaging 24 items.

The factors that affect cooperation in surveys of businesses are somewhat different. Cooperation is likely to be adversely affected by the company's staff resources, which can get strained when the economy is poor and as government filing requirements increase. Response rates in surveys of businesses have not fallen as much as they have for household surveys, but Willimack, Nichols, and Sudman (2002) stress the importance of burden reduction in order to obtain cooperation in business surveys, such as through providing alternative response modes and relaxation of time and data needs.

Reasons for Noncooperation

Some surveys record the concerns that people raise when they refuse to be interviewed. Brehm (1993, 53–56) finds that blacks are more likely to be rated as suspicious of the interviewer whereas Hispanics are more likely to raise confidentiality concerns. Higher-income people are more likely to say they are "too busy" to be interviewed, while older people are less likely to give that explanation (Brehm 1993, 59–61). People who are less interested and informed about politics are more likely to refuse at first to participate in a political survey (Brehm 1993, 61–64).

The comments that people make when being asked to participate can also provide useful information about refusals. Couper (1997) analyzed the comments that respondents made in the introductory conversations in the 1990 NES. People who had said they were "not interested" were more likely to have missing data, reply "don't know," and say little on open-ended questions, suggesting a lack of cognitive involvement in the interview. By contrast, people who said they were "too busy" were less likely to engage in satisficing behavior, with less missing data. Couper agrees with Brehm that people who are less interested in politics are more likely to refuse to participate in a political survey, so that the final sample overstates the level of political interest among the public.

Couper and Groves (2002) analyzed comments made in the introductory interactions in 101 telephone interviews. Time-delay statements (such as "I'm busy now") were common (63% of the calls), as were questions (35%), and negative statements (27%). As to be expected, people making negative statements were less likely to cooperate, with a cooperation rate of only 56%, 26% lower than the rate for people who did not make such comments. Yet that still means that most people who made negative statements still cooperated in the end. Interestingly, the cooperation rate was not lower for people who made time-delay statements or who asked questions.

Demographic Correlates of Cooperation

Much of the literature on survey cooperation looks at the demographics of cooperation, since it is easier to assemble data on demographics than on psychological correlates. Some studies use information in the sampling frame about the nonrespondents or ask others (proxies) about the nonrespondents. The demographics of participants are often compared with Census Bureau data to ascertain characteristics of nonparticipants, though it is important to realize that such comparisons combine contact and cooperation issues. Additionally, it is possible to see what types of people are most likely to drop out of multiwave panel studies in order to gain information about patterns of cooperation.

Several examples of common findings show what we know about the correlates of survey participation in the United States. Age is one common correlate, though the pattern is complicated because the age groups with higher cooperation rates have lower contact rates. Young people are more willing to participate if they are successfully contacted, but they are less likely to be at home, which results in underrepresentation of young people in some surveys. For example, a greater cooperation rate was found in government face-to-face surveys among households in which everyone was under age thirty (Groves and Couper 1998, 133–37) and in families with young children (Groves and Couper 1998, 140–41). Still, the greater difficulty in finding younger people at home led to an underrepresentation by as much as 6%–9% of people under thirty in the face-to-face surveys taken by the General Social Survey (GSS) and National Election Studies (NES) from 1978 to 1988 (Brehm 1993, 26–29). The pattern reverses for older people: they are less likely to cooperate but are more likely to be at home, which typically balances out to overrepresentation of older people in surveys. Kaldenberg, Koenig, and Becker (1994) obtained a lower response rate from a mail survey of retired people with increasing age. Older voters are also less willing to participate in election exit polls, though that difference diminishes when older interviewers are used (Merkle and Edelman 2000). Though older people were more likely to refuse to participate

(Groves 1989, chap. 5), their greater likelihood of being at home led an overrepresentation by 2%–9% of people over sixty-five in the face-to-face 1978–88 GSS and NES surveys (Brehm 1993, 26–29).

Surveys routinely underrepresent men. Brehm (1993, 30–31) finds that NES and GSS face-to-face surveys undercount men by 4%–8%. He concludes that men are less likely to be at home and more likely to refuse. Also, phone surveys tend to get a higher proportion of women than men because there are many more female-only households than male-only households and because women are more likely to be at home and to answer the phone (see also the discussion of gender and within-household selection of respondents in chapter 10). Women are also more willing to participate in election exit polls (Merkle and Edelman 2000).

The differences are less consistent on race. NES face-to-face surveys overestimate the proportion of blacks by as much as 5%, whereas its phone surveys underestimate the proportion by 2%–6% (Brehm 1993, 29–30). Brehm suggests this pattern is due to blacks' being less likely to refuse to be interviewed, along with a lower proportion of blacks having telephones. Groves and Couper's (1998, 137) multivariate analysis finds a greater cooperation rate among Hispanics in government face-to-face surveys, whereas racial differences were not significant when the age composition of the household was controlled.

Educational differences are also complicated. NES and GSS face-to-face surveys have higher proportions of people who did not graduate from high school than Census Bureau statistics show, but NES telephone surveys have lower proportions of non–high school graduates (Brehm 1993, 31–32). People with higher education are more likely to return mail questionnaires (Groves 1989, chap. 5).

In Marton's (2004) analysis of 53 monthly Buckeye State Poll surveys described above, months with higher response and cooperation rates generally had distributions of race, gender, and income that were closer to census values. The exceptions involve age and education, partially echoing the Keeter et al. (2000) result that a study with a lower response rate had an education level more similar to census values. There was not much variance between months in some of these monthly series, but this study still suggests that surveys with higher response rates are likely to be more representative demographically.

Some plausible factors have not been found to affect cooperation rates. In particular, Groves and Couper's (1998, 123–31) analysis of several federal government surveys find that cooperation rates were not affected by the number of adults in the household or amounts of discretionary time.

However, there was greater cooperation among people with lower housing costs, which suggests that higher cooperation was obtained with lower-income people.

As to Internet surveys, a 2001 telephone and Internet study by the Council for Marketing and Opinion Research (Sheppard 2001) found that males were more likely to refuse the Internet study. In both modes, males reported having refused during the previous year more often than did females. Also, those with higher educational levels and higher income were more likely to report having refused to participate in surveys in the past year, whereas older people were less likely. No differences were found among race or ethnicity categories.

Note that some of the separate correlates of cooperation discussed above overlap with one another. For example, face-to-face surveys report less success in areas with high crimes rates and high population density, in large metropolitan areas, and in housing units that are in bad repair (Groves and Couper 1998, chap. 4), situations that overlap to a considerable degree. Also, Groves and Couper (1998, chap. 4) find greater success in households with several adults, with children under age five, and with adults over age seventy, all three of which may simply be instances of greater success with households in which there are more likely to be adults at home.

Multivariate analysis is required to determine which factors have separate effects on cooperation. The correlates that have been discussed so far are important, but some may be spurious artifacts of other correlates. Two studies show that many demographic correlates survive such multivariate controls. Brehm's (1993, 132) multivariate analysis of demographic factors in the 1988 NES study found significant effects associated with income, gender, Hispanic ethnicity, race, and whether the respondent and interviewer were of different races (which depresses participation), with the participation status of 84% of the cases predicted correctly from these factors. Groves and Couper's (1998, 146–50) analysis found low housing costs, young and elderly households, and households with young children had greater cooperation in government surveys, whereas single-person households had lower cooperation, with several other factors controlled.

While there have been many studies of the demographics of nonparticipation, demographic correlates do not necessarily translate into response bias. For response bias to result from demographic imbalance, the demographic in question would have to relate to the behavior being measured. For example, an election survey that overrepresented women would be off the mark if there is a gender gap in voting, but an election survey that overrepresented left-handed people would be unlikely to produce biased results. As discussed later in this chapter, it is possible to weight the sample to correct for some demographic differences, but weighting makes assumptions about nonresponse that may be wrong. As a result, understanding the factors that lead to nonparticipation is more important than focusing on the demographics of participation.

It is worth emphasizing too that willingness to participate in surveys may differ between countries, so the extent of nonresponse error is likely to differ across countries in multicountry surveys. The correlates of cooperation and effects on response bias are also likely to vary across countries.

Panel Surveys

Panel surveys face special problems maintaining cooperation. People may refuse to be reinterviewed, depending in part on recollections of their feelings about the previous interview. The greatest threat to reinterview refusals is usually at the first attempt. For example, the 1984 Survey of Income Program Participation (SIPP) experienced a 5% nonresponse rate in the first wave, 3% more in the second wave (not counting people who moved), 2% more in each of the next two waves, and so on, until the total attrition rate not due to moving was 16% by wave 9 (Bailar 1989, 6–7). In practice, people who refuse to be reinterviewed in one wave are sometimes willing to be reinterviewed in a subsequent wave. SIPP does try to recapture nonrespondents from previous waves; for example, 5% of fifth-wave respondents had missed at least one previous interview (Kalton, Kasprzyk, and McMillen 1989). Kalton and Brick (2000) detail the issues involved in weighting panel surveys to handle wave nonresponse.

When looking at what types of people are most likely to continue cooperating on panel studies, Lepkowski and Couper (2002, 265–68) found that their cooperation was related to community attachment (with renters less likely to participate). The pattern was somewhat different on an election study panel survey (Lepkowski and Couper 2002, 268–71), with cooperation being affected by survey experience (people who were reluctant in the first wave being less cooperative in the second) and accessibility (people who moved in the past three years participating less).

Panel studies have both additional advantages and potential difficulties as regards unit nonresponse issues. Panels with overlap and/or rotation allow comparison of the remaining part of the panel to the new part of the sample. However, there are potential definitional problems with household panels, since the composition of many households changes over time as people move in or out.

Techniques for Increasing Response Rates

The most common strategies for increasing the response and cooperation rates include advance letters, incentives, tailored requests to participate, refusal conversions, and easing the technological burden. These strategies will be discussed in the remainder of this section, with discussion of how they play through for interviewer-administered surveys, self-administered surveys, and Internet surveys. Note that substitutions to compensate for noncooperation are considered undesirable for the same reasons they are undesirable as a means to compensate for noncontact.

Advance Letters

Advance letters (also called *prenotice* letters) are sometimes sent for surveys so that the respondent recognizes that the interview attempt is legitimate rather than a disguised sales attempt. Traugott (1987) finds that advanced contact increases response rate on face-to-face surveys by more than 10%. There do not seem to be backfire effects against advance letters, except for a study showing one if the letter is sent a month in advance of contact (Groves and Couper 1998, chap. 10). Although studies differ on the impact of advance letters for telephone surveys, a study in Arizona (Goldstein and Jennings 2002) found a 16% gain in cooperation rates among contacted households with letters, with a more accurate representation of older people and men. Often telephone numbers are selected randomly for phone surveys, without knowledge of the person's name or address, but advance letters can be sent by using services like Telematch, which do reverse searches to obtain names and addresses for a list of sampled phone numbers (see http://www.gannettoffset.com/html/telematch.html). Advanced notification is also useful in e-mail and Internet surveys (Cook, Heath, and Thompson 2000; Kaplowitz, Hadlock, and Levine 2004), to the extent to which it is possible to do so. Advanced contact is especially necessary for interviewing elites.

As to the content of advance letters, Dillman (2000, 156) suggests that they be short, personalized, and worded in a positive manner. The letter should give only a general description of the survey topic but should stress how important the survey is. If an incentive is being sent with the questionnaire, the letter can indicate that a "token of appreciation" will be provided.

One approach to avoid in Internet polls is bombarding newsgroups and other e-mail lists with announcements of them. This approach is considered equivalent to spamming, and it has resulted in flaming of even well intentioned researchers.

Incentives

An increasingly popular approach for dealing with noncooperation is giving respondents incentives to participate. The social exchange interpretation of an incentive is as a kindness that evokes reciprocity as well as demonstrating the thoughtfulness of the giver, so long as it is not seen as an indication that the survey will be unusually burdensome. An alternative interpretation is in terms of the cost-benefit trade-off, viewing the incentive as a payment for services, so that it may be more effective with low-income respondents. Incentives can also be targeted, given only to respondents who are deemed the least likely to cooperate. For example, Groves, Singer, and Corning (2000) show that incentives have a larger effect for people with low involvement in the topic of the survey but have less effect for people with high involvement, who have a high cooperation rate even without incentives (also see Groves, Presser, and Dipko 2004).

There is now considerable research on the types of incentives that are most productive. The usual advice is to use prepaid incentives rather than the promise of an incentive (Church 1993). Singer, van Hoewyk, and Maher (2000), for example, found that prepaid incentives increased response rates even for telephone surveys by 10%, whereas promised incentives did not increase responses. The effect was not an interviewer expectancy effect, since it occurred even when the interviewer did not know that the person received the incentive. Five-dollar incentives were cost effective in that study, reducing the average number of phone calls per case from 10.22 to 8.75. Fortunately, prepaying incentives did not decrease the chance of the person's participating in a later survey, even if that later survey did not include an incentive (Singer, van Hoewyk, and Maher 1998). There is some fear that people who participate only with incentives will not bother answering all items, but item nonresponse was not increased in this study; indeed incentives seemed to decrease item nonresponse among older people and nonwhites.

Incentives generally are $2 or $5, with the amount obviously increasing over time with inflation. An early study by James and Bolstein (1990), for example, reports a response rate in the 70% range in a mail survey with a $1–$2 incentive versus 54% when no incentive was given. Larger incentives lead to higher response rates, but there seems to be a diminishing effect with larger incentives. With small incentives, it is usually easier to include cash in the envelope than to write checks, which would have to be kept on the survey organization's books for several months if they are not cashed. Larger incentives are increasingly being used, as when the

2000 NES survey gave a $20 incentive, which was increased to $40 in the week before the election.

A large-scale study by Trussell and Lavrakas (2004) provides further detail on the usefulness of monetary incentives. Respondents for a mail survey were randomly assigned prepaid incentives from zero to $10, with respondents contacted in advance by telephone. Respondents who were given some incentive were more likely to cooperate, with higher incentives generally leading to more cooperation. Larger incentives had the greatest effects among people who could not be contacted on the phone and those who refused to participate when contacted on the phone; this suggests that incentives should be targeted to people who have not shown a willingness to cooperate.

Nonmonetary incentives can be effective for particular types of respondents. For example, movie tickets are reasonable incentives for students, whereas bottles of wine are better incentives for elites than money would be. Material incentives such as pens are sometimes used and have been found to increase the response rate (Willimack et al. 1995), but funds used for the added effort and for packaging may increase participation more if distributed as financial incentives. Promising a contribution to a charity in the respondent's name costs less than prepaid incentives, since only people who participate are rewarded, but the evidence seems to be that charity incentives are not effective (Warriner et al. 1996; Hubbard and Little 1988; Furse and Stewart 1982; cf. Robertson and Bellenger 1978).

Some surveys offer lotteries as incentives, giving participants a chance to win a large reward. Lotteries have the advantage of rewarding only actual participation, whereas some people who are sent monetary incentives for mail surveys still do not participate. Also with lotteries it is necessary to send money to only one person rather than go to the trouble of sending out money to a large number of people. If it is a large survey in which the chances of winning the lottery prize are small, however, people may not participate if they understand the small chance of winning—though this does not seem to deter people from participating in state lotteries. Still, it is not clear that lotteries are as effective as prepaid incentives for survey participation.

Incentives can be given very efficiently in Internet surveys. Participants in one-time surveys are often awarded a gift certificate number that they can use at an Internet merchant's site. People who have joined ongoing survey panels are sometimes given points that can be accumulated for large prizes. These techniques reward only people who finish the surveys, while turning over the distribution of incentives to firms that specialize in that.

Tailoring the Survey Request

In understanding cooperation with interviewer-administered surveys, it must be kept in mind that this is a very quick decision by respondents. Typically, people have no idea who is on the phone or at the door when they answer. They must quickly figure out how to categorize the survey attempt: if it is real rather than a disguised sales attempt, if the source is one they respect (such as their home state university), if the interviewer seems friendly and competent, if they have some time at the moment, and so on. This makes the introductory survey request particularly important.

The importance of how the interviewer begins the conversation with the respondent is shown by Oksenberg and Cannell's (1988) finding that 40% of refusals occurred during the first few sentences, 50% later in the introduction, and only 10% after the interview has started. Thus the first 15–30 seconds of an interviewer-administered survey should be seen as crucial. Unfortunately, the respondent selection techniques that yield the most representative samples (see chapter 10) take longest, are more intrusive, are more likely to cause suspicion, and therefore are most likely to lead to nonresponse.

It was originally felt that there should be standard introductions to surveys, using the best possible justification of participation. Few interviewer-administered surveys still adopt that approach. Even survey researchers who believe in standardization now recognize that scripting is not effective. A scripted introduction can actually lead to problems, raising concerns that would not otherwise come to the person's mind. Morton-Williams (1993) was one of the first studies to show a much higher response rate with unscripted introductions, 17% higher in that instance. Going back to the discussion of conversational interviewing (chapter 4), a Dutch study found greater success in obtaining interviews when interviewers began with a conversational approach than when they followed a standardized script in asking for the interview (Houtkoop-Steenstra and van den Bergh 2002), possibly because the interviewers sounded more natural and/or were more motivated.

The modern approach to survey participation is to "tailor" the pitch to participate to the potential respondent. Different respondents have different values, so no single strategy for achieving cooperation will be optimal for all potential respondents. The interviewer is supposed to learn to read cues about the best way to motivate particular types of people. Tailoring is particularly effective in face-to-face interviews in which the interviewer can size up the respondent and his or her household. It is also used in telephone interviews, with interviewers being told to use their natural skills to respond as the conversation develops. The tailoring approach can be taught as part of interviewer training. Groves and McGonagle (2001) report a study in which interviewers were trained to learn to identify and classify the cues of reluctance from householders. Interviewers with that training had better response rates than the control group. In their analysis of the introductory interactions in 101 phone interviews, Couper and Groves (2002) found that almost all interactions included at least one statement by the respondent that was informative; only 8% of the calls they analyzed had no comments at the introductory stage that could be used in tailoring the request to participate in the survey. The cooperation rate was higher when interviewers followed up with a tailored approach, though the number of cases in key cells is too small to prove this effect beyond question.

The most common approach is a brief introduction that gives essential information but is tailored to the respondent's concerns. Guidelines from the American Association for Public Opinion Research (AAPOR) suggest that telephone interviewers begin by identifying themselves, stating their affiliation, indicating the sponsor of the survey, summarizing its purpose, explaining how the person was selected (e.g., randomly), encouraging cooperation, and verifying the phone number. Lavrakas (1993) favors a brief introduction, establishing credibility, and quickly getting the respondent to answer the first question so that she or he has become committed to answering. As an example, the box on the opposite page gives the introduction used by Ohio State University's Center for Survey Research for one of its polls.

The social exchange theory has been blended with Cialdini's compliance principles in Dillman's (2000, chap. 1) tailored design approach for self-administered questionnaires. Dillman lists several possible ways of providing rewards to survey participants, including showing respect to them, thanking them for participating, asking for their advice, supporting their values as represented by their group attachments, giving them tangible incentives, making the survey interesting, providing them social validation by emphasizing that people similar to them have participated, and telling them that the opportunity to respond is scarce. The costs of participating can be reduced by not making the respondent seem dependent on the researcher, not embarrassing the respondent (as by making sure the questionnaire is not too hard to fill out), making participation convenient, keeping the questionnaire short with easy formats, minimizing requests for sensitive information that people may not want to disclose, and making

Introduction-Selection Sequence

GENERAL INTRODUCTION:
(PLEASE TAILOR WORDING TO REDUCING REFUSALS!!!)
Hello, my name is _____ and I'm calling from Ohio State University (in Columbus).
Your household has been randomly sampled to participate in the ___ Poll. This is a very important scientific survey of Ohio residents that we are conducting this month for professors at Ohio State and . . .
(The interview takes about 10–15 minutes to complete. It's entirely confidential, your household's participation is voluntary, and I *really need* and will appreciate your help.)
(This month's questionnaire asks a series of questions related to . . . In addition, there is a series of background questions to help the professors who will be analyzing the data.)

WITHIN HOUSEHOLD RESPONDENT SELECTION:
For this survey, I'd like to interview the person in your household who is at least 18 years old *and* who had the last (i.e. most recent) birthday. (Is that you or someone else?)

NUMBER VERIFICATION:
Before I begin the interview, may I please verify that I have reached you on _____

PROCEED TO QUESTIONNAIRE WHEN SELECTED RESPONDENT IS AVAILABLE. DETERMINE WHEN BEST TO CALL BACK IF SELECTED RESPONDENT IS NOT AVAILABLE AND RECORD INFORMATION ON CALL-SHEET.
Source: Ohio State University Center for Survey Research

the requests to participate similar to those in other surveys that the person has participated in. Trust can be established by emphasizing an authoritative sponsor, making the survey seem important (such as with professional-looking questionnaires), and giving incentives in advance.

Several approaches are used to boost the cooperation rate, in line with the different compliance approaches reviewed early in this chapter. Some studies identify their sponsorship as a means of justifying participation. Surveys from state universities usually emphasize their school's name when interviews are conducted in their state, and surveys from the nation's most prestigious universities use their school's name even in interviews outside their state. Some research shows the highest response to surveys sponsored by the federal government, next to those with university sponsorship, and lowest to those with commercial sponsorship. A related approach is using endorsements of the survey by people or organizations that respondents would recognize and value positively.

A study by Dijkstra and Smit (2002) found that personal appeals ("I would like your cooperation") were more effective in inducing respondents to participate in the survey than referring to authority (their sponsoring university) or social validation (that most people enjoy the interview).

Their results on the effectiveness of tailoring were not consistent. Repeating the person's objection diminished participation.

The evidence accumulated in a meta-analysis by Singer, Von Thurn, and Miller (1988) points to higher response rates with assurances of confidentiality, but the effect is small and holds only when sensitive questions are involved. With nonsensitive topics, they found that confidentiality assurances decrease response rates, but that effect is very small. However, mail return of the census, according to a Survey of Census Participation after the 1990 census, was significantly affected by confidentiality and privacy assurances, with an estimated 3–6 percentage point increase in response rates (Singer, Mathiowetz, and Couper 1993). The importance of privacy concerns is clear in a finding that mail return of the 2000 census was significantly less among people concerned that it might be used for law enforcement purposes, such as by the FBI to trace troublemakers and for locating illegal aliens (Singer, van Hoewyk, and Neugebauer 2003).

Telephone interviewers are usually given standard "fallback" arguments to use when respondents ask questions or give specific reasons for not participating. Fallback statements used by the Ohio State University Center for Survey Research are shown in the box on the opposite page. For example, the person who says he or she is too old to participate is told that it is important to represent the views of older people in the survey, whereas the person who asks about sponsorship will be assured that the sponsor of the survey is legitimate. Obviously an important part of the interviewer's job is to figure out the most appropriate way to tailor the request to this particular respondent. For that reason, interviewers can be seen as "request professionals" (Groves and Couper 1998, 18), and the best interviewers have learned how to request cooperation so as to succeed.

Of course even tailoring an appeal for cooperation requires maintaining the interaction of the interviewer and the respondent. In face-to-face interviews, this is known as the problem of keeping the interviewer's foot in the door. So long as the door is not completely slammed in the interviewer's face, there is a chance of convincing the person to give the interview. This problem is even more serious for telephone surveys, since it is very easy for people to hang up when interviewers introduce themselves. If the person stays on the phone, the interviewer can attempt to tailor the pitch to the respondent, but that requires the person to stay on the phone. Survey organizations that try to convert refusals instruct their interviewers to be very polite in initial encounters, so there is a possibility of a later conversion by another interviewer if the respondent initially refuses. Surveys with nonprobability sampling that permit substitutions for refusals instead seek

Fallback Statements—Use Only as Needed

PURPOSE OF THE SURVEY

This survey of Ohio adult residents is part of an important research program that Ohio State University has conducted every month since 1996. This is being done in partnership with . . . It also will help professors and students at the University who are conducting various research projects.

It is a monthly survey of Ohio residents to help the state understand what the public is thinking about and experiencing on a number of different issues.

This month, the interview takes 10–15 minutes to complete. All of the information you give is kept strictly confidential. Your cooperation is voluntary and *I'd greatly appreciate your help.*

USE OF SURVEY RESULTS

After we interview residents throughout Ohio, the OSU professors who are conducting this survey will be analyzing the findings and working with the news media and various other public and private sector agencies in Ohio to help interpret them.

For example, reporters at . . . will use some of these findings in several news stories that will be written next month, including ones focusing on the Ohio economy. This type of information should give Ohio businesses and government officials a better sense of the strengths and weaknesses of the state's economy.

I want to assure you that all of the information you give will be *kept strictly confidential.* Your cooperation is voluntary, and *I'd greatly appreciate your help.*

CONFIDENTIALITY

Although I have your telephone number on a separate sheet if paper, this sheet will never be associated with your answers to the survey questions. All of the responses will be analyzed in the aggregate and individual responses won't be singled out.

Furthermore, all of the interviewers at the OSU Survey Center have signed a "Statement of Confidentiality" which ensures that all interviews we conduct will not be discussed outside of the Center. Of course, your cooperation in this study is voluntary, but *I'd greatly appreciate your help.*

HOW DID YOU GET MY TELEPHONE NUMBER?

Your number was randomly chosen using a technique called Random-Digit Dialing. The way this works is that a computer randomly chooses an area code in Ohio and a three-digit prefix that works within that area code. Then, the computer randomly adds four more numbers to make up a telephone number, like the one I used that reached your home.

Many times we dial numbers that are not working numbers; others reach businesses, and some reach households such as yours. We do this because it's important that we interview a random sample of Ohio residents, regardless of whether they have a listed or unlisted telephone number.

WHY THE LAST BIRTHDAY?

We ask to speak with the one person in each household we reach who had the last birthday, because this further randomizes the sample of citizens who we interview. We do this so that we get a balanced number of men and women and of younger and older adults.

WITH WHOM CAN I VERIFY THIS SURVEY?

If you would like, I would be glad to ask one of my supervisors to come to the phone to speak with you.

Or you may call the professor who directs the survey at Ohio State, . . . He also happens to be the Director of the OSU Survey Center. He can be reached at . . . during regular business hours.

Source: Ohio State University Center for Survey Research

to maximize their number of interviews per hour, which means the interviewer is better off contacting the next person on the list than trying to maintain the interaction with a reluctant respondent.

Tailoring the survey request is more difficult in mail surveys. Using different approaches for different subpopulations is not feasible, since little will be known about differences between respondents before they are sent the survey instrument. At best, tailoring to subpopulations may be possible when they appear on separate lists or if side information can be obtained before mailing them the survey.

Tailoring is similarly hard to achieve in Internet surveys, but minimally it is important to appeal directly to the respondent to participate, such as by sending e-mail surveys in a personalized manner rather than as part of a mass mailing that shows e-mail addresses of large numbers of people. Additionally, Internet surveys require an inviting welcome screen so as to minimize unit nonresponse. Similarly, the first question should be one that will interest people and be easy to answer, as a way of drawing people into the survey.

The Council of American Survey Research Organizations (CASRO) suggests that online e-mail surveys have been most effective when the subject line indicates the topic (e.g. "Survey Invitation for Pet Owners"), there is an explanation of how the e-mail address was obtained, the sponsor and company conducting the survey are identified, the purpose of the research is explained, the approximate completion time is stated, the incentive for participation is described, and valid contact information is given.

Refusal Conversion

Many survey organizations recontact people who initially refuse and try harder to get them to participate. Interviewers who excel at conversion of refusals often make this second attempt. More than a quarter of the people who originally refuse to participate will often cooperate on a second attempt, which shows that the quick choice not to participate is not necessarily a permanent decision. This partially suggests that the initial interviewers were not fully effective, but it also points out that it can be useful to try different approaches on the same person. Studies show that the verbal reasons people give for refusals are not stable under alternative cues given to the person, so it is often worthwhile to try again to get the interview.

The initial interviewers can be given "refusal forms" to fill out when someone is not willing to be interviewed. The interviewers describe the type of appeal they used and the person's response, so that the conversion attempt can try a different appeal or a different way to mollify the person's concerns. The box on page 186 shows the Refusal Report Form used by Ohio State University's Center for Survey Research CATI unit. Conversion attempts are not usually made when the person was rude, mad, or obscene, or directly said not to call back ("hard refusals"), but many refusals are not that clear-cut, as when the person said they did not have time at the moment or when the person answering the phone hung up before the interviewer could ascertain whether he or she was the intended respondent ("soft refusals"). Some surveys switch to a different interviewing mode to try to convert refusals, trying a phone interview when a face-to-face interview failed, or vice versa. Sometimes a refusal to a questionnaire is followed up with a personal interview.

Survey organizations must decide how much effort to put into refusal conversions. Refusal conversions may be costly without being very successful, and, even worse, some studies find them not to be representative of all refusals (Lynn et al. 2002). It may be possible to use the initial comments during refusals to decide who is worth recontacting. Groves and Couper's (1998, 261–65) multivariate analysis shows that people who ask questions during the first attempt to survey them are more likely to cooperate in later contacts, whereas those who make negative comments or say they're too busy are less likely to cooperate eventually.

To give an idea of the stakes involved, a report on a CATI survey at the time of an election in 1996 required 7.56 calls on average to complete interviews with refusal conversions versus 4.01 for nonrefusals, with a cost 1.6 times the cost for nonrefusals (Mason, Lesser, and Traugott 2002). The refusal conversions increased the response rate from 46% to 61%, but with more item-missing data—19% on the crucial vote intention item versus the 12% level for nonrefusals and with about 25% having missing data on explanatory variables compared to 11% for nonrefusals. This example nicely illustrates the considerations in deciding whether refusal conversions are cost effective. They are costly and many do not provide full data anyhow, but they can boost the response rate considerably, and imputation can help handle the item-missing data that can result.

Instead of trying to convert refusals, another approach is to seek to conduct a very short interview (possibly with someone else in the household) to collect some demographic information that can be used for later analysis. The discussion below of postsurvey statistical adjustments shows how this information can sometimes be employed. A variant of this is to sample a subset of the refusals to get more information about them,

Refusal Report Form

Interviewer # ___   Case id ___
1. Did the person who refused have the last [most recent] birthday?
   Yes
   No
   Uncertain

2. Demographics of the person refusing
   GENDER   AGE   RACE
   Female   Child   Asian
   Male   Adult < 30   Black
   Uncertain   30–59 Yrs   Hispanic
                   60 or older   White
                   Uncertain   Uncertain

3. How much interaction between you and the person refusing has gone through?
   No interaction at all   1 2 3 4 5 6 7   Much interaction (conversations
   (immediate hang-up)   beyond the standard introduction,
                                             questions and answers, explanations, etc.)

4. Reason for refusal (check all that apply):
   ___ Immediate hang-up/uncertain of the reason
   ___ Time-related concerns/reasons/excuses
   ___ Objects to surveys/topics/sponsors
   ___ Concerned about confidentiality/legitimacy of survey
   ___ Other reasons (specify) _____

5. Strength of refusal: VERY WEAK   1 2 3 4 5 6 7   VERY STRONG
   Respondent attitude: VERY POLITE   1 2 3 4 5 6 7   VERY RUDE
                         NOT AT ALL ANGRY   1 2 3 4 5 6 7   VERY ANGRY

6. Did you tell the person:
   A. How s/he was sampled?   Yes / No
   B. The nature/purpose of survey beyond the standard intro?   Yes / No
   C. Confidentiality?   Yes / No
   D. How the data would be used?   Yes / No
   E. Verification with supervisor/sponsor?   Yes / No

7. What can you recommend, if anything, for gaining respondent/household cooperation if a conversion attempt is made?_____

Source: Ohio State University Center for Survey Research

particularly when fully studying the refusals would be expensive. Statistically, data obtained from the refusals can then be treated as a stratum (see chapter 10), using stratified sampling formulas (Kalton 1983). A related approach would be to do extensive refusal conversion in the first phase of a survey, to estimate what the optimal level of persuasions would be for the rest of the survey.

While refusal conversion is generally associated with interviewer-administered surveys, attempts should also be made to elicit responses from people who do not respond to Internet surveys (Best and Krueger 2004, 77) and those who do not send back mail questionnaires on the first mailing. Dillman's Tailored Design Method will next be explained as a method that combines tailoring with persistence in seeking survey cooperation.

The Tailored Design Method

Mail surveys are notorious for low response rates, often as low as 20%. As a result, there has been considerable focus on methods to increase cooperation with mail surveys. In particular, Dillman (2000, 27–28) claims that mail surveys can average a 75% response rate when his Tailored Design Method is followed. This method combines insights on how to obtain compliance from respondents and very practical advice gained from large numbers of studies. Five features are emphasized (Dillman 2000, chap. 4): a questionnaire with a "respondent friendly" design (see chapter 6), stamped return envelopes, personalized correspondence, token financial incentives given in advance, and five mailings as described below.[4]

Dillman's five mailings begin with an advance letter, alerting the respondent to a survey that will arrive in a few days and requesting cooperation. The questionnaire mailing includes the incentive and a cover letter explaining the importance of responding. A thank-you postcard is sent a few days later, asking for the questionnaire to be returned soon if it has not yet been. This is followed in two to four weeks by a replacement questionnaire for people who have not responded by that date. Those first four mailings are all by first-class mail, but the fifth is by a different mode, such as special delivery or Federal Express (or even by telephone if the person's phone number can be obtained).

Dillman (chap. 4) also gives detailed advice on each mailing. The cover letter should have a short introductory paragraph indicating what is being requested and a second paragraph explaining why the survey is useful and important. The letter must also stress that answers will be confidential and that participation is voluntary. Dillman recommends a casual mention of the incentive, such as "We have enclosed a small token of appreciation as a way of saying thanks for your help" (162). The cover letter should provide information on how to contact someone if the respondent has questions. Dillman urges that the letter have a real signature and a postscript that again expresses thanks, as well as explaining what to do if the person is not eligible for the study.

When the questionnaires are mailed, they are usually given identification numbers so that follow-up mailings can be employed to increase the response rate. It is usually best not to conceal identification numbers, because of the problems that could ensue if respondents realize that there is a concealed number. For example, an identification number under the return postage stamp in a survey of Michigan state legislators in the early 1970s was detected, and the story circulated immediately around the legislature resulting in bad feelings toward the sponsoring university. If sensitive data, such as on drug use, are being collected, the identification number can be put on a separate postcard that respondents are asked to send back when they return the completed questionnaire.

The message on the postcard at the third mailing should be much simpler than in the original cover letter. It is intended to remind people of the survey. The card should begin by saying that a questionnaire was sent and why, which helps in situations in which the person never noticed the original mailing. Then comes thanking people who have sent the questionnaire back and asking others to do so "today." The message should end with an explanation of how to get a replacement questionnaire if it is needed.

The replacement questionnaire in the fourth mailing should also be accompanied with a cover letter, but the letter's message at this stage should be phrased more strongly. It should emphasize that other people have responded and again highlight the importance of the survey. The fifth contact should be viewed as the final attempt to obtain cooperation, using more expensive contact methods if increasing the response rate is urgent.

Dillman (2000) gives the specific evidence on which the Tailored Design Method is based, but it is useful to mention a few findings regarding how much effect particular parts of the process have. A meta-analysis of journal articles based on mail surveys from 1961 through 1985 (Fox, Crask, and Kim 1988) finds the largest effects were university sponsorship (9% effect), advance letters (8%), and stamped return postage (6%). Smaller significant effects were found for postcard follow-up (3.5%), using first-class outgoing postage (2%), and a green questionnaire (2%). Notification of a cutoff date and the postscript requesting cooperation did not have significant effects. Incentives had effects but with diminishing returns: the added response rate increment was largest for using a small incentive and was more minimal with large incentives.

Further evidence on the size of effects comes from a series of experiments testing approaches to increase the response rate to the U.S. decennial census (Dillman 2000, chap. 9). Government surveys are not immune to the response rate problems of mail surveys, even when compliance is mandatory by law, as shown by the 75% response rate to the census in 1980 and the 65% rate in 1990. The experiments showed that response rates could be improved by using a respondent-friendly design (a 2%–4% effect), an advance letter (5%–7%), a postcard thank-you reminder (5%–8%), a replacement questionnaire (6%–11%), and a prominent reminder that response is mandatory (9%–11%). These procedures, other than the replacement questionnaire, were implemented for the 2000 census and helped bring the response rate up to 67%.

Easing the Technological Burden

A special concern for Web surveys is the extent to which unit nonresponse is due to technical problems. The respondent's computer hardware or software may not be able to handle the questionnaire. Browser problems and slow connect times may discourage survey participation. These situations may improve over the years as browsers are improved and standardized and as fast connections become more common, but researchers have to remember that many respondents do not have Internet capabilities that permit the same multimedia access that the researchers themselves have.

The respondent may also lack the technical skills that the survey assumes. Dillman and Bowker (2001) point to several such factors that lead to nonresponse in Internet surveys. First, some people do not know how to provide answers, as in how to handle a drop-down menu. People may also not understand how to change answers, since some formats require only choosing a different answer whereas others require clicking a second time on the original answer. It is awkward when not all the answer options show up on the screen at the same time so that viewing the options requires scrolling up or down the page. Also, sometimes it is necessary to take several actions to answer a question (such clicking on an answer and then having to click elsewhere to go on to the next question).

An early problem with Internet surveys was that respondents could not tell how long the survey would be. If people got the impression that the survey was endless, many would break off early. Therefore it has become common to show respondents how far along they are in the survey through progress indicators, such as a bar in a corner in the screen which darkens as the person gets further through the questionnaire.[5] Allowing people to skip questions rather than requiring each question to be answered before the respondent can go on to the next one can also decrease breakoffs.

An unusual danger of Internet surveys is that respondents seem to be especially sensitive to the survey administration. A 2001 study by the Council for Marketing and Opinion Research (CMOR) found that Internet respondents tended to be more critical of the mechanics of the survey than were telephone respondents (Business Market Research

Association 2002). Experience in being members of Internet panels can further exacerbate this tendency. One result is that people may self-select out of Internet surveys that do not satisfy their view of how such surveys should be presented.

Some of these factors involve item nonresponse as well as unit nonresponse, but respondents who get discouraged by factors such as these are probably less likely to complete the questionnaire and to participate in future Internet surveys. Skilled programmers can prevent some of these potential problems, and standards are developing for others, but differences in respondents' familiarity levels with computers will inevitably persist. Therefore it remains essential to seek to ease the technological burden for respondents in Internet surveys.

Response Bias

Nonresponse has pervasive statistical effects (Brehm 1993, 93–117). It leads to underestimates of the variance of a variable, which can lead to false statistical inferences. The bias for a regression coefficient can either be positive or negative, depending on whether the variables are less or more related to each other among nonrespondents than respondents (Groves and Couper 1998, 10–11). Nonresponse can also bias estimates of multivariate relationships, particularly when people who are extreme on one of the variables do not respond (Brehm 1993, 93–117). For example, losing people who are low on an independent variable such as political interest (so the sample is truncated) can bias survey estimates of variables affected by political interest if the relationship between interest and the dependent variable is different for people with low political interest. Sampling to exclude extreme low cases on the dependent variable (so that variable is "censored") would decrease estimates of the effects of independent variables, an attenuation effect. Studies that use variables related to survey participation as either their independent or dependent variables are particularly susceptible to these effects.

A very minimal strategy is to examine the survey results to see how big a difference there has to be between respondents and nonrespondents to reverse sample findings. If 55% of the sample is satisfied with a service provider, for example, but that is based on only a 60% response rate, the researcher calculates how much nonparticipants would have to differ from participants for the true proportion to be less than a majority. Or if positive change were found for 55% of those who change on a question between two waves of a survey, but that is based on reinterviewing only

80% of the original sample, one can calculate how much those who were not reinterviewed would have to differ from those who were reinterviewed for the true proportion changing in a positive direction to be less than a majority.

Unit Nonresponse and Response Bias

The underlying assumption of most of the research on unit nonresponse is that it is vital to make heroic efforts to maximize the response rate. A low response rate indicates that the sample size is smaller than desired (which can affect statistical significance testing), that there may be nonresponse bias, and that people will be suspicious of the study. Readers of a report on a survey with low response rates may doubt the representativeness of its sample. Sponsors may feel that a low response rate indicates lack of effort by the survey organization.

There is, however, a revisionist view regarding the importance of high response rates. As surveys began to obtain lower response rates, researchers began to test whether that really matters. A set of studies now suggest that higher response rates do not necessarily lead to more accurate results, because lower response rates do not always translate directly into bias.

In an important experiment, Keeter et al. (2000) examine the effects of vigorous efforts to increase the response rate in a telephone survey. They compared a "standard" five-day survey in which people at home were interviewed (the youngest adult male at home, otherwise the oldest female at home) with at least five callbacks and one follow-up for a household refusal to a "rigorous" eight-week survey with random selection among all household members and a prepaid incentive. The response rate was only 36% for the standard procedure versus 61% for the rigorous procedure; the contact rate was 68% for the standard versus 92% for the rigorous. The key concern is whether these differences lead to sizable differences in results, and the important finding is that they do not. Regardless of the large differences in response and contact rates, the samples obtained in the standard and rigorous approaches were very similar on most variables, with an average difference of only 2%, with no difference being greater than 9%. There were only slight differences on demographics, with the respondents in the standard version being more nonwhite (4%), less educated (5% less college), lower income (4% less high income), and more renters (6%). Also, the standard procedures obtained more political independents, which fits with Brehm's (1993) study and others showing that politically uninvolved people are harder to contact. Significant differences were not obtained on attention to media, engagement in politics, social trust, connectedness, social and political attitudes, or attitudes toward surveys.

Although the differences in results between the standard and rigorous approaches in the Keeter study were minimal, there were, of course, substantial cost differences between them. The standard approach yielded about one interview per hour compared to only 0.8 interviews per hour for the rigorous approach. The implication of this study is that we may not need to conduct many callbacks or use rigorous procedures, since the differences between the two approaches were few. Still, this study does not examine the limits to that logic—such as whether comparable results would have been obtained with only three days of interviewing instead of five days. Also, the report of this study focuses only on single-variable results, without examining any implications for multivariate analysis. Yet the result is important, because it implies that expensive procedures for increasing response rates may not be necessary. Furthermore, it is not the only study with this claim.

A similar implication derives from a study by Curtin, Presser, and Singer (2000) on the surveys on which the Index of Consumer Sentiment is based. They find that excluding respondents who required more effort to contact and to interview would not have affected monthly analysis, since nonresponse bias was constant and did not affect comparisons over time. At most, some differences appear when responses to multiple consumer sentiment surveys are aggregated, since the impact of nonresponse bias is more evident when larger sample sizes reduce the sampling error.

As mentioned earlier, Merkle and Edelman's (2002) analysis of 2000 VNS election exit polls found that there was no relationship between the response rate in a precinct and the deviation of the precinct's exit poll result from its actual vote. This result again suggests that lower response rates do not necessarily lead to higher response bias.

Perhaps the most provocative of these studies is one involving the accuracy of predictions of preelection surveys. Visser, Krosnick, Marquette, and Curtin (1996) compared the forecasting accuracy of mail and telephone surveys in Ohio. Over a fifteen-year period, they found that the mail surveys predicted more accurately than phone surveys (average error of 1.6% versus 5.2%) even though the mail response rate was only about 25%, compared to a response rate around 60% for the phone surveys. The explanation the authors give is that people who were interested enough in the election to return the mail questionnaire were showing the same type of initiative that is required to go to the polls to vote and therefore are more predictive of actual voting.

This result suggests that low response rates are not necessarily bad. On the basis of this study, Krosnick (1999) directly argues that phone surveys would forecast election outcomes more accurately if they did not aggressively seek high response rates. Further, he generalizes this to argue that high response rates are not necessary for survey representativeness, citing other evidence that correcting for demographic biases in the sample has little impact on correlational analysis (Brehm 1993) and that improvement in response rate does not necessarily alter substantive conclusions of a study (Traugott, Groves, and Lepkowski 1987).

However, there are reasonable limits to this argument. One could interpret the results of the Visser et al. study as unique to election forecasting and could view the evidence from the nonelection studies cited in this section as slim. Additionally, simply accepting low response rates would permit survey organizations to be lazy in contacting respondents, which would let response rates fall even lower than they have already fallen. Yet it is interesting to see that several studies converge in arguing that low response rates do not matter, a conclusion that completely contradicts the conventional wisdom in the field.

Statistical Solutions to Unit Nonresponse

Attempts to eliminate survey nonresponse can never be totally successful, but there are statistical means for dealing with the problem through weighting and through modeling the nonresponse process.

Weighting

Survey respondents are sometimes weighted in a differential manner so that the obtained sample better resembles the population of interest. Three weighting systems will be mentioned here: probability of response weights, weighting class adjustment, and poststratification adjustment.

The simplest weighting method is using *probability of response weights*. Observations are weighted inversely to their ease of acquisition, with people who were easiest to interview being weighted less than the hardest. For example, the Politz-Simmons method uses weights based on the number of days during the previous five days that the respondent would have been available to be interviewed at the same time of day. People who would have been home at that time every day would receive the lowest weight, whereas people who were home only on that one day would receive the highest weighting, being treated as representative of people who are hard to contact. This assumes that people who are at home only on Friday night are similar to those at home only on Monday night, for example, which can be a false assumption, but it recognizes that people who are rarely at home differ from those who are always at home.

A generalization of this procedure is to weight cases by the difficulty of obtaining them; so, for example, an interview obtained on the first mailing, visit, or phone call would receive less weight than one obtained in a much later attempt. The logic of this *extrapolation for difficulty* procedure is that people who are impossible to contact may be most like those who are difficult to contact—though some survey analysts would argue that the unreachables are not like actual respondents. In any case, the high proportion of refusals among nonrespondents makes this weighting less useful, since those who refuse are not necessarily similar to those who are difficult to survey (Smith 1983).

A *weighting class adjustment* method uses variables that were involved in the sampling design. For example, if one were sampling from lists of different types of doctors who responded at different rates, one could then weight them back appropriately. Say that one were interviewing off two lists of doctors, with 55 specialists on one list and 45 family doctors on the other list, and that interviews were taken with 33 specialists and 40 family doctors. As shown in table 8.1, the response rate for specialists would then be 60% and that for family doctors would be 89%. Weights are then calculated as the reciprocal of those response rates: $1/0.60 = 1.667$ for specialists and $1/0.89 = 1.125$ for family doctors. When the 33 interviews with specialists are weighted at 1.667, the weighted number of cases (n) for them becomes $33 \times 1.667 = 55$, and when the 36 interviews with family doctors are weighted at 1.125, the weighted n for them becomes $40 \times 1.125 = 45$. Thus the weighted n's for the two types of doctors together are back up to their numbers in the original lists that were to be interviewed.[6]

Weighting treats the interviews with the harder-to-interview group as more valuable than those with the easier-to-interview group. The basic assumption is that the data are missing at random (MAR) and that people in the category who were interviewed are like the people in that same category who were not interviewed, so they can be used to represent those who were not interviewed. That means, however, that weighting gives false impressions to the extent that nonrespondents in a category differ from respondents in that category.

The most common weighting method for unit nonresponse is called *poststratification adjustment*. It differs from weighted class adjustment in that information about nonrespondents is not available, so these weights are instead population based. The weighting is based on external informa-

Table 8.1 An Example of Weighting Class Adjustments

	Specialists	Family Doctors	Total
Population	55	45	100
Responded	33	40	73
Response Rate	33/55 = .60	40/45 = .89	
Weight	1 / .60 = 1.667	1 / .89 = 1.125	
Weighted N	$33 \times 1.667 = 55$	$40 \times 1.125 = 45$	100

tion about the number of people in the category, usually population values from the census. A typical instance involves taking a random-digit-dialing telephone sample and then having to adjust the results on the basis of sex, race, and age. Poststratification weighting assumes that the data are missing completely at random (MCAR), and it fails to the extent that the people who were not interviewed differ from those with the same demographics who were interviewed.

It is not possible to weight on variables for which the true population value is not known, and it is not advisable to weight on variables whose values are highly changeable. For example, weighting preelection polls on the basis of past party identification distribution can bias poll forecasts, since that distribution can change during the election campaign (Allsop and Weisberg 1988). Thus, in 2004 there was much concern in preelection polling about whether to weight up Democrats, since their numbers seemed to be fewer than in previous years, but in the end some of the most accurate polls were those that did not weight, because there had indeed been a drop in the percentage of Democrats among actual voters. The best rule is to weight on the basis of factors involved in selecting the sample while being cautious about weighting on other variables. In any case, it is important that weighting be disclosed in reports on surveys, so readers can judge for themselves whether the results are vulnerable to faulty weighting decisions.

Poststratification weighting works best when the poststratification variables are strongly correlated with the dependent variable of interest. However, multiple variables in a study could be used as dependent variables, which means that different variables might be useful for poststratification adjustment for different data analyses. It is common to adjust on several variables at once, using computer programs that adjust for each variable in turn iteratively until the weights stabilize (a procedure known as *raking*). When there are multiple design variables, the differences of the dependent variable within combinations of these variables should be made small while the differences between different combinations of these variables should be made large. This procedure adjusts for biases in the sample, but the variance increases, so there is a trade-off between bias and variance.

The estimates of the vote given by the Voter News Service (VNS) on election night were based on poststratification weights, using correlations between exit poll results and past election results for the sample precincts, with actual vote totals for the sample precincts substituting for exit polls when they become available (Mitofsky and Edelman 1995). The problems that VNS had in dealing with the 2000 presidential election led to reconsideration of that procedure (though the Florida debacle—in which Florida was first called early for Gore and then called in the middle of the night for Bush before the original call was retracted, sending the Florida election into recount territory—was more likely due to the differences between how people thought they voted and whether and how their votes were counted). Similarly, the overstatement of the Kerry vote in some key states in the 2004 exit polls seems to have been partially due to the use of problematic mathematical models.

Weighting is a common solution, but a caution should be emphasized: some statisticians believe that weights should be used only for univariate analysis, not for multivariate modeling such as regression analysis or factor analysis. If education affects results, this argument would be, then education should be used as a predictor in the regression equation rather than as a weighting factor. Even weighting by probability of response can attenuate effects, so that weighting leads to missing significant effects (see, for example, Brehm 1993, 119–21). In fact, weighting often makes little difference in actual practice. The results are often so similar with and without weighting that weighting is sometimes not considered worth the trouble and the added difficulty of explaining the weighting to clients. However, as Voss, Gelman, and King (1995) show, weighting sometimes does have substantial effects, with support for George H. W. Bush appearing 2%–4% higher in preelection polls in 1992 when weighting was not used.

Modeling Nonresponse

Another set of statistical procedures for dealing with unit nonresponse is *model based*, modeling the nonresponse process. Nonresponse is treated as stochastic in this approach. This approach is especially appropriate when the likelihood of responding is related to variables of interest in the study. There are actually several statistical procedures of this type, including Little's maximum likelihood estimation and Bayesian methods that incorporate prior information about the underlying distribution of the variable (see Lessler and Kalsbeek 1992, 199–207; Little 2003). These methods are controversial in that they make assumptions about the nonresponse, and they are difficult to employ because they must be estimated separately for each variable of interest rather than simply deriving a single weight as above. They require having some predictors of participation in the survey, including at least one predictor that affects participation but not the dependent variable of interest.

The model-based procedure that will be described here is Heckman's method for correcting for selection bias. It involves developing a pair of equations, a probit equation for the chances of entering the survey and a regression equation for the variable of interest in the survey, with the two equations being solved simultaneously through maximum likelihood procedures.

A useful example of the Heckman method is John Brehm's (1993) analysis in his book *The Phantom Respondent*. Brehm develops three separate models of survey participation, making use of different sources of information about cooperation. One is an administrative model, using predictors for the survey participation equation based on aspects of the field administration, such as the number of calls to the household in order to get the interview, whether or not a persuasion letter was sent, and whether or not refusal conversion was necessary. Brehm admits that the administrative model can lead to inconsistent estimates of residuals because these predictors are endogenous to the survey participation model, but he finds that these effects are minimal in practice. His second approach is behavioral, using predictors such as the respondents' relationship to strangers, to the interviewer, to the interview, and to self-image, based on respondents' reasons for refusal or compliance. In the studies he examined, the most important of these was relationship to the interview, with relationship to the interviewer also having a large effect. The third possibility would be using demographic predictors in the selection equation, but demographic information about nonrespondents can usually be obtained only while the study is being administered, at which point it is easier to use the field administration model (particularly if demographics are to be part of the second substantive equation). As a result, Brehm relies mainly on the administrative model.

Some examples of Brehm's results usefully illustrate the potential of this approach. The Heckman correction for estimating voter turnout as a function of demographics leads to a larger estimate of the effect of education on turnout, due to the censoring effect of survey participants' being more likely to participate in voting (136–38).[7] The largest change in regard to the turnout equation is a more negative constant term after correction, meaning that turnout is overestimated in surveys when a correction for nonresponse is not employed, with an effect estimated at 18% in the 1986 National Election Study. On the NES abortion question, the correction shows that the survey understates opposition to abortion by blacks, after controlling for other demographics, and a change in the constant term implies that NES underestimates the proportion of people who are adamantly against abortion (154–57). On the other hand, estimation of income as a function of demographics does not change when a Heckman model is employed. All in all, Brehm finds that although not all variables are affected, our understanding of some variables is affected by unit nonresponse, so assuming that nonrespondents are the same as respondents can lead to serious error.[8]

A final statistical approach for dealing with unit nonresponse is imputation, in a manner analogous to how imputation is used for item nonresponse (see chapter 7). Rubin and Zanutto (2002) propose a "matching, modeling, and multiple imputation" method in which matched substitutes are used to impute data for nonrespondents (modeled using predictions from background covariate variables that are available for the nonrespondents), though this assumes that appropriate covariate variables can be obtained for the nonrespondents.

Unit Nonresponse and Survey Constraints

Unit nonresponse can be minimized by aggressive attempts to get the cooperation of designated respondents. There are several methods for this. In the case of noncontact, the solution is to do repeated callbacks in the hope that contact will eventually be made. In the case of noncooperation, many survey organizations have some interviewers who specialize in refusal conversions—convincing people to participate in the survey. Another way to deal with the potential problem of noncooperation is to offer incentives to participate: money, other material items (from pens to DVDs), or a chance to participate in a lottery. Additionally, tailoring the request to participate is becoming the most common approach to securing cooperation in interviewer-administered surveys.

It is also useful to collect as much information as possible on nonresponse, especially if corrections for nonresponse are contemplated. Obviously there are limits to this possibility, but it is often possible to keep track of simple demographics (at least gender on phone surveys) and reasons given for nonresponse as well as number of callbacks and whether refusal conversion was attempted.

There is also an important choice to be made between weighting the data for nonresponse and trying to model the nonresponse, especially when multivariate analysis is planned. Survey researchers are more familiar with the use of weights, but a strong case can be made for modeling instead. Obtaining a single set of weights for adjusting for nonresponse takes less time and effort, whereas deciding on an intricate set of modeling choices for each dependent variable is more costly. Menard (2002, 41) summarizes that "weighting may be preferable when nonresponse rates are low, while modeling nonresponse may be preferable when nonresponse rates are high (e.g., over 30%), but in this latter case, neither is likely to be entirely satisfactory."

Minimizing unit nonresponse in a survey is not an inexpensive task. The trade-offs are clear here, since spending more money on avoiding unit nonresponse directly takes away money that could otherwise be spent on increasing the sample size. Academic survey organizations tend to be more likely to do large numbers of callbacks and to have conversion specialists on their interviewing staff, while inexpensive commercial phone banks are least likely to do callbacks or bother trying to get conversions. Survey clients will sometimes require a particular level of response rate, which survey organizations will factor into their bids for the project.

Academic researchers usually desire to maximize their response rates, expecting that this will yield more valid data and that it will minimize the chance of journal reviewers' rejecting their work on the basis of having too small a response rate. Market research operations that require timely results are least likely to emphasize response rates, particularly if they need daily data on topics such as satisfaction with services provided. Survey researchers who employ the total survey error approach tend to take the middle route, doing some callbacks and conversion attempts but emphasizing to their clients that doing either extensively would take away too much from the total sample size that could otherwise be afforded. As long as this does not become a justification for low response rates, it is certainly advisable for researchers to be realistic about response rates.

The most challenging controversy is whether response rates really matter. Low response rates raise credibility problems, since poll consumers will assume that samples with low response rates are not representative of the target population. Studies are now finding that higher response rates do *not* necessarily lead to less response bias. Yet the credibility problem remains: will a survey be believed if its response rate is low? The public is already somewhat skeptical of surveys, and reporting surveys with low response rates could further exacerbate that skepticism. Clients and journal reviewers similarly will often not believe survey results unless the response rate is reasonably high. Accepting lower response rates would lower survey costs, but caution should be exercised so that survey credibility is not sacrificed.

Appendix: Response Rate Calculation

Response rate statistics for surveys are often stated, without explaining how they were calculated. As a result, the American Association for Public Opinion Research (AAPOR) has prepared a standard set of definitions for response rates, cooperation rates, contact rates, and refusal rates (AAPOR 2004). This appendix will summarize those definitions, which are available on the AAPOR Web site.

Disposition Classification

Since surveys can involve thousands or even tens of thousands of contact attempts, it is important to give a clear code as to the disposition of every interview attempt. The AAPOR standard definitions distinguish four results of interview attempts: (1) interview, (2) eligible cases that are not interviewed (unit nonresponse), (3) unknown eligibility, noninterview, (4) not eligible.[9] Each is divided into various types, as follows (see fig. 8.1, p. 202).

1. Interviews can either be complete interviews (I) or partial (P). Some survey organizations include partials as a successful interview in their calculation of response and cooperation rates, whereas other organizations are more conservative and do not consider them successful interviews. Both categories are counted when one is looking at refusal and contact rates. Breakoffs, which occur when the respondent terminates the interview so early that less than half of the questions were answered, are usually instead treated as eligible, no interview.

2. Unit nonresponse ("eligible, no interview") consists of three situations. The first is refusals (R), including household refusals before identifying the proper respondent and breakoffs. The second is noncontacts (NC). For phone interviews, this occurs when the interviewer never talked to the selected respondent. For in-person household interviews, this occurs when the interviewer is unable to gain access to the building, no one was reached at the housing unit, or the respondent was away or otherwise unavailable. For mail surveys of specifically named persons, noncontacts occur when there is notification that the respondent could not fill out the

questionnaire during the field period and when the questionnaire was mailed back too late for inclusion. The other (O) category consists of situations where the designated respondent had died or was physically or mentally unable to do the interview, or there was a language problem, plus literacy problems for mail surveys. All eligible, no interviews count as failures when the response rate is computed.

3. The unknown eligibility includes two subcategories. One is when it is unknown if the household is eligible (UH). In telephone surveys this occurs when the number is sampled but not dialed, is always busy, never answers, has an answering machine that does not make clear if the location is residential, has a call-screening or call-blocking device, or in the case of technical phone problems such as circuit overloads. For face-to-face household surveys, the unknown household eligibility situation occurs when contact was not attempted, when the interviewer was unable to reach the household (as because of weather) or an unsafe area, and when it was not possible to locate the address. For mail surveys, this occurs when the questionnaire was never mailed and when nothing was ever returned so there is no information about the outcome.

The second subcategory of unknown eligibility is a broad "unknown, other" (UO) category. It includes instances when it is not known whether an eligible respondent resides at the household. For phone and in-person interviews, this is the case when it is not found out whether there is an eligible respondent and when there was failure to complete the needed screener. There are several additional situations included under this category for mail surveys, partly because the U.S. Postal Service (USPS) puts a variety of different explanations on returned mail, such as "illegible address," "insufficient address," "no mail receptacle," and "no such address." Mail that is returned with forwarding information is also treated in this category.

Different survey organizations treat the unknown eligibility, noninterview category differently in computing the response, refusal, and cooperation rates. The most conservative approach is to count each of these as a failure, whereas the survey effort appears more successful when they are totally omitted from the calculation.[10]

4. The last disposition category, not eligible, does not count in computing response and related rates but is included so that every interview attempt is coded in some manner. This category occurs when the person lives outside of the designated geopolitical boundary. For phone samples, it also occurs when the phone number is a fax/data line, is a nonworking (not yet assigned) or disconnected number, or is a beeper or cellular/mobile

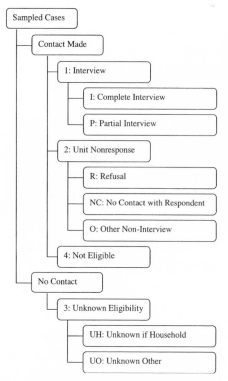

Figure 8.1 Possible results of attempts to contact sampled cases
Source: Adapted from Marton (2004, figure 3.1)

phone (if not the household's only phone). For in-person interviews, it also occurs when the housing unit is vacant as well as when there are other security measures such as a doorman precluding entry. Not eligible also includes contacts with nonresidences (such as business or government offices,) institutions (such as prisons and sanitariums), and group quarters (such as military barracks and phones in the hallway of a college dormitory, fraternity, or sorority). Not eligible also includes situations when there is no eligible respondent at the household, as when all residents are under age eighteen or when no one passes the screening questions for sampling subgroups. Some samples have quotas for particular subgroups (such as a given number of older men), so a household is categorized as not eligible if the quota for that subgroup is filled by the time that household was contacted. Finally, not eligible includes duplicates in large mailings of mail surveys. The frequency of noneligibles is sometimes reported to clients, as when the client wants to know the proportion of people screened out in a survey of grandparents because they have no grandchildren.

Response Rate Formulas

AAPOR recognizes a variety of formulas for response rates (RR), cooperation rates (COOP), contact rates (CON), and refusal rates (REF), depending on how particular subcategories are treated. These formulas vary in
how conservative they are, depending on whether partial interviews
are treated as interviews and how unknown eligibility is considered.
Academic and government surveys sometimes require using the most
conservative formulas, whereas commercial organizations likely prefer
reporting to clients higher response, cooperation, and contact rates and
lower refusal rates. The formulas will be summarized here, using the category number above as part of the notation. That is,

I = Complete interview
P = Partial interview
1 = Interviews = I + P
R = Refusal
NC = Noncontact
O = Other eligible nonresponse
2 = Unit nonresponse = R + NC + O
UH = Unknown if household eligibility
UO = Unknown other eligibility
3 = Unknown eligibles, noninterview = UH + UO

The response rate indicates the proportion of eligible reporting units
that led to interviews. The most conservative version is $RR1 = I/(1 + 2 + 3)$,
whereas the highest response rate is obtained with the formula $RR6 = 1/(1 + 2)$. CASRO uses $RR3 = I/(1 + 2 + 3e)$, where e is an estimate of the proportion of category 3 that is eligible.[11]

The contact rate is the proportion of cases in which some member of
the household unit was reached. The different versions of the contact rate
formula differ only in whether to count category 3 as relevant to the cal-
culation. The most conservative contact rate is $CON1 = (1 + R + O)/(1 + 2 + 3)$, whereas the highest contact rate is $CON3 = (1 + R + O)/(1 + 2)$.[12]

The cooperation rate is the proportion of all eligible units ever contacted that were interviewed. Two choices are whether to count partial
interviews as cooperation successes and whether the other subcategory of
eligible, noninterview counts as a failure. The most conservative house-
hold-level cooperation rate is $COOP1 = I/(1 + R + O)$, whereas the highest
cooperation rate is obtained with the formula $COOP4 = 1/(1 + R)$.[13]

Finally, the refusal rate is the proportion of all potentially eligible cases
in which the respondent refuses to be interviewed, including breakoffs. As

with contact rates, the different versions of the refusal rate formula differ only in whether they count unknown eligibility, noninterviews as relevant to the calculation. The most conservative refusal rate is REF3 = R/(1 + 2), whereas the lowest is REF1 = R/(1 + 2 + 3).[14]

Table 8.2 illustrates the calculation of the conservative versions of these formulas for the 1998 The Ohio Political Study (TOPS) conducted by the OSU CSR.

The AAPOR Web site provides a calculator for these response rates: http://www.aapor.org/default.asp?page=survey_methods/response_rate _calculator.

Table 8.2 Response Rate Calculations for 1998 TOPS Survey

1I	Completed interview	510
1P	Partial interview	24
2R	Household refusal	171
2R	Respondent refusal	317
2NC	R never available	91
2NC	Residential answering machine	31
2O	Incapacity	46
2O	Non–English speaking	12
3UH	Always busy	29
3UH	Never answer	59
3UO	Phone problem	18
4	Nonworking number	423
4	Disconnected number	24
4	Number changed	28
4	Nonresidential number	327
4	No adult household member	6
4	Not Ohio resident	4
	Total	2120
	SUBTOTALS	
1total	Interview	534
2R	Refusal	488
2NC	Noncontact	122
2O	Other eligible nonresponse	58
2total	Eligible, no interview	668
3total	Unknown eligibility	106
4total	Not eligible	812
	Total	2120

Response Rate #1 = I/(1 + 2 + 3) = 510/(534 + 668 + 106) = 39.0%
Contact Rate #1 = (1 + R + O)/(1 + 2 + 3) = (534 + 488 + 58)/(534 + 668 + 106) = 82.6%
Cooperation Rate #1 = I/(1 + R + O) = 510/(534 + 488+58) = 48.3%
Refusal Rate #1 = R/(1 + 2) =488/(534 + 668) = 40.6%

Source: Ohio State University Center for Survey Research

9

Coverage Error
SAMPLING FRAME ISSUES

This chapter will cover several relatively obscure types of survey error that are associated with the *sampling frame*, the actual set of units from which the sample will be taken. *Frame error* is the error that can arise when the elements in the sampling frame do not correspond correctly to the *target population* to which the researcher wants to make inferences. The most important frame error is *coverage error*, the mathematical difference between a statistic calculated for the population studied and the same statistic calculated for the target population. A simple example of a coverage problem involves sampling from phone books, because they exclude people with unlisted numbers who are certainly part of the target population. Sampling frames and coverage error will be explained in detail in the first section of this chapter.

The second section of this chapter deals with a frame error problem of *ineligibles* that arises when some elements in the sampling frame are not part of the target population. For example, random-digit-dialing (RDD) methods for telephone surveys call some business phone numbers even when the target population consists only of households. There is often a trade-off between coverage error and eligibility, since increasing the coverage may also bring in cases that are ineligible. This often necessitates finding an efficient sampling frame with a level of coverage that is high but not so high as to require contacting many cases that prove to be ineligible.

The third section describes two additional types of frame errors. *Clustering* happens when there is one element in the sampling frame for

Table 9.1 Sampling Frame Problems

Coverage error	Sampling frame omits elements from population
Ineligibles	Sampling frame includes nonpopulation elements
Clustering	Groups of units are listed as one in sampling frame
Multiplicity	Case appears more than once in sampling frame

several elements in the target population, as when multiple people share the same phone line. *Multiplicity* occurs when there are several elements in the sampling frame for one element in the target population, as when there are several phone lines in a single house. Table 9.1 summarizes these four types of frame error.

The Theory of Sampling Frames

The *sampling frame* in a survey is the list from which the sample is selected. It is the actual population of units to which sampling is applied, as distinguished from the target population that the investigator wishes to study.[1] The sampling frame has been described as the operational definition of the population (Henry 1990, 50). The extent to which a sampling frame covers all units in the target population can be viewed as its "comprehensiveness" (Fowler 2002, 12).

Sampling from a population may sound fairly easy, but it is often difficult to obtain an appropriate sampling frame. For example, there is no central listing of eligible voters in the United States; this makes it necessary for election pollsters to devise reasonably efficient sampling frames. Given the lack of lists of U.S. residents, most surveys of the general population are actually samples of households, because it is easier to obtain plausible sampling frames of households. It is also difficult to derive a sampling frame for businesses in the United States; some government agencies maintain business registers, but legal restrictions sometimes preclude their availability. By contrast, government agencies in many other countries are responsible for maintaining comprehensive lists of eligible voters, residents, and businesses, thus providing better sampling frames than are available in the United States.

Magnitude of Coverage Error

Coverage error occurs when there is bias due to the omission of noncovered units. For example, coverage error occurs in telephone surveys if people who do not have phones differ systematically from those who do, so that studying only people with phones leads to biased estimates on the variables of interest. This is also sometimes referred to as noncoverage, undercoverage, or incomplete coverage.

The effect of coverage error is to bias statistical estimates. The coverage error for means is equal mathematically to the product of the proportion of people not covered in the sampling frame and the difference between the means of the sampling frame and those omitted from it. The mean for those omitted from the sampling frame is, of course, unknown, which makes it difficult to estimate the actual extent of such bias.

Coverage error for mean = (% of population not included in the sampling frame) × (mean for sampling frame − mean for those omitted from sampling frame).

According to this formula, the extent of coverage error depends on the interaction between two factors. First, the potential bias depends on how much of the population is not included in the sampling frame. There is a large potential bias when a large proportion of the population is excluded from the sampling frame. For example, the coverage error can be substantial when an unadjusted Internet survey is used to represent the U.S. population, since at least a third of the population are not Internet users as of this writing. By contrast, there is little potential bias if the sampling frame covers nearly the entire target population. Thus, omitting Alaska and Hawaii from the sampling frame of national surveys for cost-saving reasons is usually thought to introduce little bias, given that they together constitute only 0.65% of the U.S. population according to the 2000 census.

Second, whether there is actual bias on the variable of interest depends on whether the missing cases resemble those that are interviewed. Within the constraint of the proportion of the population excluded from the sampling frame, the coverage error is large if the mean for those in the sampling frame differs greatly from that for those who were omitted, whereas the coverage error is small if the two means are similar. These differences may be small in surveys on some topics but may be large in surveys on other topics. Indeed, the differences between the sampling frame and population can vary by variable in a study, with large coverage error on some variables even when there is little coverage error on other variables. For example, omitting residents of Hawaii is unlikely to affect most attitude questions but may bias results on topics on which they would have distinctive answers (such as whether respondents have ever been to a luau).

Although the coverage error problem is usually described in terms of the means of variables, it also affects more complex statistics. In particular, it can lead to incorrect regression coefficients (Groves 1989, 90–95). Since coverage error is a selection problem, this raises the possibility of doing a Heckman selection model (see chapters 7 and 8) if it is possible to develop an equation for the likelihood of being in the sampling frame.

Noncoverage Problems in Surveys

Coverage exclusions are actually quite common in surveys. For example, face-to-face interviewing of the mass public typically excludes the institutionalized population, such as people living in hospitals, nursing homes, college dormitories, prisons, and military bases. Presumably the proportion of the population that is institutionalized is small enough that this exclusion should not have a major effect on most survey results. Furthermore, the exclusion of some of these groups will not affect particular surveys. For example, noncitizens and felons cannot vote, so their exclusion from election surveys is fairly benign, though their exclusion is responsible for part of the overestimate of voting turnout that is common in the National Election Studies surveys since they are included in Census Bureau turnout reports (McDonald and Popkin 2001). Language creates an additional coverage issue in that most interviews in the United States are conducted only in English, whereas the 2000 Census found that 11.9 million adults were "linguistically isolated," living in households in which no person aged fourteen or over speaks English.

Coverage error is sometimes a problem because of out-of-date or inaccurate sampling frames. Such a problem would occur if a researcher used unadjusted phone numbers from a telephone book for sampling, because phone books are out of date by the time they are published, omitting people who have moved into the area and obtained new phone numbers since the phone books went to press. This problem can sometimes be handled through collection of more up-to-date information during the survey itself. For example, face-to-face household samples sometimes have interviewers list all dwelling units in the sampled area (e.g., the sampled block in most small and medium-sized cities) to take into account the possibility that available lists of dwelling units on that block are old or inaccurate. Modifying a list in this way in the process of drawing a sampling is known as *list-assisted sampling* to distinguish it from list sampling on the one hand and random sampling on the other.

The size of the population is sometimes not even known in advance of the sampling, which makes it necessary to develop sampling frames that work well regardless. As a result, as seen in chapter 10, the most common list-assisted methods for sampling for national face-to-face and telephone interviewing do not require knowing the total number of population elements in advance.[2]

A survey, by definition, cannot provide any information about its own noncoverage, since cases outside of the frame are not contacted. Estimating coverage error requires side information, which is sometimes obtained by using auxiliary information from surveys employing different sampling frames, as by asking people in face-to-face surveys whether they have a telephone, whether they use the Internet, etc.

Coverage error is usually considered at the unit level (as in missing certain types of households) but can also occur within the unit (as in missing certain types of people within households). As an example of the latter, Martin (1999) analyzes a 1993 Living Situation Survey in which households were interviewed twice. She finds that asking people to list residents of their households misses an estimated 4 million people in the United States, especially people who are not continuously present and when the informant has incomplete information. Martin interprets the undercoverage pattern as one of confusion, whereas Tourangeau et al. (1997) ascribe it to concealment. There seems to be some coverage error for adult males in male-headed households in phone surveys; other coverage issues will be mentioned in chapter 12 in the discussion of demographic differences between surveys conducted through different modes.

Telephone Surveys

Telephone interviewing necessarily excludes people who do not have telephones, a coverage problem that precluded telephone interviewing until the 1970s, when the percentage of U.S. households with telephones finally went over 90% (U.S. Bureau of Census 1984).[3] By March 2000 telephone penetration had reached 94.6% of households and 95.2% of civilian noninstitutionalized adults, the highest levels to date (Belinfante 2000). However, phone books still do not provide good coverage. By one estimate, there were 96.7 million household phones in the United States in 1998, of which 30% were unlisted (Survey Sampling Inc. Web site). Additionally, about 20% of American households move in a year, so about 12%–15% of the residential numbers in a phone book are disconnected.

Even with more people having phones, there is bias to having one, with undercoverage of people with low incomes, nonwhites, young people, and people in rural areas. Coverage error could thus be a serious problem for phone surveys on particular topics, such as studies of welfare recipients or crime victimization. In 1976–79 studies (Groves 1989, 117–20), people without telephones had victimization rates more than twice as high as those with phones (73% versus 31%). While that difference is stark, only about 7% of the total sample did not have phones, so the total victimization rate (34%) was only 3% higher than would be found by interviewing only respondents with phones. Thus, even in this case the coverage error has a limited effect on survey results.

Smith (1990b) reports on the attitudinal and behavioral correlates of telephone ownership as measured in face-to-face General Social Surveys between 1973 and 1987 that included questions about telephone ownership. Non–telephone owners had the profile of alienated social isolates, skeptical of human nature, residentially mobile, not reading newspapers, not voting, not attending church, and not belonging to social organizations. Smith judged them as so distinctive that their exclusion could bias some surveys, such as ones on unemployment.

The 90+% coverage rate of phones in the 1980s and 1990s made phone surveys very popular, but, unfortunately, changing technology has once again made the coverage problem an important issue for telephone surveys. The change that is likely to have the greatest effect eventually is the use by many people of only a cellular phone, without any landline phone in their household. Though this may eventually change, mobile phone companies in the United States currently charge users on the basis of minutes used, so people are not thought to be willing to be interviewed on cell phones given that they would essentially be paying to be interviewed. As a result, survey organizations avoid calling the banks of phone numbers assigned to cell phones, which leads to a problem of coverage error in phone surveys if that is the household's only phone.[4] Anecdotal claims are that young people are particularly likely to use a cell phone as their only phone, so surveys that exclude cell phones have a built-in age bias. According to a Yankee Group survey (*USA Today*, July 28, 2001), 2% of wireless customers use wireless as their only phone, but that percentage is certainly growing. The base of that survey was wireless customers rather than households; a face-to-face household survey asking that question would produce more definitive results. In any case, this is a new coverage problem for conventional telephone surveys that exclude cellular phones. As Groves et al. (2004, 80) predict, "Eventually telephone surveys will sample mobile phone numbers, and this will require movement away from the household as a frame and sampling unit." The ability to keep one's phone number when one moves weakens the traditional relationship between area codes and residence locations. The developing of technologies such as VoIP for phoning over the Internet may further complicate phone interviewing.

Before the 2004 U.S. election there was considerable concern regarding whether excluding people who use only cell phones was skewing polling results. Bush had a small lead in most polls, but it was possible that Democrats were more likely to have cell phones. Exit polls after the election estimated that the 7% of voters who had only cell phones broke too narrowly for Kerry for cell phones to have thrown off preelection polls. The proportion of people with only cell phones is projected to increase rapidly, however, so it will be important to handle this potential issue better in subsequent elections.

Other important technology changes affecting coverage in phone surveys have been answering machines, caller ID, and other devices to screen unwanted phone calls (such as a "privacy manager" marketed by some local phone companies that requires callers who come up on caller ID as "unavailable" or "blocked" to identify themselves). To the extent to which people are unwilling to accept calls from people or phone numbers they do not know, the set of people available for phone interviewing does not correspond to the desired population. One estimate is that 50% of households have answering machines, 25% caller ID, and 6% voice mail (Piekarski, Kaplan, and Prestengaard 1999), and nearly all cell phones include caller ID. A face-to-face survey in 2000 found that 33% of households are frequent screeners, of whom 78% use answering machines and 45% caller ID, with some obviously using both (Tuckel and O'Neill 2001).

What matters for survey coverage is whether people are using answering machines and the like to avoid intrusive contact from the outside world (which Tuckel and Feinberg 1991 label "cocooning" behavior) or to maximize their accessibility even when not at home ("connecting" behavior according to Tuckel and Feinberg). Current views are that most people with such devices use them to avoid unwanted calls at particular times (such as during dinner) rather than to avoid those calls completely, since studies find considerable success in phoning back to numbers with answering machines and/or caller ID. Indeed, the interview completion rate for phones answered by an answering machine on the first call is not lower than for phones that were busy or had no answer on the first call (Tuckel and Feinberg 1991). However, studies in South Carolina (Oldendick and Link 1994; Link and Oldendick 1999) show that people using answering machines and caller ID for screening calls tended to be college educated and younger, have a higher income, and have children in the household; this raises a potential for coverage error.

Encountering an answering machine usually is at least a signal that the phone number is a working residential number. As a result, some survey organizations give special effort to phoning back numbers that were initially answered by answering machines. For example, Piazza (1993) reports on a 1990 California Disability Survey in which a response rate above 80% was necessary and callbacks to phones with answering machines made the difference. That study would have saved 9% of the cost of its 330,839 phone calls by stopping calling numbers after 10 machine answers, but the extra phone calls increased the response rate from 80% to 83% so that the required response rate could be achieved. Xu, Bates, and Schweitzer (1993) find that leaving messages on answering machines increased the participation rate from 33% to 46%.

As another example of coverage problems due to technology change, in 2003 a TeleZapper device, which tries to remove a person's phone number from sampling frames, was being marketed. Many telephone interviewing operations eliminate nonworking numbers by having computers first phone sampled numbers (without actually ringing) to see if they emit a particular signal that distinguishes working from nonworking numbers. When a working number with the TeleZapper attached is phoned, that device sends a tone that tells the software that the number is disconnected, after which the software hangs up and removes the phone number from the database. This device would diminish the accuracy of sampling frames, but it is difficult to determine whether it is as effective as its manufacturer claims.

The coverage issues for telephone surveys vary across countries. In some developing countries, telephones are still relatively rare, making phone surveys impractical. In other countries, cellular phones are more common than conventional wired phones and incoming calls are free; phoning cell phones would provide a good sampling in those nations. The telephone density in many nations, however, is too low for random phone surveys to be cost-effective. Thus, whether phone surveys are practical must be decided on a country-by-country basis.

Household Surveys

It might seem that coverage issues would be less for face-to-face household surveys, but there are also coverage problems in such surveys. One set of problems involves incomplete lists of dwelling units. Even when lists of dwelling units in a city are available, those lists may miss newly constructed units. Interviewers are usually told to include any units between listed units in the area in which they are sampling,[5] but there can be boundary interpretation errors when it is not clear whether a unit is or is not in the area to be sampled. A second set of problems involves access to dwelling units. Interviewers may not be permitted access to gated communities or to locked apartment and condo buildings.

Additionally, the definition of "resident" in a household can lead to some coverage problems. Some surveys define residents as people who slept in the dwelling unit the previous night, while others define them in terms of people who "usually live" there. Even under the latter definition, it can be difficult to decide whether to include people whose occupations involve travel, such as truck drivers and flight attendants, as well as whether to include young children who live with one parent some nights and another parent the other nights. The person first contacted in a household often neglects to think of such residents, thereby excluding them from the sample. Also, illegal residents are likely to be missed, as when a woman receiving government welfare payments does not list a male resident whose residence would make her ineligible for benefits. Several of these considerations often lead to undercoverage of younger males.

Internet Surveys

Coverage error in Internet surveys can be estimated with available polling data from face-to-face and telephone surveys. For example, a National Telecommunications and Information Administration survey in late 2001 found that two-thirds of U.S. households had computers but only 54% were using the Internet. The coverage differences correlated with income, race, and rural-urban residence. Both computer ownership and Internet access have increased since this study was conducted, but it still would be appropriate to be concerned about the coverage correlates. Handling the coverage problem by weighting underrepresented groups would assume that rural low-income people with Internet access are like rural low-income people without that access, which would be a risky assumption. Furthermore, many people with computer access and even Internet access may not have the skills needed to complete a survey on the computer, or at least may be too concerned about their skill level to try to complete one. As a result of these considerations, Couper (2000) views coverage error as the biggest threat to Internet surveys, though Taylor (2000) emphasizes that this may not be problematic forever.

There are two strategies for Internet polling that try to avoid this coverage problem by using a telephone sample frame to generate Internet panels. The first uses probability-based telephone sampling methods, such as random-digit dialing, to recruit people with Internet access for an Internet panel. People who agree to participate are sent e-mail requests to fill out particular surveys, with access controlled through passwords. The second approach to avoid the coverage problem in Internet surveys is to offer free Web access to respondents who are originally contacted using telephone sampling procedures to obtain a probability sample of the full population. As described in previous chapters, Knowledge Networks provides its panel participants both WebTV units and Internet access. For both of these approaches, the initial probability-based phone survey obtains background demographics to permit appropriate weighting of the Internet survey. The remaining coverage errors are those from incomplete telephone coverage.

Another strategy for handling coverage issues is to supplement the Internet poll with a telephone survey on the same topic. Harris Interactive uses this approach, with propensity weighting of the samples. A logistic regression equation is estimated, with the dependent variable being whether the interview is from the Internet or the phone survey and with standard demographic variables and the person's online usage pattern as independent variables. The cases in the Internet sample are then weighted so that probability deciles have the same portion of cases in the two samples. Each person's probability of being in the phone sample is computed on the basis of the equation, and this propensity score is then used to weight the responses of Internet respondents (Miller and Panjikaran 2001). This use of propensity scoring is controversial. Schonlau, Fricker, and Elliott (2002, chap. 4) regard the Internet sample as nothing more than a convenience sample, regardless of the propensity weighting. Best and Krueger (2004, 20) emphasize that propensity scoring must adjust for differences in the relationships, which will not happen if the decision-making process of Internet respondents differs systematically from that of the general population, as is the case when computer users rely on the Internet for their information more than do those who are not computer users.

Election Exit Polls

Election exit polls have unique coverage problems relating to time issues (Merkle and Edelman 2000). Exit polls exclude absentee voters, whose votes can sometimes reverse the results of a close election. They also exclude early voting (as in Texas, where voters can cast ballots for a period of time before election day) and mail-in voting (which is now how Oregon votes and is used also extensively in Washington). In order to estimate voting outcomes, the Voter News Service (VNS), which conducted exit polls for the major news media in national elections in the 1990s, commissioned telephone surveys in states with early and mail-in voting.

Another potential time issue for exit polls relates to when they are tallied. The VNS exit polls were conducted throughout the day of the election, but results were phoned in only at 9:00 a.m., at 3:00 p.m., and just before the polls closed. There would be a coverage problem if only the early poll results were used, as occurred in the 2004 elections, when midday exit polls released on the Internet showed Kerry leads in Florida and Ohio that would have given him the election; these leads had vanished, however, by the time the polls closed. News organizations planned their coverage on the basis of the early polls, which would lead to faulty analysis if late voters differed from early voters. However, there was no time-related coverage bias in the final exit polls. The advent of notebook computers connected to cell phones and wireless personal digital assistants (PDAs) might permit real-time transmission of results from sample precincts to a central tally facility throughout Election Day, fully removing this time-related coverage problem.

The 2000 VNS exit polls encountered an additional coverage issue: only 84% of its sampled precincts were staffed with interviewers (Konner 2003). Hiring interviewers to work only one day had serious practical problems, since they had little incentive to show up if a better opportunity came their way. This coverage error could lead to serious prediction errors, depending on the distribution of the unstaffed precincts.

Methods for Dealing with Coverage Problems

There are a few possible ways of dealing with coverage problems. Postsurvey adjustments are often employed. The procedure for poststratification weighting described in chapter 8 simultaneously corrects for nonresponse and coverage error. For example, if poststratification weighting is used to get the gender balance in a survey to match census parameters for the same geographical area, that automatically corrects for any tendency of either gender to be disproportionately excluded from the sampling frame.

An intriguing approach for telephone surveys is to give extra weight to hard-to-obtain interviews. The basis of this approach is Keeter's (1995) report on using panel surveys to examine "transients" who had phone service in one panel wave but not another wave. In the Current Population Survey conducted by the Census Bureau in 1992 and 1993 (with only people who did not move during that year being included in the reinterviews), he found that 3.4% of households were transients compared to

2.6% that did not have a telephone at either point. The transients were more similar to those without a phone than to those with phone service at both time points, as, for example, only 69% of transients and people without phone service had health insurance versus 87% of those who had continuous phone service. This leads to the suggestion (Brick, Waksberg, and Keeter 1996) of asking people if their household had experienced an interruption in phone service in the last year and weighting on that basis, so that households with longer service interruptions are given greater weight so as to represent people without any phone service.

Another approach to minimizing coverage error is using *multiple frames* in the same study in the hope that the second frame will counter the coverage limitations in the first frame. For example, the federal government's Current Population Survey employs a multiple frame design, using lists of building permits to supplement addresses from the last decennial census of population and housing. Telephone surveys can use two sampling frames, as when Traugott, Groves, and Lepkowski (1987) demonstrated cost saving from using a random-digit-dialing (RDD) sample to complement a list frame from a telephone directory that would have had severe coverage issues if it were the sole list.

The multiple frame strategy is particularly valuable when the frame with coverage problems is less expensive to interview, while using only the other frame would be overly expensive. For example, in a survey of a rare population (such as Jews), a list sample from organization membership rosters would have coverage problems since people who are not organizationally active would be missed, but an RDD sample with extensive screening to locate people in that rare population would be very expensive. Using both frames can be more efficient in that situation. Another example of a dual frame study is combining a telephone survey and a more expensive face-to-face survey, with respondents in the latter asked if they have a phone to see if they were covered in the other frame. Estimation of the proportion of the population that is in both frames allows weights to be assigned so that the two frames can be combined (Groves 1989, 125–26; Groves et al. 2004, 88). Multiple frame surveys can also be used to determine the bias from using a single frame for later surveys. Harris Interactive's use of a phone survey to supplement its Internet survey is another example of this approach. Lessler and Kalsbeek (1992, 83–88) provide more detail about the use of multiple frames in surveys, including optimum sampling fractions based on sampling and data collection costs for each frame.

Ineligibles

Another frame problem involves ineligibles—the inclusion of some elements in the sampling frame that are not part of the population of interest. This is also called overcoverage, blanks, empty listings, duds, foreign elements, or out-of-scope elements. Examples are the inclusion of aliens in a voting survey and the inclusion of businesses in a household survey. This problem is actually quite common, since foreign elements often happen to fall in sampling frames. A household survey will inevitably contact some noncitizens, and a telephone survey will inevitably phone some nonresidential numbers. In the United States, telephone prefixes (the first three digits of a seven-digit phone number) do not correspond to political boundaries, so RDD phone surveys unavoidably phone some people outside of the intended geographical area.

Interviews with ineligibles would bias the results if they were included in the analysis. The solution, of course, is simply to discard them, as when voting surveys drop interviews with noncitizens and telephone surveys drop calls to businesses. The bias introduced by accidental inclusion of data from ineligibles depends on how many ineligibles are included as well as how different they are from the eligibles. There would be no bias for the mean if the mean for the eligibles equals that for the ineligibles. Following Lessler and Kalsbeek's (1992, 68) notation, let v = the ratio of the mean for ineligible cases to the mean for eligibles and let Q_0 be the proportion of elements in the frame that are ineligible. The relative bias is then $Q_0(v - 1)$.

Conducting interviews with ineligibles (and even just contacting them to ascertain their eligibility) increases the cost of a survey. Surveys must estimate the proportion of ineligibles and then plan to contact enough extra cases to compensate for the expected ineligible rate. The extent to which a sampling frame avoids extraneous units can be viewed as its efficiency (Fowler 2002, 13). The most efficient sample, in this sense, would be one in which all contacts lead to eligible respondents.

There is often a trade-off between the comprehensiveness and efficiency of a sampling frame, since a larger frame would include more units in the target population (and therefore have less coverage error) but also more units outside of that target population. It is common, for example, that the prefixes of phone numbers that are operational in a county include some with many phones in the county's main city and some with phones mainly in the suburbs. Randomly selecting numbers from all the prefixes would yield an inefficient sample if one were trying to sample the main city, since many calls would go to ineligibles in the suburbs. Phoning only prefixes with large numbers of phones in the main city would be more efficient, but it would not be a comprehensive sample since it would exclude city residents whose prefixes were primarily in the suburbs. In this situation, the researcher must choose a tolerable trade-off between coverage error and ineligibles, realizing that both cannot be minimized simultaneously.

Screening Ineligibles

A common method of dealing with ineligibles is to begin the interview with a set of *screening questions* designed to check whether the person or household belongs to the target population. A survey of vote intention in an upcoming mayoral election would begin by checking if the person lives in the city. Preelection polls often use *likely voter screens*, such as by also checking if the person is registered, voted in the last election, and/or is likely to vote in the next election; the call is terminated if the person responds negatively to any of these questions. Similarly, an RDD survey of Catholics would have to use screening questions on religion.

Developing appropriate screening questions can be difficult. For example, predicting who will turn out to vote in an election is hard—particularly for primary elections—so likely voter screens often do not work well. People who do not vote may have different preferences from those who do, so polls with poor screens can be off the mark in their forecasts. Thus, screening may be necessary but still may not correctly eliminate ineligibles.

Good screening questions are designed such that the person contacted cannot figure out how to opt in or opt out of the survey by giving particular answers. For example, a survey of the Columbus Jewish Federation conducted by the Ohio State University Center for Survey Research did not begin by directly asking people if they were Jewish but instead began with an innocuous question and then asked them their religion.

Screening questions can be controversial. In summer 2001 a controversy developed regarding a CBS News/*New York Times* survey showing declines in presidential approval for George Bush. The *Washington Times* (Lambro 2001) criticized the poll for not screening on likely voters. Pollster John Zogby is quoted as saying that the *New York Times* polls are tilted toward the Democrats because all adults include "larger percentages of minorities or poorer voters and voters even in the $25,000-to-$50,000 income range, all of which lean to the Democratic side. When you screen for voters, you screen out a substantial percentage of Democratic individuals who do not vote." However, Zogby has a reputation as a pollster for political conservatives (Mooney 2003). In fact, the use of all adults is standard in presidential approval polls in nonelection years, following the approach used by George Gallup starting in 1935, so the CBS News/*New York Times* nonscreening was appropriate for comparing Bush's popularity with that of previous presidents at that point in their term. Still, this does demonstrate that the choices involved in screening can affect results and therefore are debatable.

Purchasing Sampling Frames of Eligibles

An alternative to screening ineligibles is to purchase sampling frames that have been refined to avoid ineligibles. This has become very common for phone interviews. The sampling frame problem is important in telephone interviewing because pure RDD would inevitably lead to phoning many nonworking phone numbers. There are firms that use devices that dial phone numbers (hanging up before ringing) and check whether the number gives the type of signal that nonworking numbers usually give. The list vendors, such as Marketing Systems Group (Genesys) and Survey Sampling, Inc. (SSI), sell the resultant lists of working phone numbers to survey organizations that are willing to pay to avoid the expense of having their interviewers phone nonworking numbers. Ideally, the sample is drawn first, after which a sample screener is used to identify disconnected numbers. This system is not perfect, but the companies claim that it can identify about half the disconnects, so that about 10% of the sample numbers need not be called by interviewers. This approach is intended to increase the efficiency of the sample while maintaining the integrity of random-digit dialing.

Lists purchased from these vendors also try to handle the geography of a sample, so that one can purchase phone numbers from a particular metropolitan statistical area, whereas phone books do not cover those geographical areas exactly. However, there is never a perfect match between geography and phone numbers. For example, if one wanted to cover a particular zip code, one would phone several telephone prefixes (the first three digits of a seven-digit phone number), but some of those prefixes would also include phone numbers outside of that zip code. Firms that sell samples can tell users what proportion of phone numbers from a particular prefix are in the desired zip code (the hit rate). The survey organization must then decide on the desired trade-off between full coverage and hit rate.

These firms are able to provide some additional useful services when they sell sets of phone numbers. For example, they can withhold numbers that were assigned to samples in the previous six months, so as to reduce the likelihood of respondents' complaining about being overinterviewed. Also they can give out phone numbers as "replicates," with each replicate being a representative sample. Survey organizations would give out these replicates to interviewers one at a time. If the full sample consists of ten replicates, for example, and if using the first eight replicates yields enough completed interviews, then the sample is representative without dialing the numbers in the final two replicates.

Purchasing a sample is particularly attractive as an alternative to screening for small groups in the population. Several firms sell "low incidence" lists of people in small groups in the population. They develop these lists of home addresses, phone numbers, and/or e-mail addresses of people based on product warranty cards, mailing lists, participation in previous surveys by that firm, and other sources. Among the hundreds of categories of targeted groups are dog owners, musicians, motorcyclists, people interested in weight management, large-print book buyers, business owners, full-figured clothes buyers, seniors, college students, child care workers, acne medication buyers, chess players, people who buy gospel music, Democrats, people who fish, and people interested in exotic vacations. These rare population samples are not true probability samples, but they are much more cost effective than the extensive screening required in pure random sampling. Alternatively, some firms maintain an Internet panel of people willing to participate in surveys. Information about those people is kept in a database, so that it is possible to purchase a sample based on personal characteristics such as state, age, gender, occupation, education, income, race, and ethnicity.

Other Frame Mismatches

Clustering

A different problem involving sampling frames involves clustering, when groupings of units are listed as one in the sampling frame. This situation is actually common in face-to-face and telephone surveys of the general public. These surveys are usually household surveys, since the unit being sampled is actually the household, even when the researcher really wants to interview only one person in the household. (Procedures for selecting the respondent in a household sample are reviewed as part of the sampling discussion in chapter 10.)

The solution for the clustering problem involves weights. Consider, for example, the situation where multiple people share the same phone line.

A phone number can actually be shared by multiple households. If there are two households on a phone line, for example, those two households individually have only half as much a chance of being selected as a household with its own phone line. Proper weighting would entail giving such households a weight of two, the reciprocal of their chance of selection. More generally, if clusters are of different size, the cases can be weighted proportionately to the reciprocals of the cluster sizes to equalize chances of selection.

Returning to the common example of sampling households, clustering occurs whenever there are several people in the same household. The chance of selecting any particular person is the reciprocal of the number of adults in the household (Kish 1965, 400), so the proper weight for respondents is proportional to the number of adults in their household. If p_i is the number of adults in household i, the weight would be $p_i/\Sigma(p_i)$. Whether this weighting has much effect depends on the distribution of household size as well as whether the variable of interest is related to household size. There was little effect for most variables in the 1950s, when most households had two adults (Kish 1965, 400). The increase of single-adult households in the 1960s and 1970s led to some effects for demographic variables but not for attitudinal variables (Hess 1985, 53–55). The weight for household size particularly affects estimates of the proportion of people who are married (or partnered), since there are at least twice as many adults in married households as in single-adult households, leading to an 8% lower estimate of the married proportion of the population in 1976, when weights were not used (Hess 1985, 252). As Fowler (2002, 27) concludes, "Any estimate of something related to marital status will be distorted if the data are not weighted." Thus there should be caution in calculating person-level statistics using household probabilities of selection when the variable of interest is related to household size (Lessler and Kalsbeek 1992). This is usually not a serious problem, but it may make correction for household size appropriate when one is looking at variables related to marital status (Weisberg 1987).

Multiplicity

The final frame problem is multiplicity: having a case appear more than once in the sampling frame, thus giving it a higher chance of selection than other cases. This problem is also known as *duplicates*. It often occurs in list samples when the sampling frame is based on several different lists, as when using lists from multiple environmental organizations that have partially overlapping membership, giving those people a higher chance of selection. It would also be a problem in sampling from lists of e-mail addresses, since it is common for people to have multiple e-mail addresses. Multiplicity occurs with some frequency in phone surveys, since some households have more than one telephone number. One estimate is that a quarter of households had multiple lines in the late 1990s (Piekarski, Kaplan, and Prestengaard 1999), with 15% of households having a dedicated computer/fax line (Tuckel and O'Neill 2001).

Multiplicity problems can also result from "repeats," such as repeat business customers. For example, when customers of a company are being sampled, a person who has bought products from several branches of that company would generate a multiplicity problem. Sampling airline flyers is complicated because most flyers are on multiple flights within even short periods of time. Similarly, it is difficult to estimate the number of people visiting doctors, since many people visit multiple doctors. It is often easier to estimate how many events occurred during a time period, such as the number of doctor visits, in order to avoid the repeat problem.

The most straightforward solution to the multiplicity problem is to put the sampling frame into an appropriate order in order to notice and remove duplicates. Names can be listed in alphabetic order, for example, after which extra occurrences of the same person are removed—if it can be told with certainty that it is the same person. That solution, however, assumes that a full list of the sampling frame is available in advance, which is often not the case.

When it is not possible to remove duplicates in advance, the proper solution for the multiplicity problem after the data have been collected involves weighting by the reciprocal of the case's multiplicity. For example, if a case appears twice in a list, it has a double chance of selection, so data for that case should be assigned a weight of one-half. Phone interviews frequently ask people how many nonbusiness phone lines they have, so that an appropriate weighting adjustment can be made at the data analysis stage. For example, a household with two lines has twice as high a chance of being selected randomly as a household with a single line, so it should be given a weight of one-half, the reciprocal of its number of telephone lines.[6] This example also usefully illustrates how information about the sampling frame can be collected during the interview itself. The most difficult situation occurs when it is not possible to determine multiplicity, in which case bias results. Lessler and Kalsbeek (1992, 73–76) provide the bias formulas as well as a technical discussion of weighting for various multiplicity problems (90–102).

Frame Error and Survey Constraints

Figure 9.1 is a graphical interpretation of the relationship between the different types of frame error.

Frame error can obviously be avoided if one simply uses a sampling frame that has a one-to-one correspondence with the population. That advice is, however, often simply not practical. Obtaining a reasonable sampling frame can be very difficult, and the best available sampling frame for the problem at hand may have each of the problems discussed in this chapter. That is certainly the case for the best sampling frames that have been devised for national household face-to-face and telephone surveys in the United States. The sampling frame problem sometimes seems less serious when an organization provides its membership list to a survey organization, but even then there are usually problems such as missing new members (coverage error) and accidentally including ex-members (ineligibles).

Obtaining a useful sampling frame is partly a matter of cost. Purchasing a good sampling frame can be much more expensive than using a less adequate one. However, there is a trade-off between the costs involved in extensive screening to avoid ineligibles and purchasing a pre-screened list requiring fewer contacts with ineligibles.

This trade-off can also be seen in terms of time required to do the interviewing. For example, in the Columbus Jewish Federation survey mentioned in this chapter, a long time was needed to conduct interviews based on random-digit dialing because of the large number of phone calls required to get enough respondents who passed the screening questions. The second half of that study involved phoning people on organizational

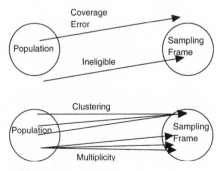

Figure 9.1 Frame problems

lists supplied by the sponsor, and those interviews were taken quickly since nearly everyone on these lists passed the screening questions. Yet the information from the RDD sample was invaluable to the sponsors of the survey, since they wanted information about people who were not currently being served by community organizations.

A point to emphasize is that information about frame problems can often be gained as part of the interview. Multiplicity can be ascertained through questions in the interview, such as how many phone lines there are in the household, so it is important for researchers to include questions to check on multiplicities. This is a way of controlling survey error without increasing costs.

The ethical issue involved in frame error is that people reading survey results often do not notice frame problems. If the report on a survey does not emphasize a lack of correspondence between the actual sampling frame and the real population of interest, most readers of the report just assume that the results can be generalized safely to that population. As a result, the writer of a survey report has a responsibility to point out explicitly where there are discrepancies between the sampling frame and the population and in what ways those discrepancies may limit generalizations.

10

Sampling Error
THE NEED FOR REPRESENTATIVENESS

The best-known source of survey error is sampling error: the error that arises when only a subset of the population is included in a sample survey. It occurs since sampled units inevitably differ from the full population. Even when interviewing a sample, the survey researcher normally wishes to be able to generalize beyond the people who were sampled, sometimes even to people in other locations and at other time points. That is possible only when the sample is representative of the larger population of interest.

Understanding the issues involved in sampling requires beginning with a review of the theory of sampling error. The second section of this chapter presents a wide variety of sampling methods, both probability methods underlain by elegant mathematical theory and nonprobability methods that many researchers still choose because of practical considerations. Issues involved in selecting a sample size are then discussed in the third section.

The Theory of Sampling Error

How likely a sample is to be representative of a larger population depends on the sample size and on the variance of the variable (whether its values are dispersed or concentrated). As would be expected, the larger the sample, the more representative the sample will be. The sample will also be more representative if the variable's variance is smaller. It is useful to begin with some intuitions on these two points.

Figure 10.1 shows the distribution of respondents on two numerical variables that range in value from −100 to +100. The values on the variable are shown on the horizontal axes, and the height of each represents the number of people giving the corresponding answer. The same number of respondents answered each variable (24), and both variables have means of zero. There is much more variance on variable 1, with the answers tending to be at the two extremes, than on variable 2, which elicited more centrist answers. Now imagine that you were taking a sample of five of these people. There is a chance of getting an extreme sample on variable 1, such as just getting people who answered −100 and −80. It would be impossible to get

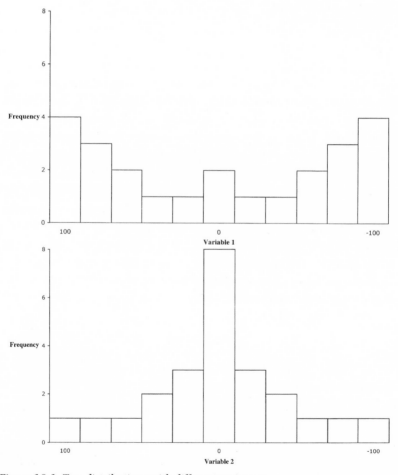

Figure 10.1 Two distributions with different variances

that extreme a sample on variable 2, since most people are concentrated near the center. Thus, for a fixed sample size, you are more likely to get a representative sample for a variable with a smaller variance.

As to the effect of sample size, imagine instead taking samples of size 9 from these two distributions. You could still get an unrepresentative sample, but not as unrepresentative as could occur with samples of size 5. For example, an extreme sample of size 9 on variable 1 would consist of people whose scores were -100, -80, and -60, which is not as extreme as was possible for a sample of size 5. Thus the sample mean is more likely to be near the population mean (here 0) when the sample is larger.

Two descriptive statistics are relevant to this discussion: the variance and standard deviation of a variable. The *variance* measures the dispersion of the variable X around its mean ($\overline{X}$). The variance is the average of the squared deviations of the individual values from their mean: Var $(X) =$ $\sum_i (X_i - \overline{X})^2/(n-1)$, where n is the number of cases in the sample.[1] The variance is an awkard statistic, since it is measured in squared units. The variance of a set of incomes, for example, would be in squared dollars, which is not a very intuitive metric. To get back to the original unit of measurement, statisticians make extensive use of the *standard deviation*, which is the square root of the variance: $s_x = \sqrt{\sum_i (X_i - \overline{X})^2/(n-1)}$.

Sample surveys seek to generalize from the sample to the relevant population. This requires moving to inferential statistics. The most important statistical result for inferential statistics is the *central limit theorem*, which states that for large sample sizes the distribution of sample means (known as the *sampling distribution of means*) is normally distributed. The central limit theorem also states that the mean of this sampling distribution is the true population mean, while its standard deviation (known as the *standard error*) is the variable's standard deviation divided by the square root of the number of cases: $s_{\overline{x}} = s_x/\sqrt{n}$. The normal curve table can then be used to establish confidence intervals for the mean. It is conventional in most survey fields to use the 95% confidence interval, which extends 1.96 standard error units on either side of the sample mean for a normal distribution. That is, 95% of confidence intervals that extend from the sample mean minus 1.96 times the standard error to the sample mean plus 1.96 times the standard error cover the actual population mean (μ), leading to this probability statement: $\Pr(\overline{X} - 1.96s_{\overline{x}} \leq \mu \leq \overline{X} + 1.96s_{\overline{x}}) = 0.95$. This 1.96 times the standard error is commonly referred to as the *margin of error* for the variable and is often reported as such in newspaper reports on surveys. Obviously, the ideal is to have a small standard error and a small margin of error so that one can make precise statements about the population from the sample survey. This can be seen in figure 10.2, where the normal curve with the smaller standard error has higher probability of centrist values and lower probability of extreme values.

To illustrate, if a report on a survey indicates that 54% of the public favors some proposed policy and cites a 3% margin of error, the 95% confidence interval extends from 51% (54% − 3%) to 57% (54% + 3%). This would allow us to conclude with high confidence that a majority of the population supports the policy. If, on the other hand, only 52% of the sample is found to support the policy when the margin of error is 3%, the survey does not show whether or not a majority favors the policy. The 95% confidence interval would extend from 49% to 55%, so it is still possible that a majority does not favor it. Such a result would best be termed "too close to call."

The standard error is actually affected by four factors. First is the heterogeneity of the sample. The standard error is smaller to the extent that there is less variance on the variable. Second is the sample size. One would expect to be more certain of results when bigger samples are employed, and that is the case. Since the sample size is in the denominator of the standard error formula, the larger the sample size, the smaller the standard error. More precisely, the square root of the sample size is in the denominator of the standard error formula, so quadrupling the sample size is required to cut the standard error in half. Some of the large national social science surveys use a sample size sufficient to obtain a 3% margin of

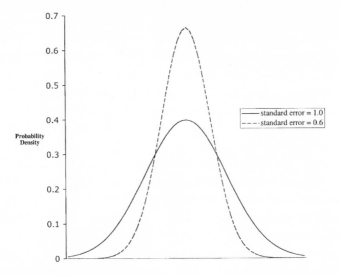

Figure 10.2 Normal distributions

error, feeling that it is not cost effective to quadruple the sample size to cut the margin of error to 1.5%. That would nearly quadruple the interviewing cost, while the other sources of error discussed in other chapters of this book probably amount to at least 1.5%. Advocates of the total survey error approach would argue that the money required to quadruple the sample size would better be spent in trying to reduce, or at least measure, those other types of error.

Third, the standard error is affected by the sample design. Particular types of sampling, such as stratifying and/or clustering as described later in this chapter, directly affect the standard error. The formulas given above assume a *simple random sample* (SRS), as when names are drawn randomly from a hat. This very pure type of sampling is not generally used in large surveys. Yet most general-purpose statistical computer programs treat the data as if they were generated by a simple random sample, without adjusting for the possibility that the actual sample design yields a higher standard error.

Finally, the standard error is affected by what proportion of the total population is sampled, known as the *sampling fraction*. If we let small n represent the sample size and capital N the population size, the sampling fraction $f = n/N$. Technically, the full formula for the sampling error includes a term based on the sampling fraction. The *finite population correction factor* is $\sqrt{1-f}$, and it is a multiplier for the rough version of the standard error formula given above. That is, the full formula for the standard error of a mean is $s_{\bar{x}} = \sqrt{1-f} \times (s_{\bar{x}}/\sqrt{n})$. Yet this technical correction can often be ignored, for reasons that are explained next.

Intuitively, there should be less error when a larger fraction of the population is being sampled. After all, a survey that samples 90% of the population should give more accurate results than a study that samples only 2% of the population. In fact, however, the sampling fraction has little effect on the standard error, unless the sampling fraction is large. Sampling fractions for surveys of the mass public are generally very small, usually less than 0.0001, so the finite population correction factor is around 0.9999. Multiplying any number by a value that near 1 basically gives back the original number, so the correction factor has a negligible effect. For example, a survey of 1,500 people leads to a minuscule sampling fraction if the population consists of millions of people, in which case the finite population correction factor is near 1 and does not have any real effect on the standard error calculation.

As an example of this point, consider the question of whether a survey of 1,500 people taken from the U.S. population is more accurate than a survey of that same size from your state. Most people expect that the survey of their state would be more accurate than the national survey, but in fact there is not a meaningful difference between the two. Both involve sampling fractions well under 1%, so that the corrected standard error is virtually the same as it would be if the sampling fraction were ignored. Most household surveys (other than very large ones conducted by the federal government) and most telephone surveys of the public are thus able to ignore this correction in computing their standard errors.

The sampling fraction has a large effect on the standard error by the time that a majority of the population is sampled. If a random sample of 50 is taken from a 100-person legislature, for example, the sampling fraction must be taken into account. A common rule (Fowler (2002, 34; Biemer and Lyberg 2003, 326) is that the sampling fraction can be ignored when less than 10% of the population is sampled.

Note that standard error is calculated for specific variables, not overall for a survey. The standard errors depend on the variance of the variable, which is variable specific. Also, some of the technical aspects of the survey design, specifically clustering effects when cluster samples are used, are variable specific. How then can newspaper reports on a survey report a single margin of error for a study? The common practice is to report the maximum margin of error for a proportion, such as the proportion of the public supporting a policy. The variance of such a proportion can be expressed as $p(1-p)$, where p is the proportion, so the variance of an item with 50% giving each answer is 0.25. Notice that the variance of a proportion would be lower if the proportion giving an answer were different from 50%, since $p(1-p)$ would be less than 0.25. Thus the variance of a proportion is maximal when 50% of the sample gives each answer. The standard deviation of a proportion is the $\sqrt{(p(1-p))}$, which has the value of 0.5 when $p = 0.5$ and is lower when $p \neq 0.5$. The standard error is $\sqrt{(p(1-p))/(n-1)}$. The maximal margin of error for a proportion using a simple random sample of a small fraction of a large population is then $(1.96 \times 0.5)/\sqrt{n-1} = 0.98/\sqrt{n-1}$, or approximately $1/\sqrt{n}$. This is what newspaper stories on polls usually report as the margin of error of a survey, though sometimes these values are adjusted for the survey's design effect.[2]

It bears repetition to emphasize that the standard error formula does not incorporate the types of error discussed in other chapters of this book. Sampling error is neat to work with since it can be computed mathematically; this makes it seductive to focus on sampling error and ignore the other types of survey error. However, as the rest of this book should make clear, sampling error is only the tip of the iceberg, and it would be a serious mistake not to take the other sources of error into account in interpreting survey results.

The final point to make in discussing sampling error is that it can be calculated only for probability samples. It cannot be calculated for other types of samples, such as when people volunteer to participate in a survey. Some reports on nonprobability samples provide an indication of what the margin of error would be for a simple random sample of the size of their survey, but that estimate has no statistical basis. Citing a margin of error for a nonprobability sample is merely an attempt to cloak the survey with a scientific status that it does not have. The sampling error would be near zero for a survey of a million people, but that does not mean that inferences to the total population from a nonprobability sample that large are scientifically valid.

Sampling Methods

The most basic distinction in choosing a sampling approach is between nonprobability and probability sampling. Probability sampling is sampling when there is a known chance that every element in the sampling frame will be selected for the sample, whereas that chance is not known in nonprobability sampling.

Nonprobability Samples

Nonprobability sampling is considered problematic because there are potential biases and because the sampling error cannot be estimated. Without a random mechanism, it is not clear how representative the sample is of the larger population. For those reasons, sampling statisticians argue strenuously against using nonprobability sampling, but such sampling is still fairly common. The term *external validity* is used to refer to the extent to which results of a study can be generalized. Nonprobability samples have serious external validity problems.

There are several types of nonprobability samples, and the terminology for them is not fully standard. Three types will be described here: availability, purposive, and quota samples.

Availability Samples

The *availability sample* (also known as a *convenience sample*) studies cases that are readily accessible. One example is when instructors have their classes fill out questionnaires. The problem with such samples is that it may not be possible to generalize the results to a larger population. How serious this problem is depends on the purpose of the study. A college student survey is not likely to be representative of the full adult population if the researcher is trying to estimate the level of public support for Social Security reform, but it might be more justified if the researcher is trying to study the decision processes of people with high levels of education.

Using volunteer subjects should also be seen as availability sampling. Volunteers are often not representative of the larger population of interest. This problem plagued the early surveys on sexual behavior. This topic was long considered to be too sensitive to ask about in regular surveys, so the original Kinsey Report (Kinsey, Pomeroy, and Martin 1948) was based on volunteers who answered advertisements to participate in a survey. This study produced the first empirical estimates of the proportions of the American public who performed particular sexual activities that had previously been assumed to be rare. However, those estimates may be invalid, since people who were willing to volunteer in a 1940s survey of sexual practices were probably more liberated in their sexual practices than people who would not choose to volunteer for such a study. A comparable bias is considered to have occurred for the later Hite Report (1976) on sexual satisfaction of women. Hite distributed her questionnaires through women's organizations. Women who were less satisfied were probably more motivated to fill out the questionnaire (Moore 1995, chap. 1), which would have produced a bias in the study toward overreporting dissatisfaction. The use of volunteer subjects is always problematic, though obtaining such a sample can be so inexpensive that availability samples will always remain common.

Another example of availability sampling comes from the early political polls. A mass circulation magazine called the *Literary Digest* conducted polls preceding presidential elections starting in 1916. Using its subscriber list and lists of owners of automobiles and telephones, it mailed out very large numbers of postcards on which people were supposed to record their intended votes. Millions of people sent in their postcards, and it was thought that survey results based on millions of replies could not be wrong. The *Literary Digest* poll correctly predicted several presidential elections, until it incorrectly called Republican Kansas Governor Alf Landon the winner of the 1936 presidential contest. One problem is that this sampling procedure led to a middle-class and therefore Republican bias during the middle of a severe economic depression when many people could not afford to own an automobile or even have a telephone, let alone subscribe to a literary magazine.[3]

A current example of availability sampling involves the approach used by the Centers for Disease Control (CDC) for assessing the success of programs for disseminating information about HIV/AIDS risks. They station interviewers outside of gay bars and other places that at-risk populations are likely to frequent and ask people leaving those locations questions (such as whether they have seen particular flyers warning against high-risk activities). Intercept polls interview people who are available, so they are successful to the extent that the people at those locations are representative of the population of interest.

One of the practical ways for sampling from rare populations can also be considered an availability sample. Sometimes respondents are chosen for low-incidence surveys out of the pool of people who have self-identified with that group in prior surveys of the general population. For example, the 2003 Annual Survey of American Jewish Opinion sampled from the set of people who self-identified as Jewish in a previous consumer mail panel survey conducted by a marketing research firm.[4] This might be considered a probability sample if the original samples were probability samples, but it also clearly has an availability aspect.

Phone-in surveys by radio and television stations and click-in surveys on the Internet are also basically availability samples. The people who participate are essentially volunteers, which makes it impossible to know how to generalize from these samples to the larger population. Even worse, there is nothing to stop people from participating multiple times in such surveys, making them even less valid. As an example of the bias involved with using volunteer subjects on the Internet, Walsh et al. (1992) show that self-selected subjects were more positive toward scientific computer networks and electronic communication in general than was a random sample, with an average bias of 22%.

Couper (2000) differentiates three types of nonprobability Web surveys that are all basically availability samples: Web surveys as entertainment, self-selected Web surveys, and volunteer panels of Internet users. The ongoing entertainment polls, like CNN's QuickVote, do not portray themselves as scientific but simply try to have people respond to a "question of the day." The unrestricted self-selected survey is usually a full-blown survey on a particular topic that is advertised on other popular Web sites and generally claims to be giving valid results. Couper (2000, 480) cites the example of a survey on the National Geographic Society Web site, which claimed to have received "twice the minimum [number of responses] required for scientific validity" though the type of person participating in that survey was not necessarily representative of the larger population. Alvarez, Sherman, and VanBeselaere (2003) report that their recruitment of Internet survey subjects through banner ads had a yield rate of only 0.02%, with respondents being younger and more educated than those in a parallel phone survey. They also recruited subjects through a subscription campaign: people registering for some services on the Web were given the opportunity to check a box indicating their interest in participating, after which they were asked for their e-mail address. That approach provided many more subjects at about one-sixth the cost per respondent. Still, self-selected Web surveys may be one of the most practical ways to survey rare populations, such as gays and lesbians (Koch and Emrey 2001) or people with particular rare diseases. Best and Krueger (2004, chap. 3) give some practical suggestions for recruiting samples on the Internet.

Couper's third type of nonprobability Web survey is the volunteer opt-in panel. People see ads on popular Web sites asking them to volunteer for the panel. Respondents provide demographic information when they first register, after which they are periodically sent e-mail requests to participate in specific surveys, with a password provided for access to the survey in order to prevent multiple entries by the same respondent. The volunteer opt-in panel is still essentially an availability sample, but some survey organizations try to compensate for this by weighting their data. In particular, Harris Interactive seeks to make its volunteer Internet panel results valid by "propensity weighting" (chapter 9). This involves comparing willingness to share information and participation in social activities to a smaller scientific telephone survey, which allows Harris Interactive to estimate probabilities of participation for weighting purposes.

Internet polls have the potential advantage of being able to generate very large numbers of respondents. Telephone surveys of tens of thousands of respondents are unusual, but it is possible to obtain that many respondents in Internet polls. The large number of respondents has the effect of making sampling error calculations virtually irrelevant, though it should increase concern about the representativeness of a sample based on volunteer respondents (Dillman and Bowker 2001).

At the extreme, surveys of one's own country can be considered to be instances of convenience samples. Social scientists are generally trying to develop generalizations that are not country specific, but they usually take surveys only in their own country because of costs as well as limited funding possibilities. Thus many surveys that appear to be probability samples are in this sense actually availability samples.

Purposive Samples

A *purposive sample* involves the intentional selection of particular cases to study. One example of a purposive sample is the study of *typical cases*. The problem with this approach, however, is that it is not obvious how to judge what is typical. Further, what is typical in one respect may not be typical in another. We know that the typical American is white because that is the majority race, Protestant because that is the plurality religion, and so on, but that does not mean that people with those characteristics have "typical" attitudes on policy questions (cf. Scammon and Wattenberg 1970).

Purposive sampling is sometimes used for substantive reasons. Two common approaches to case selection are to select cases that are either maximally similar or maximally dissimilar. As explained in Przeworski and Teune (1970), both of these approaches are thought to have particular advantages. Studying Australia and the United States as most similar cases, for example, would guarantee that any differences found between them would not be due to language or British heritage. Studying Britain and Haiti as most dissimilar cases would guarantee that any similarities found would not be due to their language heritage or their economic development. Similar strategies could be used in deciding which U.S. states to study if a research project had funds to survey only a handful of states. These case selection strategies are sometimes appropriate, but because they are purposive, they do not permit estimation of sampling error.

Another variant of purposive sampling is the use of *critical cases* (Henry 1990, 21). Researchers often feel that some cases are particularly important for them to study because they provide critical tests of the theory they are testing. If the hypothesis of interest does not hold for a particular set of respondents, the researcher may be able to argue that it is not able to hold anywhere, so that those respondents provide a critical test.

A more specialized type of purposive sample is a *snowball sample*, in which respondents are asked to suggest more respondents. Snowball samples are common in studies of elites in cities, with a researcher starting by interviewing some people who are thought to be among the influential decision makers in the city, asking them who else is influential, and then interviewing those added people. Snowball samples are nonprobability samples, since it is impossible to know the probability with which any person in the larger population ends up in the sample. Such studies can shed valuable light on community influence, but they are nonprobability samples so their sampling error cannot be estimated.

Quota Samples

The final type of nonprobability sample to be discussed here is *quota sampling* (also known as *model-based sampling*). Interviewers are assigned quotas of types of people to interview, but the interviewers decide themselves whom exactly to interview within each of those categories. For example, an interviewer might be told to take 12 interviews each day, with quotas such as 6 with men, 6 with women, 10 with whites, 1 with an African American, 1 with a Hispanic, and so on. The interviewer would decide whom to interview within those quotas; this freedom allows the interviewer to be as efficient as possible. The problem, though, is that this selection is nonrandom, so it can be biased. In particular, interviewers are only human, so they are likely to choose people who seem to be the easiest and safest to interview, which often means that they choose people who are like themselves.

Quota sampling became prominent in the United States after the 1936 *Literary Digest* debacle. While the *Literary Digest* predicted the wrong winner, more accurate forecasts were made that year by some smaller face-to-face surveys using quota sampling, most notably by a young George Gallup after he had successfully polled for his mother-in-law, Olga Miller, when she ran for Iowa secretary of state in 1934 (Moore 1995, chap. 2). That led the polling industry to adopt quota sampling. Unfortunately, quota sampling hit its own crisis in 1948, when the Gallup Poll, the Roper Poll, and all the other large commercial polls predicted that Republican New York governor Thomas Dewey would be elected president. The interviewers were apparently most likely to choose to interview middle-class people, who were more likely to vote Republican, leading to a Republican bias in the surveys. One survey that was correct in 1948 was taken by the University of Michigan's Survey Research Center, which used probability sampling rather than quota sampling (Mosteller et al. 1948). As a result of studies of the poll failures in 1948, many survey organizations changed to probability methods. However, because of their lower cost, quota samples are still used in many instances, particularly in market research.

Quota sampling is the main approach used for preelection polls in Great Britain. The election period there is just three weeks long, making it impossible to do enough callbacks to obtain a high response rate. Instead the British pollsters choose areas to survey on the basis of probability techniques, including stratifying, with interviewers being assigned very small areas so as to minimize their discretion (Moon 1999, 46–64). Quotas are usually established on the basis of gender, age, social class, and working status, sometimes interlinked so that interviewers are given quotas for various age-class combinations and various gender-working status combinations. The claim is that if the quotas are set properly, a quota sample does as well as a probability sample and is more reliable than a probability sample with a low response rate (Moon 1999, 40). However, Jowell, et al. (1993) argue that part of the failure of the British polls in 1992—when a Conservative win of 7.6% over the Labour Party was not detected in preelection polls—was due in large part to quota sampling without callbacks, since a postelection survey found a 6% pro-Labour bias at the first call.

Groves (1989, 247–48) cites as an American example a NORC quota sample in which counties, cities, and blocks were chosen in probability terms, but interviewers were then given quotas based on gender, age, and employment status. The researchers knew the probability of people's being available within those categories. If the people in the same quota group are homogeneous, this method is considered unbiased. Regardless of the potential difficulties with the quota approach, it can reduce costs as well as nonsampling error. In this instance, the NORC sample cost only a third as much as a probability sample would have, making it very attractive to the extent to which those quota groups were homogeneous.

This review of nonprobability sampling has been lengthy so that several points could be seen. First, the usual statistical advice is to avoid nonprobability samples. Second, many surveys either explicitly use nonprobability sampling or do so implicitly (as in studying only the United States while trying to generalize more broadly). Third, the balancing of survey errors and survey costs can sometimes justify nonprobability sampling, regardless of the usual textbook injunctions against it. Probability sampling would be universal in an ideal world, but research is never conducted in an ideal world.

Probability Samples

As stated above, probability samples are ones in which each element of the sampling frame has a known chance of selection. This makes it possible to generalize from such samples to the larger population from which they are drawn. Additionally, it is possible to obtain the sampling error for probability samples, so that statistical inference and confidence statements can be drawn.[5]

Simple Random Sampling

The simplest form of probability sampling is *simple random sampling* (SRS), in which each element has an equal chance of selection, as when names are drawn randomly from a hat. Most standard statistical analysis computer programs use significance tests that assume simple random sampling.

Simple random sampling is, however, actually uncommon because of the requirements it imposes. First, one needs to sample from a list of all possible cases in order to give each element an equal chance of selection, and such lists often do not exist. Second, simple random sampling can be very expensive, as it would be if a national face-to-face sample required transporting an interviewer to an isolated town to do a single interview. Similarly, purely dialing random ten-digit phone numbers would be very expensive, since many of the numbers would turn out to be nonassigned or business numbers rather than residential. After all, one estimate is that only about 45% of phone numbers in 1999 were residential (Brick, Montaquila, and Scheuren 2002). Because simple random samples require lists of possible cases and can be very expensive, other types of probability samples are more common.

Simple random sampling is possible in Internet surveys sampling from a list of e-mail addresses but is not possible for surveying the general public on the Internet, since there is no list of e-mail addresses for the whole U.S. population. It is feasible in organizations that have a high level of Web access, as in sampling university faculty. Typically people are sent requests to participate through e-mail and are given a PIN code to a Web site in order to preclude people's filling out the questionnaire multiple times. These surveys often attain reasonable response rates, though there is a problem when potential participants do not check their official e-mail address frequently enough to receive the invitation to participate.

Simple random sampling is the most basic probability sampling approach. There are two important adjustments to simple random sampling: stratification and clustering. *Stratified sampling* involves taking separate samples from subgroups, which has the advantage of generally reducing sampling error. *Cluster sampling* entails sampling clusters rather than individual cases, which increases sampling error but which has the advantage of decreasing costs since complete lists are not required. *Systematic sampling* is sometimes used as a practical means of doing simple random and stratified sampling. Furthermore, complex sample designs often involve *multistage sampling*, such as cluster sampling followed by sampling of individuals within the selected clusters. These procedures will be detailed in the sections that follow.

Stratified Sampling

Stratified sampling—taking separate samples in subgroups—requires sampling from lists. Sampling college students from separate lists of seniors, juniors, sophomores, and freshmen would be an example of stratified sampling. As just stated, its advantage is cutting sampling error. It reduces sampling error because the right proportions of people are sampled within the categories chosen for the stratification. There would be no reason, for example, to take the chance of obtaining a random sample with too few seniors when it is possible to get the right proportion by stratifying.

Stratification is commonly used in national household surveys, since it is easy to stratify on region of the country. In face-to-face national household surveys, one may as well stratify on region, since that is known in advance. Similarly, area codes are nested within states, making it easy to stratify by region in telephone interviewing. Stratification works, though, only if it is possible to get lists by appropriate subgroups. A researcher wanting to stratify a phone sample of the general public by religion would probably not be able to do so, since there are not separate lists of the phone numbers of Catholics, of Protestants, and so on.

In *proportionate stratification*, the number of cases sampled from each stratum is proportional to the relative size of that stratum. The standard error for the mean in proportionate stratified sampling is the $\sqrt{\sum_k w_k \times Var_k(X)/n_k}$, where w_k is the weight for stratum k, $Var_k(X)$ is the variance of X for that stratum, and n_k is the number of cases in that stratum. The *design effect* (deff) of such a sampling method is the variance of the variable divided by the variance of a simple random sample of that same size. The square root of that design effect shows the gain in precision from stratifying. That gain depends on the relative size of the variability between strata and the homogeneity within strata. That is, there is greater gain from stratification when the variability within a stratum is small and the differences between strata are large. Thus stratifying by year in college would be useful if seniors were fairly similar on the variable in question, as were juniors, as were sophomores, and as were freshmen, while there were large differences between seniors, juniors, sophomores, and freshmen. As this example suggests, stratification can be useful when it decreases sampling error for variables because they are correlated with the stratification variable.

In *disproportionate stratification*, the number of cases sampled from a stratum is not proportional to the relative size of that stratum. Disproportionate stratified sampling is useful in several situations. One is to get a smaller sample where a group is unlikely to live, such as not bothering to use a large stratum for Eskimos in Arizona. Conversely, a larger sample can be obtained where the group is more likely to live. For example, a redesign of the National Crime Survey in the 1980s defined strata on the basis of FBI crime indexes, with oversampling of high crime areas to increase the selection probability for crime victims (Skogan 1990). Additionally, disproportionate sampling can be used to oversample strata with low data-collection costs. As an example of these considerations, Lavrakas (1993) advocates determining in advance the number of phone lines with each prefix (224 in 224-5555), so that the sampling can concentrate on prefixes with the most lines in the desired area.

Another situation where disproportionate sampling can be useful is to get larger samples for subgroups of interest, such as an extra-large sample of Asian Americans, a group that might not otherwise be large enough in a sample to be able to make estimates for it separately. Finally, oversampling of strata with higher variance is recommended. However, the variance within strata will be different for different variables, which means that the optimal allocation for disproportionate sampling can vary between variables. This calls attention to a reason to be cautious in choosing to stratify a sample. Stratification is usually used to reduce sampling error, but disproportionate stratification can instead increase sampling error if poor guesses are made about the proportions to use, as would happen if more cases were selected from strata with less variance on the variables of interest.

Technically, differential sampling weights should be used in data analysis when disproportionate stratification is employed, so as to compensate for the unequal probabilities of selection.[6] If Asian Americans are double sampled so there are enough respondents for reasonably precise estimates for that group, for example, each Asian American in the sample would be given a weight of one-half (the reciprocal of the uneven chances of selection) when the whole sample is studied together.[7] Note that weights would not be necessary in this instance in multiple regression analysis if the stratifying variable were one of the predictors.

Stratification may seem reminiscent of quota sampling, but there are meaningful differences. Respondents in stratified sampling are chosen through a probability mechanism, whereas the interviewer chooses respondents in quota sampling (though within the quota rules). The difference, in other words, is the difference between probability and non-probability sampling, and that difference permits estimation of sampling errors for stratified samples while it opens up the possibility of bias in quota samples.

Systematic Sampling

A technique known as *systematic sampling* can be used when one is sampling from a written list. Every i-th case on the list is chosen after a random start. Say that there are N elements on the list but that one can afford to sample only n of them. The *sampling interval* is then $i = N/n$, and the random start is taken from the first i elements, after which each i-th case is used. Consider, for example, a situation in which the list consists of 900 names but you can afford to interview only 45 cases. That is a 1 in 20 sample (45/900), so the sampling interval i is 20. One of the first 20 names on the list is selected randomly (say the 4th), after which each 20th subsequent name is used (the 24th, 44th, 64th, and so on), which will produce 45 people to be interviewed.

Systematic sampling has potential problems associated with patterns in the list. In particular, cycles in the list (such as every tenth house on a block being a corner house) can lead to biases. Alphabetic lists have unusual properties since particular last names often have the same ethnic origin, which can again lead to problems in systematic sampling from lists. Furthermore, systematic sampling is similar to simple random sampling in that a list of the population to work from is still needed. Many lists are computerized nowadays, in which case having a computer program do simple random sampling would generally be preferable to systematic sampling. However, when one is sampling from a list that is not computerized, systematic sampling can be easier and less expensive than simple random sampling.

Intercept polls, when an interviewer is instructed to interview every i-th person at a downtown street corner or leaving a polling place, can be seen as following a systematic sampling approach to the extent that the sampling locations are chosen in a probabilistic manner. The systematic approach is often used in mall intercept surveys, with interviewers told to interview every i-th person, in order to reduce interviewer bias (Sudman 1980). Intercept polls are also used on the Web, asking every i-th visitor to a site to fill out a survey. The usual problems with sampling frames for Web surveys are alleviated, in that the sampling frame is the set of people who have visited that Web site. Some Internet merchandisers use this approach to measuring customer satisfaction with their online purchase experience.

The response rates in these surveys tend to be low, though some Internet intercept surveys use cookies to track the person's browsing behavior for a period of time so as to obtain information that can be used for weighting. Still, nonresponse bias can be considerable, since people who are willing to participate may differ considerably from those who do not.

Systematic sampling is also advantageous if the list happens to be arranged in a useful order that provides a de facto way of achieving stratification in sampling, as discussed in the preceding section. As an example, if a list of college students is sorted by the number of credits each student has taken, then systematic sampling would guarantee getting the proper proportions of seniors, juniors, sophomores, and freshmen. Because systematic sampling can provide a de facto means of stratification, it is often used in choosing phone numbers for RDD interviews. For example, working banks (292-64 in 292-6446) of phone numbers in a selected county might be ordered according to the number of residential phone numbers assigned in each, after which systematic sampling is used on that list to choose banks to use in the sample.

Cluster Sampling

Cluster sampling—sampling clusters of cases instead of individual cases—is useful for reducing costs in interviewing. For example, taking several interviews on the same city block entails less transportation cost than taking the same number of interviews scattered through the same city. An additional gain comes from the interviewer's ability to go to another house on the same block if no one is at home on the first house and then to return to that first house later to check whether someone has returned home. Unlike stratifying, where items are sampled from each strata, in clustering some clusters are sampled and others do not fall in the sample at all. Instead, the elements in the clusters that are sampled serve to represent elements in other clusters that do not end up in the sample. If one were sampling students in a high school, for example, one might randomly choose a set of homerooms as the clusters from which to select students to interview, after first stratifying the homerooms by class year (sophomore, junior, or senior) to get the proper representation of each. Thus there would be homerooms from each stratum, but some homerooms would not be chosen in the clustering part of the sampling.

Cluster sampling is often used with *multistage sampling*. A set of clusters is sampled in the first stage, after which cases are randomly sampled within those clusters. After a city block is sampled in the first sampling stage, houses on that block are listed by the interviewer, who then samples several of them for the actual interviews.

Cluster sampling has the further advantage of requiring less extensive lists than simple random sampling does. A full list of the entire population is not required for cluster sampling; as in the example of sampling blocks above, all that is needed is a list of potential clusters, followed by lists of all the cases in the selected clusters if multistage sampling is used. Furthermore, in face-to-face surveys, the interviewer can check the accuracy of the listing of households on the block before sampling some of them.

Cluster sampling is inefficient, however, to the extent that people in the same cluster tend to be similar to one another. Most U.S. cities have residential zoning laws that result in people with similar social-economic status living on the same block. As a result, people on the same block are likely to be more similar than an equivalent number of people selected randomly throughout that city. That problem has led to the development of measures of the efficiency of cluster sampling.

The design effect in cluster samples depends on the differences between the cluster means and the overall mean, on the heterogeneity of the clusters, and on the number of clusters selected. There are greater differences in the means if the clusters are homogeneous. The more the clusters differ internally, the less precision there is. More clusters also lead to more precision. One common statistic used to describe cluster samples is the *intracluster correlation* (rho), which measures the covariation of pairs of people in the same cluster, calculated as deviations from the overall mean. Correlations near +1 represent inefficient samples, since the information from different interviews taken in the same cluster is redundant; negative correlations show that the clustering is more precise than a simple random sample of that size would be, which would be an unlikely situation. The optimal allocation strategy is to take large numbers of sample elements from homogeneous clusters and small numbers from others (Groves 1989, 263), but the variance within clusters can be different for each variable, and wrong choices can increase the sampling error. Groves (1989, 271–79) also shows that the increase in standard errors from clustering is usually largest for the mean of the full sample, smaller for subgroup means, and smaller yet for differences between subgroup means. One piece of advice is that sampling 6 items per cluster is attractive if the expected rho's are small (below 0.10; Groves 1989, 266).

The design effect for a cluster sample is deff $= 1 + (\rho\,(\bar{n} - 1))$, where ρ is the intracluster correlation and $\bar{n}$ is the average number of items sampled in a cluster.[8] The deft, the square root of the design effect, is useful for gauging the effects of the sampling design on the standard error of variables. The 2000 NES survey provides a useful example. The 2000 NES reports a mean deft of 1.098 for its total sample of about 1,800.[9] The standard error for a percentage around 50% with a simple random sample that size would be about 1.18%, whereas the corresponding standard error for this sample is 1.29%. The margin of error (1.96 times the standard error) for NES would be 2.53%, versus 2.31% for a simple random sample, leading to confidence intervals one-tenth wider than ones calculated under simple random sampling. Reports of national samples often make this correction in giving their maximum margin of error, but users of standardized computer programs, such as SPSS, rarely make this adjustment, so their significance tests are often done incorrectly.

Another way of looking at this is that just as precise results could have been obtained with a smaller simple random sample. Compute the reciprocal of the deft, in the above case $1/1.098 = 0.911$. The precision of the cluster sample is then the same for a simple random sample 91% of the size of the cluster sample. A random sample of size 1,639 would have been as precise as the NES cluster sample of size 1,800. This computation highlights the trade-off involved in clustering: cutting the survey costs by clustering required taking 161 extra interviews. Again, the design effect differs by variable, being larger for variables in which elements in the same cluster are similar and smaller for variables that vary as much within clusters as between clusters. Thus the design effect would be relatively high for candidate vote by precinct in exit polls, since some precincts vote cohesively for one party. On the other hand, there would be no design effect on gender in most cluster samples, so no adjustment is required in significance tests involving gender.

Cases can be sampled within clusters either equally or unequally. An important approach is the *equal probability of selection method* (epsem sampling), which is required for some government and legal surveys. This can be achieved in multistage cluster sampling, with the probability of selection of a case being the product of its selection at each stage. For example, if a sample of one-in-twenty schools in a large city is used with a sample of one-in-thirty students in each school, then the overall probability of any student's being selected is an equal one in six hundred ($1/20 \times 1/30 = 1/600$). Another epsem approach would be to use *probability proportionate to size* (pps) sampling with an equal number of cases from each primary sampling unit. For example, say that there are enough funds to interview in 30 organizations that have to be chosen from a population of 100 organizations that have 40 members each and 100 that have 20 members each. Since there are twice as many members of larger organizations, two-thirds of the organizations sampled should be large ones. Randomly choose 20 of the larger organizations and 10 of the smaller ones, and then randomly select 5 people in each organization to interview. The chance of someone in a larger organization being selected is $(20/100) \times (5/40) = (1/5) \times (1/8) = (1/40) = 0.025$, and the probability for interviewing someone in a smaller organization is $(10/100) \times (5/20) = (1/10) \times (1/4) = 0.025$, so that equal probability of selection has been achieved. Epsem sampling can also be done with phone samples, using an equal probability sample of all possible ten-digit phone numbers in banks with at least one listed telephone number. Even if this approach is sometimes required, it is an inefficient system since business numbers are not excluded, so there will be multiple phone calls to banks with few residential numbers. An advantage of epsem sampling is that the sample is self-weighting.

Sampling within Households

The discussion of clustering provides a useful opportunity to deal with the choice of whom to interview in a household. Since most surveys of the mass public are really household surveys, a decision must be made as to which member of the household to interview. Just interviewing the person who happens to answer the door or the phone would lead to bias, since young unmarried adult males are less likely to be at home than others. Interviewing the most willing participant instead would overrepresent younger and better-educated people (Lavrakas 1993, 107). In either case, the probability basis of the sample would be compromised, by availability considerations in the former case and volunteer subjects in the latter.

Sampling specialists have devised several different procedures to choose respondents within households in an unbiased manner. These procedures vary in their ease of administration and how intrusive they seem to the person with whom the first contact is made at the household. The best procedures are hard for interviewers to administer, and people often terminate the contact early because these procedures are seen as intrusive. This has resulted in using suboptimal procedures with which people are more willing to comply.

The classic procedure for selecting respondents in a household is through the use of *Kish tables*, named for Leslie Kish, who developed them in the 1950s. Interviewers ask people to list for them all the household occupants, including their genders and ages (a procedure known as *rostering*). The interviewer is randomly given one of eight different Kish tables (see table 10.1) to use for a particular household (different tables for different households, chosen by chance). These tables show the interviewer whom to interview as a function of the number of adults in the household. One table might tell the interviewer that the designated respondent in a household with two adults is the oldest adult female, if there is one, and otherwise the youngest adult male, whereas a different table might tell the interviewer that the designated respondent in such a household would be the oldest adult male, if there is one, and otherwise the youngest adult female, and so on. This procedure worked very well in the 1950s, but its success has diminished over the years as people have become more suspicious of strangers asking them to provide such private information, so other procedures have been developed that are less intrusive. A reliability check by Groves and Kahn (1979) found 9% of households providing inconsistent listings of the number of adults, generally involving counting young adults.

The Troldahl-Carter-Bryant procedure (Troldahl and Carter 1964; Bryant 1975) asks people only two questions: how many adults there are over age eighteen in the household and how many of these are women. Interviewers then randomly use one of four versions of a matrix to choose the designated respondent (see table 10.2). This system undersamples middle-aged people and men; Troldahl and Carter (1964) estimated the noncoverage at 2%–3%. Traugott (1987) showed that the success of these randomizing methods depends on the persistence of interviewers in contacting the designated respondent. He found that interviewing the ini-

Table 10.1 An Example of a Kish Table

If the number of adults in the household is	Interview the adult numbered
1	1
2	2
3	3
4	3
5	3
6	5

Instruction to the interviewer: List all adults in the household, males first and then females, listing them in order of age. Number the oldest male #1, number as #2 the next oldest male if there is one or otherwise the oldest female, and so on. Then use the table above to choose the designated respondent.

Table 10.2   An Example of a Partial Troldahl-Carter-Bryant Matrix

	1 Adult	2 Adults	3 Adults	4 Adults
0 female	sole male	younger male	2nd youngest male	2nd youngest male
1 female	sole female	sole male	older male	2nd youngest male
2 females		younger female	younger female	younger male

Note: The full table would include more columns and rows to allow for larger households.

tial contact would have led to an overstatement of the Democratic vote in the 1984 presidential election.

Since even that system asks personal questions that many people are nowadays unwilling to answer, the Hagan-Collier system was developed in which no questions on household composition have to be asked (Hagan and Collier 1983). Instead, the interviewer is randomly given one of four possible sets of instructions to identify the designated respondent in a household. In some households, for example, the interviewer will automatically ask to speak to the oldest female, with a fallback as to which male to interview if there are no adult females in the household (see table 10.3). In other households, the interviewer instead asks to speak to the youngest female over age eighteen, the oldest man, or the youngest man over age eighteen, in each case with appropriate fallbacks if there is no such person in the household. Hagan and Collier found that the refusal rate with their system was 5% lower than with the Troldahl-Carter system. Lavrakas (1993, 115–16) suggests an even simpler system for inexperienced interviewers, in which there are again no questions on household composition. The interviewer is randomly told to interview one of two types of people in a household: either the youngest male or female, with the emphasis on the youngest so as to avoid the problem of young people's being at home less than older people.

The most popular system nowadays for determining the designated respondent in a household is the "birthday method." The interviewer simply asks to interview the adult with the most recent (or, for some survey organizations, the next) birthday (Salmon and Nichols 1983). The idea behind this system is that this question does not make people suspicious, so the compliance level is higher than with the other systems. One study that is somewhat dated (O'Rourke and Blair 1983) found that only 1.8% of phone calls were terminated when the most recent birthday system was used, versus 4.1% refusals for the full Kish system. There is some evidence that people do not sort their memories so as to be able to recall immediately who has had the most recent (or will have the next) birthday, so this

Table 10.3 An Example of a Hagan-Collier Instruction Set

"For this survey I need to speak to the oldest woman in your household, if there is one."
(If no woman in the household), "Then can I speak to the youngest man (over age 18)?"

system is prone to error. Groves and Lyberg (1988) point out that, strictly speaking, this is not a probability sampling method, as birthdays are not assigned randomly within households. The improvement that it achieves, by the way, may well be due to interviewer expectancy rather than respondents. Interviewers are often more hesitant in asking the household composition questions, which could reduce cooperation, whereas they are more willing to ask the birthday question.

To measure the accuracy of responses to the birthday method, a study using that method asked respondents at the end of the survey, after rapport with the interviewer was developed, to list all adults in the household and their birthdates. The researchers later checked the accuracy of the respondent designation and found 20%–25% inaccuracy (Lavrakas, Bauman, and Merkle 1993). The most common error seems to be choosing oneself, regardless of whether that person had the most recent (or next) birthday; since women are more likely to answer the phone, this tends to lead to a higher proportion of women in samples. In spite of that evidence of inaccuracy, birthday methods are commonly used, especially for telephone surveys, because people seem to be so willing to accept them.

Rizzo, Brick, and Park (2004) have suggested a very practical variant. People are first asked how many people eighteen and over live at the household. If there is only one such person, the respondent selection process is over. Otherwise, a random procedure is instituted. If there are N people in the household, the screening respondent is given a $1/N$ chance of being selected; if that person is selected, the selection process is done. If the screening respondent is not chosen and there are just two adults in the household, the interviewer simply asks to speak to the other adult. If there are more than two adults in the household, then the screening respondents can be asked which adult other than themselves had the most recent birthday. Since 85% of U.S. households contain no more than two adults and since the screening respondent is chosen by chance in some households with more than two adults, the birthday question need be asked in only about 12% of calls. Furthermore, the birthday question is used in such a way that the screening respondent cannot self-select into the sample.

Multistage Sampling

Actual large-scale surveys often employ complex multistage sampling. Complete examples of techniques for sampling for face-to-face interviews, telephone surveys, and exit polls illustrate this complexity.

NATIONAL FACE-TO-FACE SURVEYS. One example for face-to-face interviews is the method that the University of Michigan's Survey Research Center used for its national household survey after the 1990 census (Heeringa et al. 1994, National Election Studies Study Implementation at http://www.umich.edu/~nes/studyres/datainfo/implement.htm). The sampling frame was households, excluding Alaska and Hawaii and also military bases and institutional living units such as prisons, dorms, and institutional homes for adults. A three-stage sampling method was used for selecting households. The first stage involves sampling 84 "primary sampling units" (PSUs)—counties, Standard Metropolitan Statistical Areas, and standard consolidated areas. Of these 84, the 16 largest SMSAs are "self-representing," so that New York City, Chicago, Los Angeles, etc. are in every sample. The other 68 are sampled using stratification by region and size, ending up with one medium-sized southern city to represent that type of location, one midwestern rural county, and so on. These 84 PSUs were used for all the face-to-face interviews taken during that decade, which allowed interviewers to be hired in those areas on a long-term basis.

The second stage of the Michigan SRC sample involves sampling area segments within those PSUs—census blocks in urban areas or enumeration districts in rural areas. This requires extensive maps, but only for those 84 PSUs rather than for the entire country. Sampling at this stage involves stratification by geography, size, and, where available, median income, with probability proportionate to number of occupied housing units to yield at least 6 segments, and up to 25 in the largest 6 PSUs. Third, several households are randomly cluster sampled within those segments by interviewers after they list all housing units in the segment. Finally, once the household is chosen, the interviewer uses the Kish table method described earlier to select the designated respondent in each household.

As another example, the Eurobarometer surveys in Europe use a two-stage sample separately in each nation. At the first stage, a sample of primary sampling units is taken based on the national resident population, stratified by metropolitan, urban, and rural areas in each administrative region of each country, using proportional-to-population size sampling to handle differences in population density. At the second stage, a cluster of addresses is selected from each PSU, with systematic sampling used to choose addresses within those clusters and then a random method (such as a birthday method) to choose one respondent per household.

TELEPHONE SURVEYS. Cooper (1964) originally suggested the idea of random-digit dialing for random selection of phone numbers. However, pure RDD is very wasteful, since many combinations of ten-digit numbers do not exist. There is no need to call numbers with nonexistent area codes. Further, it is possible to purchase lists of the operational prefixes. For example, 614-292-xxxx is an operating series of phone numbers, but 614-999-xxxx may not be. Avoiding nonoperational prefixes is a way of saving costs without any increase in sampling error. However, it is worth seeking a system that is more efficient yet, since only 16% of numbers in operational prefixes are working household numbers.[10]

In a local telephone survey, a telephone book can be sampled to obtain residential phone numbers in the working ranges. Although sampling from a telephone book has disadvantages, Groves and Couper (1998, 298) also stress that it has some advantages. Fewer calls will be made to non-working phone numbers, and, more important, the availability of the street address makes it possible to link the household with census and other aggregate information about the neighborhood. Additionally, street addresses make it possible to mail advance letters to the sampled house-holds.

Most phone samples, though, do not call the actual numbers in a phonebook. The "add-a-digit" system has interviewers phone not the number in the phone book but that number plus one. If 614-292-5555 is obtained in the sample from the phone book, the number that is actually dialed is 614-292-5556. (Alternatively, randomization on the last digit or two of the phone number is common.) This system has the advantage of allowing unlisted numbers and newly assigned numbers to be included in the sample, which would not be the case if the actual numbers in the phone book were used. Add-a-digit is considered a "list-assisted" sampling design, since it uses information in a list to obtain numbers, including some that are not on that list.

The Mitofsky-Waksberg method (Waksberg 1978) is an important multistage sampling technique for phone surveys. It uses random-digit dialing to identify phone numbers for seeds in the second stage, where the seeds are essentially clusters. Specifically, a randomly generated phone number is called to see if it is a working household number. If it is not, it is dropped and no more calls are placed to its "100-bank" of numbers (292-64 in 292-6446). If it is a working household number, the second stage consists of replacing the last two digits of the number with multiple pairs of random numbers. For example, if 292-6446 is a working residential number, then several pairs of random digits are generated and substituted for the 46 at the end of that number, so that calls might be made to 292-6413, 292-6485, 292-6439, and so on. Lavrakas (1993, chap. 2) recommends using at least 11 pairs of random last two digits per bank. The Mitofsky-Waksberg method reduces costs, since businesses and residences are usually assigned different banks of phone numbers. Further, many banks of phone numbers contain no residential numbers, while other banks will consist of many working residential numbers, so this method guarantees many calls to clusters of phone numbers that are found to include residential numbers.[11] The Mitofsky-Waksberg method leads to a self-weighting sample if the same number of households is interviewed in each cluster, but that is not always possible since some banks may not have enough households.[12]

A variant (known as *one-plus listed telephone banks*) that is used in many government surveys is to use as the sampling frame all 100-banks (such as 292-64xx) with at least one listed telephone number in the telephone directory (Casady and Lepkowski 1993), with systematic sampling within that frame. This list-assisted frame has coverage bias since households in banks without any listed numbers are not included in the frame, but those banks have a very low probability of containing working residential numbers. Brick et al. (1995) estimate the coverage error at 3.7% for this sampling frame. This sampling procedure is generally more efficient than the Mitofsky-Waksberg method, particularly when few numbers are selected within each 100-bank, since Mitofsky-Waksberg requires many calls to determine clusters to retain. Mitofsky-Waksberg is more efficient only in situations in which the same 100-banks can be used for more than one survey, so that the effort involved in determining productive clusters is amortized over multiple surveys. Brick et al. (1995) find only a small bias in the National Household Education Survey from using this truncated, list-assisted sample frame, compared to the larger coverage biases resulting from sampling only households with telephones. Tucker, Lepkowski, and Piekarski (2002) show that this list-assisted approach was still more efficient in 1999, with the largest gains in precision for short interviews.

Another issue in phone samples is how to give out phone numbers to interviewers. Tracking polls seek to assess change over the time of the polling, but differences could be due to differences in the phone numbers that happen to be called a particular day. A *rolling cross-section design* handles this problem by dividing the sample into a series of "replicates," each replicate being a miniature of the total sample design. One replicate is given to the interviewing staff daily. Looking only at interviews from the replicate that were taken the assigned day would bias the sample to people who are the easiest to contact. Instead, the rolling cross-section approach is to combine the interviews that are actually taken the same day, which is effectively a random event, probably omitting the first few days in which there were not enough callbacks to generate a full set of interviews. Results from interviews taken the same day can then be combined to yield a time-series that can be examined for change in attitudes (Johnston and Brady 2002).

EXIT POLLS. The election exit polls conducted by Voter News Service (VNS) in the 1990s used a two-stage sampling design. The first stage is a sample of voting precincts in which interviewing would take place, chosen on the basis of a stratified systematic sample of precincts in each state, proportionate to the total number of votes cast in the precinct in a prior election. Second, within those precincts, interviewers chose which respondents to interview using a systematic sample, approaching every i-th person as they left the polls, with the sampling interval based on that precinct's expected turnout given a goal of about 100 interviews per precinct. There could be a sizable design effect, due to the clustering among certain subgroups, for example, when people of the same racial background tend to cluster in the same neighborhoods and therefore precincts. For example, the design effect for the 1996 VNS survey was estimated to be about 1.7, leading to 30% more sampling error than under simple random sampling, but with the design effect being larger for some variables (Merkle and Edelman 2000).

After the 2004 U.S. presidential election, there was a very lively debate in Internet blogs about apparent discrepancies between exit polls and official results, nationally and in the key battleground states of Florida, Ohio, and Pennsylvania. Exit polls on election night showed John Kerry with clear leads in those states, yet he won Pennsylvania only narrowly, lost Ohio even more narrowly, and lost Florida. Conspiracy-minded bloggers argued that the exit polls showed that Kerry had really won and that the official election results had been manipulated. Pollsters replied that the more than 2,000 exit poll interviews taken in each of those states were clustered in only about 50 precincts per state, so the sampling error had to be adjusted for a very large design effect (80% more sampling error than with simple random sampling), which makes the differences between exit poll and official results nonsignificant. Furthermore, they argued that the election results were consistent with preelection polls, so the exit polls were the outliers. This debate does not negate the value of exit polls for examining how particular social groups voted in an election, but it implies that exit polls are not useful for election-night calls of close races.

WEIGHTING. Because of the clustering, the sampling error with multi-stage designs is often at least 1.25 to 1.5 times that of a simple random sample with the same sample size. The sampling variability can be estimated in several ways. *Bootstrapping* uses repeated half samples that maintain the original sample design and then averages the sum of the differences between the full sample estimate and the half sample estimates. A second possibility is *jackknifing*, dropping one primary sampling unit at a time, reweighting the other units in the stratum, computing the statistic to estimate the contribution of that stratum to the sampling variability, and then summing over all strata to estimate the overall sampling variability. Both of these methods are considered to be *resampling* techniques, using the actual sample to create artificial samples to permit assessing the variability of the sample design. Fortunately, there are computer programs for estimating sampling variability for complex sample designs. WesVar and SUDAAN are two of the more common stand-alone programs used for this purpose, and there also some procedures in SAS and Stata that can be used.[13]

There are some special weighting issues that arise in multistage household sampling. First, when only one person in a household is interviewed, each individual in a large household has a smaller probability of being selected than each individual in a small household. As a result, it is common to weight samples by the person's chance of selection—the reciprocal of the number of people in the household (see chapter 9). At the same time, however, it is easier to contact large households, since there are more people who might be at home when the household is contacted. Gelman and Little (1998) argue for also using poststratification weighting to weight households up to the census estimates of households of that size. That leads to using iterative weighting procedures so that both inverse probability weights and poststratification weights can be used at once.

The final point to emphasize in the discussion of multistage sampling is that there is controversy as to whether the sample design should be reflected in estimating analytic statistics (see, for example, Groves 1989, 279–90). Data analysts often ignore weights because they feel that the values of the variables are the result of a random process, quite apart from the sampling method. Sampling statisticians, however, feel a need to be able to generalize to larger populations, so they want to use weights to incorporate differences in sampling, at least for descriptive purposes.[14] This controversy is often handled by using weights for descriptive purposes in summarizing the distribution of each variable but not using weights for analytic purposes when doing complex modeling.

Determining Sample Size

A crucial decision in planning a survey is determining the desired sample size. There are several considerations involved in making that decision, cost and time being two of the most important of these. As seen throughout this chapter, sample size directly affects the size of the standard error and therefore of confidence intervals, so larger samples are invariably more precise. With too small a sample, the standard error will be large, and there is considerable risk of committing a Type II error of not rejecting a null hypothesis when it is false. However, the relationship is a square root relationship: quadrupling the sample size only halves the standard error. As a result, there is considerable gain in precision from taking 300 more interviews when the original thought was only to take 100, whereas the gain in precision from adding 300 more interviews to a sample of size 2,100 is much more negligible.

It is also important to consider what degree of precision is required. If the effects being examined in the survey are likely to be large, not much precision is necessary. If the expectation is that a school tax levy in a city is going to pass by a large margin, a small sample should be sufficient to test whether that expectation is correct. However, a large sample is necessary if it is important to be able to detect a small effect. If the expectation is that the school tax levy vote is going to be close, a very large sample would be needed to tell whether it is ahead or behind. Also, it may be necessary to take a larger sample in order to detect a relationship between variables if the measures to be used are of limited reliability.

Although one consideration in setting the sample size is the amount of precision needed for the overall sample, it is even more important to consider the precision needed for describing subgroups. If, for example, a survey of the general population also needs to be able to discuss Catholics and their behavior and attitudes, then it is essential to choose a large enough sample size to be able to make fairly precise statements about Catholics, which requires either devising some method for oversampling them or increasing the size of the total sample. An even larger sample is required if it is important to be able to discuss Catholic men and women separately, and larger yet if those groups must be subdivided.

The size of the sampling pool that is needed depends not only on the desired sample size but also on the incidence rate when screening is required (such as in a sample of Catholics), the expected completion rate (removing unit nonresponse), and, for some survey modes, the expected hit rate (such as the proportion of phone numbers called that will be working residential numbers). For phone interviews, Scientific Sampling, Inc. (SSI) indicates on its Web site that the hit rate varies from 0.30 to 0.70 for random-digit-dialing surveys, with 0.85 being a reasonable estimate for list sampling. They estimate completion rates from 0.23 for long interviews on sensitive topics to 0.35 for short interviews on nonsensitive topics. The needed size of sampling pool is (final desired sample size)/((hit rate) × (incidence rate) × (expected completion rate)). Lavrakas (1993, 58) recommends adding 10% to this figure to be conservative. Applying this formula, if 600 completed interviews are desired, with a working phone rate of 0.70, an incidence of 0.85, and a completion rate of 0.33, a total of 3,056 phone numbers should be used, or 3,361 if the 10% overage rule is adopted. A general rule of thumb for phone samples is that it usually takes 4 or 5 phone numbers to obtain a completed interview, taking into account the working phone rate, incidence, and the likelihood that the person will agree to be interviewed.

Not all the considerations involved in choosing a sample size are statistical. In particular, it is necessary for a sample to be large enough for the survey to be credible. If a random national sample of 50 people found that 75% were satisfied with a company's products, that sample size is large enough to show that a significant majority of the public is satisfied. However, few companies would believe a sample of just 50 people, even if statisticians assure them that the result is statistically significant. Large sample sizes are required to establish the credibility of government surveys and surveys to be used as evidence in courts. Indeed, governments and courts sometimes require interviews with the full population. In any event, credibility is often as important a criterion in determining sample size as the statistical criteria discussed above.

Intriguingly, there are some suggestions in the literature that sample size does not affect the accuracy of polls. In particular, Lau (1994) finds that errors in estimating presidential candidate support in trial heat polls before the 1992 election were not significantly affected by sample size, after other factors in the poll administration are taken into account; Crespi (1988) obtained similar results beyond the national level.

The credibility issue raises one final concern: should polls be considered valid even if you yourself were not surveyed? This question often comes up when people are deciding what to make of election polls. Consider the mathematics. There are now 200 million adults in the United States. If the average sample size is 1,000, only one person in 200,000 will be included in a particular poll, so it would take 200,000 polls of 1,000 people to call everyone once, assuming that no one is called twice. Thus the chances of being phoned in an election survey are still small, even with the large number of election polls that are taken. For example, if there were 1,000 election polls of size 1,000 and no person was called more than once, only 1 person in 200 would be surveyed, so there is less than a 1% chance of any given person's being surveyed. Thus it is not surprising that most people are not surveyed, even in election years.

Sampling Error and Survey Constraints

Sampling is employed when the constraints of money and time prohibit collecting data from the entire population of interest. Furthermore, sampling can allow a researcher to spend scarce resources to achieve better-quality data from the people who are interviewed. Obviously there is no reason to sample when one is studying a small population, such as a seven-person city council. The U.S. Supreme Court has decided that sampling is unconstitutional when it comes to estimating the size of the population for apportionment of the federal Congress, and obviously elections require a full vote rather than a sample survey. In most other instances, though, a sample is the most practical way to study a large population.

The very act of sampling, however, creates sampling error, since inevitably the results for a sample will not correspond exactly to those for the full population. It is desirable to minimize sampling error, but that can be expensive since interviewing more people to reduce sampling error directly increases costs as well as time for completion of the survey. The costs of extra interviews, depending on the survey mode, can include interviewer wages, transportation costs, refusal conversion costs, follow-up costs, and costs for processing completed survey instruments. It takes four times as many interviews to cut the sampling error by half, and often that extra money would be better spent trying to measure and/or reduce some of the forms of nonsampling error. As seen in this chapter, sampling error can also be decreased in surveys by judicious choices in the sampling, such as a thoughtful decision on how to stratify.

Table 10.4 lists the several types of nonprobability and probability sampling methods that were discussed in this chapter. The probability methods are considered preferable because they are less biased and they permit

Table 10.4 Types of Samples

Nonprobability Sampling	Probability Sampling
Availability samples	Simple random samples
Convenient cases	Stratified samples
Volunteer cases	Proportionate stratification
Purposive samples	Disproportionate stratification
Typical cases	Systematic samples
Critical cases	Cluster samples
Snowball samples	Multistage samples
Quota samples	Epsem samples

mathematical estimation of the sampling error. However, they are also more complicated, often requiring a sampling statistician to work through sampling procedures and subsequent weighting of the data. As a result, nonprobability methods are often less expensive as well as requiring less set-up time.

Two ethical constraints are involved in sampling. One is the importance of including in a survey report not only a description of the sampling procedure but also a reminder that nonsampling errors must be considered when the validity of any conclusions based on survey evidence is assessed. The other is the need to remember that the sampling method is more important than the sample size. In the era of Internet surveys, it is easy to fall into the same trap as the *Literary Digest* surveys of the 1930s: imagining that a large number of responses can substitute for a probability sample in yielding representative results. The results of an Internet survey are not necessarily valid even if millions of people participated in the survey. Regrettably, this is a lesson that seems to have to be learned again with every innovation in data collection modes, whether for mail surveys, call-in polls, or Internet surveys.

PART IV

Survey Administration Issues

11

Postsurvey Error
THE DATA EDITING STAGE

Postsurvey error is error that occurs after the interviews are conducted. It was not included in some early treatments of the total survey error approach (Groves 1989), probably because it is after the actual interviewing. However, postsurvey error is important since it directly affects the accuracy of survey results. Postsurvey error at the data processing and the survey reporting stages will be described in this chapter, along with the effects of postsurvey error and ways to minimize such error.

Additionally, this chapter will discuss the important postsurvey process that has become known as *data editing*, trying to locate and correct the individual errors in survey data. Federal government surveys, in particular, devote a great deal of attention to data editing because of their need to produce accurate official data. For example, the Bureau of Labor Statistics uses surveys to determine the nation's unemployment rate, and given the importance of that measure, they first examine the raw survey data to look for anomalies. Unless they are found and corrected, problems such as entering respondents' answers incorrectly into the computer can undermine the analysis of survey data. However, the proper amount of data editing has become a controversial topic, with some concern about the possibility of "overediting."

The data editing stage raises the question of what the appropriate standards are for survey data. Unfortunately, data standards in the survey field are relatively underdeveloped.

Data Standards

The most important standard for survey data is that the data, survey questions, and procedures all be preserved in such a manner that they can be reexamined after the project is ended. This is termed a *replication standard*, since it implies that the details of the study should be retained in such a manner that it would be possible to replicate the steps taken in the study. After all, an important part of science is producing results that can be replicated. If another researcher performed the identical steps, the same results should be obtained, at least within the range of random error. An important corollary is the need to keep extensive notes regarding procedures followed in doing research; replication is possible only if subsequent researchers are able to see exactly what was done in the original study.

The difficulty involved at this stage can be evident even in analyses of survey data in academic journal articles. People trying to replicate published analyses often have difficulty figuring out exactly how the original results were obtained. They may not be able to duplicate how missing data were handled or how a variable was coded. Whether for academic, government, or commercial surveys, there are a large number of decisions that are made in processing a study, and clients should expect as much care in the making and documenting of those decisions as in the original data collection.

How serious is this problem? An article by Dewald, Thursby, and Anderson (1986) sought to replicate the results in a year's worth of articles in the *Journal of Money, Credit, and Banking*. They found that "inadvertent errors in published empirical articles are a commonplace," though "correction of the errors did not affect the authors' conclusions in most studies" (587–88). This can be read as the glass being half empty or half full; some see it as a serious problem, whereas others see it as a minor flaw (see the extended debate in *P.S.: Political Science and Politics*, Sept. 1995). Still, it points out that the care taken in the survey stage is often not continued at the postsurvey stage.

Replicability is crucial at the postsurvey stage, since it provides a set of standards for quality research. At the same time, there are no universal standards. What is important is developing standards that promote replicability while not creating additional hurdles that are onerous to researchers. Three separate aspects of replicability will be discussed here, and it is important to recognize that they are separate—a survey researcher might want to adopt one, even if another is deemed inappropriate.

One useful approach is to keep logs of decisions that are made in processing the data, especially the judgment calls. Take the simplest case of a self-administered questionnaire in which it is hard to read whether the person checked the box for being a male or a female. This is likely to be a very rare situation, but it is best to keep track of it and how it was resolved. Doing so allows someone to go back to the questionnaire later to see if the correct decision was made or if there might be some information in answers to other questions (such as giving an answer of "housewife" to the occupation question) that might help show what the person's gender is. Minimally, keeping track of judgment calls allows the researcher to see the relative extent of problems that occurred with each question.

Second, replicability should be a crucial consideration when reporting on a survey. It is essential to describe the key procedures that were employed in the survey process and in the data analysis. Without such information, others will be unable to know how the research was conducted and how it could be replicated. Disclosure standards will be discussed further in chapter 13.

A third consideration is making data available to other researchers for secondary analysis so they can attempt to verify and/or replicate the work. Standards differ on this between fields. Commercial clients generally keep their data private, whereas depositing data in public archives has become standard in some academic disciplines. Some national media political polls deposit their data in academic archives, such as the Inter-university Consortium for Political and Social Research (ICPSR), while some pollsters deposit their data at archives they support, such as the Roper Center at the University of Connecticut. Many federal government grant agencies, including the National Science Foundation, require their grantees to deposit data collected under their grants in an archive. Some academic journals require articles to include information about the availability of the data analyzed in the article. At the same time, there are property rights issues, and researchers who have collected the data often believe that they deserve exclusive access to their data for a lengthy period, at least long enough to be the first to publish from the survey. Further complications ensue when archiving data that are confidential, especially when it is possible to identify elite respondents from identifying material in the data set.

These different possible aspects of replication should be kept in mind as the postsurvey process is discussed in the remainder of this chapter. Furthermore, what happens after the data are collected differs considerably between studies, depending largely on the survey mode. Computerized surveys, in particular, often build some of these processes into the data collection stage. Even so, every survey designer faces decisions on data processing and reporting, and those are the steps to be discussed next.

Data Processing

After the survey data are collected, it is necessary to prepare it for computer analysis. There are a number of separate steps involved in processing the data.

Initial Data Edits

When pencil-and-paper questionnaires are processed, the first stage (sometimes termed the *scan edit* stage) is to go through them and prepare them for entering the data on the computer. Typically human editors go through the questionnaires and write down on them the numbers that are later to be entered into the computer. This is seen as more efficient and less error prone than doing the editing during the data entry.

In going through the questionnaires, editors handle many different types of problems. If the respondent was told to check a box for gender, the questionnaire editor writes a number by that box corresponding to the number to be entered into the computer. Sometimes the respondent did not write down an answer, such as her age, in the appropriate blank, so it is rewritten where the data entry person will expect to find it. When a question was open ended, such as occupation, the editor writes in the correct numerical code for the person's answer. If the respondent answered questions that were to be skipped, the editor crosses out those answers so they are not entered into the data file.

Additionally, the person doing the data editing is expected to handle responses that are difficult to read, resolve internal consistency problems, and recode "other" answers into appropriate answer categories. There should be a predetermined set of edit decision rules to apply, and the person doing the editing is often instructed to document in a log any decisions going past those rules. Some operations have two people separately edit some of the questionnaires so that the editing quality can be checked. More data quality checking is done later in the processing stage.

Coding Data

Coding is the process of putting answers to open-ended questions into numeric categories. Coding routine open-ended material can be part of the editing process, or it can be separate if complex coding is necessary.

The first stage is constructing the code. Coding schemes must be set up so that categories are mutually exclusive and exhaustive. Categories are *mutually exclusive* when every separate answer fits into just one category. Categories are *exhaustive* when there is a code category for every possible answer. Coding schemes are often made to be exhaustive by using an extra category called "Other" for responses that do not fit neatly into the substantive codes.

Additionally, when the variable is ordinal, the coding scheme must be set up so that the categories are *properly ordered*. For example, the common question on party identification asks people if they are Republican, Democrat, or independent, in that order, but correlational analysis requires first reordering the categories to make "independent" the middle category: Republican, independent, Democrat.

The coding scheme should be as clear to apply as possible, but some subjectivity is inevitable in coding schemes of any complexity. Imagine, for example, asking people what the essence of democracy is. Separate codes could be developed for answers that stress elections, individual rights, laissez-faire economics, and so on. However, many answers might blur these distinctions, so that coding becomes a difficult judgment call. As an extreme example of coding difficulty, Mathiowetz (1992) finds that experienced coders of occupation using the Current Population Survey (CPS) coding scheme agreed with employer records for detailed three-digit occupation classifications only 52% of the time, with disagreements on the basic one-digit grouping (such as managers and administrators, laborers, etc.) in one-quarter of the cases.

It is best to have multiple coders code some of the answers independently so the reliability of the coding can be checked, determining *intercoder reliability*. Intercoder reliability can be calculated as the percentage of the cases that the coders code identically, or the correlation between codes given by pairs of coders. It is a measure of the reliability of the coding process, but it does not verify that the codes correctly assess the variables of interest, which is instead a validity concern. Coder disagreement is sometimes in the 20% range; this means it is important that researchers carefully work through their coding instructions so that they are clear and easy to follow. Note that coder error can be correlated if some coders interpret the codes differently than others do, in which case it would be appropriate to estimate the correlated coder variance and the design effects in a manner similar to that presented for interviewer effects in chapter 4.[1] There are computer programs for coding open-ended survey answers automatically according to rules that the researcher specifies. The ability of these programs to code text accurately has improved over the years, but human coding of survey answers is still more common than automated machine coding.

Data Entry

Data entry (or data capture) is the stage of entering data from paper-and-pencil questionnaires into the computer. This is a tedious process that is consequently error prone. Data entry error is usually random, so it will not bias results, but it does decrease the reliability of the data.

The best method to minimize data entry error is to use a double-entry system (often termed *verification*). The data values are keyed in twice, independently, after which the two are compared in order to catch entry errors. However, double entry doubles the personnel costs for data entry, as well as requiring a computer program for comparing the two versions as well as time to make the corrections. To keep costs down and speed up the process, many survey organizations instead just enter the data once, accepting the resultant risk of processing error. Studies have found low rates of data entry error prior to verification: 0.1% (Jabine, King, and Petroni 1990) to 0.6% (U.S. Bureau of the Census 1993) to 1.6% (Henderson and Allen 1981).

One way to minimize processing error for written questionnaires is to have answers marked directly onto optical-mark recognition (OMR) forms. There are some costs associated with using and reading such forms, but this is usually less expensive and faster than double entry of data would be and may make data editing unnecessary. As scanning and character-recognition technology has improved, optical character recognition (OCR) is becoming feasible even for handwritten answers. Dillman (2000, 427) is especially optimistic in assessing the future of scannable technologies, predicting that most mail questionnaires will soon be moved to scanning and imaging.

The most serious example of data entry error occurred in the 2002 Voter News Service (VNS) election exit polls. Public attention was focused on VNS that year, after its problems in calling Florida in 2000. The VNS began to revise its system, but in 2002 it found that the data being sent to them by interviewers were inconsistent. As a result, VNS did not report its results on election night, and consequently it went out of business a few months later when its corporate sponsors withdrew.

Quality Review

The next step should be an assessment of the data quality. *Data editing* (also known as *data cleaning*) is the process of looking for and correcting data errors so as to minimize the possibility of wrong values' being entered into the survey data set. An occasional inconsistency in the data is usually an instance of random error that will lower reliability but not bias results. Systematic inconsistency can, however, lead to biases, so it is important to try to find such errors. Several types of data editing are employed in processing data for major surveys (Pierzchala 1990), with many logical data checks now being automated.

The simplest form of data editing is looking for incomplete surveys. *Completeness error* occurs when forms are blank or so unreadable as to be unusable. When incomplete surveys are found, they can be treated as missing data (unit-missing data if the whole form is unreadable or item-missing data if only some items are blank), or, if feasible, the respondents can be recontacted if it is necessary to ask a few crucial questions again.

Domain error (or *range error*) exists when illegitimate values have accidentally been entered for a variable; the error is so named since the values fall outside the allowable range of answers. Domain errors can be found by *wild-code checking*. For example, if a variable is supposed to have only two legitimate codes, say 0 and 1 (plus a missing data code), the "marginal" distribution of the variable is examined to see if any other codes have been accidentally entered. When wild codes are found, the original survey protocol is checked so that the correct code can be entered. Domain errors can also be found by looking for *outliers*, values that are implausibly small or large.

Routing error or *skip pattern error* occurs when respondents are asked a question that was to be skipped or when a question was skipped that was supposed to be asked. For example, people who say they did not graduate from high school should not be asked which college they went to. Surveys often include complicated skip instructions, in which the answer to one question determines which set of questions is asked next. Respondents often have trouble following these skip instructions correctly in self-administered questionnaires, and interviewers sometimes get them wrong in interviewer-administered surveys. This should not be a problem with computer-assisted interviewing, but errors in programming a complex survey often lead to such errors even in computer-assisted interviewing. Questions that were not supposed to be answered can be dropped, but questions that were skipped by mistake must be handled as missing data.

Consistency error occurs when recorded answers to different survey questions are not logically compatible. For example, it is truly rare to interview a ninety-seven-year-old woman who has a three-year-old daughter. That answer combination is clearly inconsistent. Sometimes it is possible to look for such unusual answer combinations and edit the answers to make them more consistent. If there is no obvious way to resolve the inconsistency from the interview, one could seek other information about the respondent or if need be recontact the respondent to find which answer was incorrect. There probably is a way to find out if the ninety-seven-year-old woman is really much younger, or if the child is her granddaughter rather than her daughter, without recontacting her, but recontact should be able to resolve the matter. If it is not still possible to resolve the problem, one of the missing-data imputation methods described in chapter 7 can be applied to the inconsistent entry.

Computer-assisted interviewing might be able to avoid many of these situations if the software is programmed to do so, as by not asking people who have not graduated from high school about their college education. If inconsistent answers are given, some CATI software can be programmed to require that the inconsistency be rectified before the interview can proceed, though it can also be programmed to permit accepting apparently inconsistent answers if the respondent insists they are correct or if the interviewer feels that asking about the inconsistency would impair the rapport of the interview.

If these problems are not avoided in the data collection, consistency checking at the postsurvey stage is advisable, though it will never be possible to check for all possible inconsistencies. Computer software has been developed for data editing that can identify inconsistencies, such as by looking for outliers in data or for unusual over-time patterns in monthly surveys. (Specific computer programs are discussed in Bethlehem and van de Pol 1998.) However, poorly designed consistency checks in a computer-assisted survey can produce rather than solve problems, and incorrect data edits can introduce rather than remove error. Therefore, it is important that data editing be done accurately when it is performed. There is some concern about overediting (Granquist 1995; Bethlehem and van de Pol 1998), particularly because of the potential for introducing incorrect checks. Instead of focusing on "microediting" of individual questionnaires, some researchers prefer using "macroediting" techniques that edit on the basis of potential impact on aggregate survey results, stopping when fixing errors would not affect survey estimates.

One of the few published examples discussing consistency error in an important data set involves Indians and teenage widows in the 1950 census. Coale and Stephan (1962) found an unusually high number of widowed and divorced teenage males, for example 1,670 widowed males of age 14 compared to only 565 widowed females of that age and only 85 expected on the basis of their frequency in the 1940 census. Also, there were an unusually high number of Indians aged 10–14 and 20–24, even in parts of the country where few native Americans usually live, and there were more widowed and divorced Indian males aged 14–24 than would be expected. The detective work of Coale and Stephen shows that each of these anomalous patterns could be explained if variables had accidentally been entered one column off in the computer cards that were used in those days. Unfortunately, the data entry had not been entirely verified, since that would have doubled the data entry cost and thereby reduced funding for other parts of the data collection. Coale and Stephan conclude that the error rate was 0.5%. This example also demonstrates the trade-offs involved in avoiding processing error—even a systematic error can lead to such a low error rate that it is not always cost effective to try to minimize processing error.

Validation

Some surveys include validation exercises as part of their postsurvey operation. It is sometimes possible to validate factual survey answers, though obviously not attitude questions. As an example, the National Election Studies has validated turnout reports of its respondents for some elections, ascertaining from official records whether the person did vote—though, of course, not how they voted (Katosh and Traugott 1981; Bernstein, Chadha, and Montjoy 2001). Validation studies show that people overreport voting and church attendance, whereas they underreport income, abortion, and illegal drug use. Administrative record checks have also been conducted on crime victimization studies and hospitalization surveys. Record checks generally are best at detecting underreporting, since they can find instances the person did not report in the survey. Overreporting is harder to determine with certainty, since overreports can occur because of confusion about the time frame in the question, errors in the official records, or other problems. Unconfirmable claims that people voted in a particular election, however, are usually interpreted as exaggeration of socially desirable behavior.

Validation of survey data is an important concern because of the difficulty in guaranteeing the quality of survey results. There have been several survey validation studies over the years. A meta-analysis of thirty-seven published validation studies from 1944 to 1988 (Wentland with Smith 1993) examined the correlates of accuracy. It found that accessibility (the ability to recall the necessary information) was the most important predictor of accuracy, implying that respondents try to be accurate when they can be. Answers were more accurate on yes/no questions. Answers on nonbinary questions were more accurate when the information was fairly inaccessible, possibly because the added categories aided recall. Answers seemed to be more accurate when more time was spent answering them, which suggests that asking questions that take longer to answer can be useful in getting accurate answers. Answers were less accurate on sensitive questions, but people were more accurate on those questions when there were more than two answer categories. Additionally, people are more likely to exaggerate in answering "yes" when that is the socially desirable answer than when the socially desirable answer is "no." The meta-analysis found less exaggeration toward socially desirable answers on written questionnaires than in face-to-face interviews. There are not consistent differences between questions asking about descriptive information, deviant behavior, and financial matters, except that questionnaires yielded more accurate answers about deviant behavior than did face-to-face interviews. People who performed more of the behavior being asked about were more accurate with descriptive information and finances but less accurate when asking about deviant behavior.

Another means of gauging the accuracy of factual questions is to reinterview respondents and ask them the same questions (Biemer and Lyberg 2003, 291–301). Some surveys try to reconcile differences between the two interviews, as a means of estimating the measurement bias and/or to evaluate interviewer performance. Other surveys simply use the reinterview to estimate response variance and reliability. This approach is susceptible to a conditioning effect, in that the previous interview may affect how people respond the second time. Still, this reinterview approach can be a useful aspect of assessing survey quality.

Validation is sometimes conducted at an aggregate level rather than at the individual level, comparing survey-based estimates of behavior with other data sources. Examples include comparing estimates of hospitalization from surveys with annual data from hospitals and comparing survey-based estimates of crime victimization with official crime statistics. However, coverage issues sometimes throw off results from such validation attempts. For example, surveys on hospitalization face a coverage error if nursing homes are excluded from the sampling frame, since older people are sometimes released from hospitals to nursing homes. The extreme case is comparing estimates of murder from surveys with annual data from government sources, since murder victims are not included in the sampling frames for surveys.

Data File Management

The final data processing step is preparing the data for a statistical analysis program and/or a database management computer program. Each program has its own particular requirements, but there are several common elements. Preparation of the data set usually includes choosing names for each variable and assigning labels to values of some of the variables. Most statistical analysis programs permit missing-data codes to be defined for each variable.

Derived variables are then added to the data file. This includes calculating weights for each case, based on sampling, nonresponse adjustment, multiple telephone adjustments for phone surveys, adjustments for household size when that is appropriate, and any poststratification adjustments to the census values or other target population information (see chapters 8–10). The final weight that should appear in the data set is the product of these separate weights. *Adjustment error* is error due to faulty weighting.

Some surveys add other contextual information to the respondent's record, such as census block information on race, ethnicity, income, and so on. Election surveys, for example, sometimes merge the individual data with aggregate data for the governmental unit in which the respondent resides. Global positioning system (GPS) identifiers can be used to coordinate survey data with appropriate information about the area in which the interview was taken. Geocoding information about the interview location is now routinely included in some surveys.

Additional steps in preparing a data set include imputing missing data (chapter 7) and making sure that respondents cannot be identified from their responses (chapter 14).

The most difficult data management problems come in dealing with panel surveys. It is essential to match records accurately in order to follow the same case across time, since errors would mean that one person's data at one point is being combined with a different person's data at a later wave. Also, it is necessary to have an effective database to manage records in complex panels.

Data management error is error that occurs in the preparation of variables for data analysis. For example, sometimes it is necessary to reorder categories of a variable to make the variable ordinal or to recode the variable in some way. Take questions that are asked in a branching format, as by asking people first if they agree or disagree with a statement and then asking if that is strong or weak agreement/disagreement. For analysis purposes, it is necessary to combine these questions into a scale from "strongly agree" to "strongly disagree." Accidentally combining these questions in a different way would constitute data management error. The best practice is to retain the original variables in the data set as well as any new ones that are constructed, in case the data analyst wishes to combine the original variables differently.

Another common source of data management error has to do with handling missing data. When data analysts receive a data set, they often assume that missing data have been handled appropriately. They may not realize that missing data have not been defined for some variables, and they may thus accidentally analyze an 8 code for a "don't know" response as if it were at the high end of a 1–5 scale. As this example illustrates, data management error leads to validity problems: the variables being analyzed are not correctly measured, so statistical results are meaningless.

A useful way to look for remaining data errors is to produce a statistical summary of each variable. Weighted and unweighted frequency and percentage distributions should be compiled for all variables. Additionally, means and variances should be computed for continuous variables. These statistics should be examined for each variable, in a final check for mistakes. Finding a mean greater than 100 on a 0–100 thermometer scale, for example, usually shows that codes such as 888 or 999 are being used for "don't know" or "not ascertained" but have not been defined as missing data. Some errors inevitably occur in large data sets, so this added check is an important stage.

The best practice is to prepare a separate codebook for the data, giving information about the survey and about each variable. The survey information should include presentation of the sampling details as well as information about the response rate. The exact wording of each question should be given, along with the numerical codes for each possible answer and indication of which codes are treated as missing data.

An extra step in cross-cultural research is *harmonization* of the different versions of the same question used in different nations. This involves transforming different formats for different countries into a variable that is comparatively equivalent across those countries. For example, if education was measured with different categories in variants of the same survey that were administered in different nations, then it is necessary to develop a set of comparable categories that can be used for all the countries so there can be a single education variable for the study.

Reporting Surveys

The reporting of surveys is the other important part of the postsurvey process. This includes both preparing a final summary of the project and analyzing the data.

The Survey Report

A final summary of the project is usually prepared for large surveys. The report should describe the survey methodology, including the target population, sample selection details, how the sample is being weighted, and response rates. This is important both for disclosure and to permit replication. The survey procedure should be summarized, including the dates of data collection, the interview procedure, quality control procedures, and data cleaning procedures. Modern survey reports often go through each of the different types of survey error treated in this book (measurement error, nonresponse error, coverage error, and sampling error) and analyze how serious each might have been in the study. It is also appropriate to review the efforts of the survey organization to achieve survey quality, including special efforts taken in this survey as well as how the structure of the organization is set up so as to achieve quality. Some large government surveys now do this in the form of a quality profile, summarizing the quality control and quality assurance procedures that were used.

Documentation for the data can be included in the survey report, with a data dictionary that indicates the variable name, labels, format information, and sometimes question wording. Then a summary of survey responses is given, possibly in the form of the frequencies of each answer for each question. Statistical analysis required by the client is included in the final report, and a copy of the final questionnaire is usually appended.

Reporting error occurs when the results of a survey are reported incorrectly or incompletely in academic reports or in the mass media. This can happen when full details of the survey are not included. Minimally, it is important to give the response rate on a survey and the number of cases, explain any weighting procedure that was used, and describe the sampling procedure. Incomplete reports imply that the survey findings are stronger than they actually are, such as when survey results are reported as factual without admitting that the response rate was low. Disclosure rules are discussed further in chapter 13.

Reporting error is a problem for the mass media because of the complexity of conveying the uncertainties associated with surveys. Some of the better newspapers include with their survey reports an indication of the margin of error along with a brief explanation of polling procedures. As seen in the last paragraph of the reprinted article in the box, the *New York Times* in its reports of its polls now even indicates that the sampling error calculation does not take into account other possible sources of error such as question wording.

Another reporting problem is how to characterize polling results. For example, should 51% favoring a policy be reported as majority support? Given the statistical margin of error in surveys, 51% does not necessarily demonstrate majority support. It is only natural for the media to interpret 51% as majority support, but as seen in chapter 10, a statistician would not view 51% as an indication of majority support in a poll of 1,500 people. Sometimes the phrase "clear majority" is used to indicate that the support level for a policy is above the statistical margin of error, but there is no consensus on that phrasing.

Data Analysis

The final stage in working with a survey is analyzing the data. *Estimation error* occurs when the wrong statistical techniques are applied to the data.

How the Poll Was Conducted
New York Times, January 27, 2002

The latest New York Times/CBS News Poll is based on telephone interviews conducted Monday through Thursday with 1,034 adults throughout the United States.

The sample of telephone exchanges called was randomly selected by a computer from a complete list of more than 42,000 active residential exchanges across the country.

Within each exchange, random digits were added to form a complete telephone number, thus permitting access to both listed and unlisted numbers. Within each household, one adult was designated by a random procedure to be the respondent for the survey.

The results have been weighted to take account of household size and number of telephone lines into the residence and to adjust for variation in the sample relating to geographic region, sex, race, age and education.

In theory, in 19 cases out of 20, the results based on such samples will differ by no more than three percentage points in either direction from what would have been obtained by seeking out all American adults.

For smaller subgroups the margin of sampling error is larger.

In addition to sampling error, the practical difficulties of conducting any survey of public opinion may introduce other sources of error into the poll. Variation in the wording and order of questions, for example, may lead to somewhat different results.

For example, use of linear regression analysis on ordinal or nominal data is strictly speaking inappropriate, though some researchers prefer linear regression when it seems to be giving similar results as more appropriate techniques. Using techniques that assume normally distributed variables when the variables are not normally distributed is also an estimation error. Statisticians use the term *specification error* when the omission of relevant independent variables from the analysis of a dependent variable results in biased estimates of the effects of the predictors that are included in the model. Estimation error is a statistical problem; lay readers of survey results would not be able to detect such error.

Researchers can best handle estimation error by working with a statistician in the data analysis stage, particularly when multistage sampling was employed in the survey. The statistics that are learned in a first-term statistics course are often insufficient for correct handling of complex survey data, and it is important to recognize when expert statistical advice is required.

Postsurvey Error and Survey Constraints

Table 11.1 summarizes the various types of postsurvey error discussed in this chapter. The topic of postsurvey error emphasizes the importance of care in research even after the data are collected. If sampling error is only the tip of the iceberg of survey error, then postsurvey error is truly at the bottom of that iceberg, often not even included in discussions of total survey error, but care in collecting the data is wasted if these postsurvey operations are not conducted with equal effort.

Table 11.2 lists some of the major types of data errors that require data editing checks in the postsurvey process. It is never possible to eliminate all data errors in a large and complex survey, but it is important to do some checks on data quality.

Table 11.1 Types of Postsurvey Error

Coding error	Verbal material is put into numeric categories incorrectly
Data entry error	Data are entered incorrectly into computer file
Adjustment error	Incorrect weighting of data
Data management error	Variables are created incorrectly from survey answers
Estimation error	Wrong statistical techniques are employed
Reporting error	Survey results are presented incorrectly to an audience

Table 11.2 Types of Data Editing Error

Completeness error	Surveys are not filled out fully
Domain error	Illegitimate data codes appear for a variable
Skip-pattern error	Instructions for skipping questions were not followed
Consistency error	Answers to different questions are not logically compatible

However, there are constraints even on the extent to which postsurvey error should be minimized and on the amount of data editing to do. Hiring high-quality specialists in data management will minimize these types of error, but their salaries will increase survey costs. Overnight polls are particularly vulnerable to these problems—the lack of time to recheck data for errors makes it essential that the original computer program makes editing checks.

It is expensive and time consuming to perform a large amount of data editing, whether during the interview stage in computer-assisted interviewing or at the postsurvey stage. Editing has been known to take 40% of the budget for a survey, with more than 20% of the budgets for many federal government surveys spent on editing (U.S. Federal Committee on Statistical Methodology 1990). The better survey organizations hire specialists who are trained in information technology and statistics to deal with the data management and analysis stages; the organization then claims that it will provide a higher-quality product to prospective clients than would organizations that do not hire degreed professionals. However, Biemer and Lyberg (2003, 228) complain that "a large share of resources specifically devoted to editing could be better spent improving other components of the survey process." In practice, the greatest effort to minimize postsurvey error is expended by large research projects that are designed to collect survey data that many subsequent researchers will examine. Most establishment surveys (of business firms) conducted by or for the federal government are in this category, as are some other ongoing projects such as the National Election Studies (NES) and the General Social Survey (GSS) surveys and surveys taken by large commercial firms like Gallup and Mathematica. Smaller organizations, however, are likely to do only minimal checks for postsurvey error, which makes it important that the researcher watch for incongruities in the data.

Coding open-ended materials (including occupation) also adds to the cost of a survey. The temptation is to have interviewers put open-ended responses into the code categories (a practice known as *field coding*), but that adds one more complex task to the job of already overburdened interviewers. Hiring coders for "office coding" requires one more set of workers and adds to the time of the survey processing but produces higher-quality results. In either case, it is important to spend the time to train the people doing the coding in how each code is to be interpreted, as well as to develop codes that are easy to understand.

Finally, it is important to keep track of data editing rules and decisions so that the research is replicable. This also allows errors to be discovered later in the process of survey analysis. The ethical issue that is raised at this stage is how public the data should be. Surveys are often conducted for a client that expects to keep the data proprietary. However, it can still be useful to keep track of study statistics for proprietary surveys, such as response rate, number of callbacks attempted, and sampling method, so that aggregate characteristics of survey administration can be analyzed.

12

Mode Differences
DEALING WITH SENSITIVE TOPICS

Different survey modes can obtain different results. Survey modes are not considered a type of survey error, probably because there is no objective standard against which results from different modes can be compared. Still, it is important to understand the effects that the choice of survey mode can have.

Several studies have explicitly compared different modes. Some of these studies were intended to determine whether it was safe to move from face-to-face to telephone interviewing, and more recently whether it was safe to move from telephone to Internet surveys. Many studies check whether particular modes are better for dealing with sensitive topics. Additionally, some studies advocate using multiple modes, partly to overcome coverage problems that may be inherent in any single mode.

Mode comparisons are difficult, however, because surveys in different modes often differ in many respects at once. In particular, differences in survey administration confound differences in response. As an example of this problem, when Schuman, Steeh, and Bobo (1985) looked for racial differences, they found more mode differences for whites in the South than in other parts of the United States. As they point out, however, there were important confounding differences in that study. The interviewers in the face-to-face survey were people who lived in the community, often neighbors of the respondents or at least people with similar southern accents. The interviewing for the telephone survey was done from a centralized facility at a northern university, with staff of that university conducting the interviews. That raises the possibility that the differences Schuman and his colleagues found would not have been obtained if the interviewers for both modes had been drawn from the same pool of employees. This example usefully illustrates the difficulty of demonstrating that mode effects are real rather than due to confounding factors. Furthermore, this suggests that when one is looking at any survey result, it is appropriate to think through whether the results would likely have been different if a different mode had been used.

One point to bear in mind is that many of the differences that have been found between modes are based on very few studies. Often a single comparison study reporting a difference becomes part of the lore regarding differences between survey modes. Studies of differences between survey modes are difficult to set up, since they require doing parallel studies using similar questions in the same period of time, so there are few such studies. It is important to be cautious in accepting claimed differences, since there has generally not been robust testing. In particular, a difference that one survey organization may find between its face-to-face and its telephone operation, or its telephone and its Internet operation, may be due to how that organization happens to run those operations. Other survey organizations with somewhat different procedures for their operations may not obtain the same differences between survey modes.

The Theory of the Survey Mode

Mode effects in surveys are hard to study because they are contaminated by multiple sources of differences between modes. In regard to differences between face-to-face and telephone interviewing, for example, the mode effect actually involves the interplay between differences in sampling frames, interviewer selection, interviewer supervision, types of questions that are employed in the two modes, callback rules, refusal conversion approaches, and, often, use of computer-assisted interviewing in one mode but not the other (Groves 1989). Comparisons between other pairs of survey modes would similarly involve complex interactions among many differences between the techniques. Even though the differences that are found can be due to these many factors, they are still important since they show ways in which different modes get different results. These differences will be examined in this chapter, including differences in response effects and in costs.

Tourangeau, Rips, and Rasinski (2000, 305–12) posit three main factors in mode differences that can have psychological effects: whether the mode is impersonal (as in self-administered surveys), how it affects the perceived legitimacy of the survey (which is difficult for respondents to ascertain in telephone surveys), and how much of a cognitive burden the mode places on the respondent (such as needing to learn how to move on to the next question and how to change an answer in Internet surveys). They also mention three other potential factors: the pace of the interview, the order of processing questions (respondents to self-administered questionnaires can skip ahead if they choose), and the respondents' mental model of the survey.

Different survey modes can be thought of as eliciting different mindsets in the respondent. Groves (1989, 510) puts this into the framework of the "script" elicited by the mode. A cognitive script is an ordered set of memories about an episode in one's life, often a routine event such as the memories that one develops as to what to expect when walking into a medical office.

Groves suggests that the interviewer in face-to-face interviews in people's houses evokes a "guest" script. The interviewer is a guest in the person's house, and there are particular responses, such as politeness, which are appropriate for the respondent. Additionally, the interviewer gives off visual and nonverbal cues in face-to-face interviews, and the enthusiastic and committed interviewer may give off cues that keep the respondent engaged in the survey.

By contrast, in phone interviews the interviewer is thought to evoke a "solicitation" script. The interviewer is not actually soliciting funds, but respondents are used to dealing with anonymous phone callers who do so. Phone interviewers must make clear that they are not soliciting, but they are indeed soliciting the respondents' time, asking them to contribute it for free to the research effort. Every script has particular appropriate responses. Whereas the guest script evokes politeness, the solicitation script does not. Instead, the solicitation script leads to wariness on the part of respondents, so it is essential that interviewers make respondents understand that they are legitimate and not part of a sales effort. Furthermore, it can be harder for telephone interviewers to keep respondents engaged in a survey without visual and nonverbal cues.

Other survey modes can be similarly understood in terms of scripts. Self-administered questionnaires may call to mind the script of taking a multiple-choice test. Mail questionnaires would seem to induce a junk-mail script in nearly the same way as phone interviews evoke a solicitation script. People have developed a routine response to most junk mail, and thus the researcher must work to distinguish the survey from junk mail so that it is not immediately trashed. Surveys sent to people by e-mail might call up a spam script, again very similar to a solicitation script. By contrast, surveys that respondents find on the Internet would seem to evoke a Web surfing script, which is more positive than a standard solicitation script. People may be wary of surveys they find as part of Web surfing, and the usual response is to stay on a Web site for only a very short period of time, but the curiosity factor may get people to examine the survey and possibly become intrigued enough to take it. The other script that may be evoked by Internet surveys is a game script, since many people are used to playing computer games. The game script poses a challenge to survey designers, since it emphasizes the need to keep Internet surveys enjoyable.

A corollary of the different scripts is that respondents may treat surveys given in different modes differently. In face-to-face interviews, respondents not only have to answer questions but also have to deal with the uninvited guest. They may need to return to other activities, but they will generally pay attention to the interviewer. Also, in many cases there is another person in the room at the time of the interview. As indicated in chapter 5, there have been some studies of the effects of having other people present during an interview, especially spousal presence effects.

The respondents' mental sets for phone interviews are quite different. Their lives have been interrupted, usually without any prior notification. Even respondents who are willing to cooperate may not fully drop the activities they were performing and the concerns they had before the phone call. Respondents may be multitasking, continuing to do what they were doing before the phone rang, particularly if they are using a cordless phone that allows them to wander around their house while talking to the interviewer. They may need to get off the phone quickly, perhaps to get back to the television show they were watching, to leave the house, or to put their children to sleep. Someone else may be in the room as they answer questions, possibly providing a distraction from the phone call. Thus there are several possible attention problems that can affect phone surveys.

Respondents are usually even more negative to self-administered questionnaires, unless the topic is of intrinsic interest to them. They often view such surveys as something to complete in order to get onto a more enjoyable activity, and they do not take most mail surveys seriously enough to answer them unless the sponsor has some special legitimacy. Similarly, Internet surveys could be seen as intrusions into the person's time on the computer.

Counterbalancing the above considerations, interviewer-administered surveys raise a potential problem in terms of the impression respondents want to leave with the interviewer. Because interviews resemble conversations, people naturally want their conversation partner to have a good impression of them, or at least not to have a bad impression. That can limit the candor of respondents on topics such as racial prejudice and drug use. Self-administered questionnaires do not raise this problem, as long as the respondent feels that the answers will be anonymous. Self-administered computer-assisted surveys fall in between on this continuum, since people are sometimes not sure whether the computer can somehow trace their identity.

This script analysis is useful to the extent that it gets researchers to think through how potential respondents will view their effort. For example, realizing that surveys often evoke a solicitation script should make researchers consider how to set their surveys apart and make them interesting enough for respondents to complete.

Mode Differences

There are several types of mode differences that are possible, including differences in response levels, the demographics of respondents, question formats, answer quality, and—most important—results, especially involving social desirability effects and sensitive topics.

Nonresponse

There are typically differences in unit nonresponse between different survey modes. Response rates are higher in face-to-face surveys than telephone surveys. An NES report estimates this as a 15% difference (NES Ad Hoc Committee 1999).

Given the move toward computerized surveys, it is particularly important to examine the effects of computerization on response rates. The limitations of computer-assisted self-interviewing are apparent in a study by Couper and Rowe (1996) in which participants in a computer-assisted personal interview (CAPI) survey in the Detroit area in 1992 were asked to self-administer some questions on a notebook computer. Only 79% of the sample were able to complete the computerized self-interview themselves, 7% entered the answers after the interviewer read the questions, and the interviewer completed the items for the remaining 14%. Multivariate analysis showed that nonwhites and people who did not graduate from high school were less likely to self-administer. Those groups also had more missing data, as did older people and women. Vision problems and a lack of previous computer experience also led to less willingness to complete the computerized part of the survey. Familiarity with computer use has increased since 1992, but this study usefully indicates the limits to expecting respondents to answer questions on a computer even when computer access is not a problem.

The survey mode can also affect the rate of item nonresponse. The early literature found more item nonresponse with phone surveys than face-to-face surveys, but that problem may have been due to inexperience with telephone interviewing; a well-functioning telephone operation should not necessarily obtain higher levels of missing data than its counterpart that still does face-to-face interviews. However, there is more item-missing data on self-administered questionnaires, due to failures to follow skip instructions and the absence of an interviewer who would minimize missing data. In a comparison of e-mail and mail surveys of university faculty, Schaefer and Dillman (1998) found that the e-mail survey had less item nonresponse and more complete answers to open-ended questions. Of course, university faculty members generally have more familiarity with computers than the public as a whole, so this result may not generalize to the broader population.

Types of Respondents

Some differences have been found in the types of people who participate in different types of surveys, particularly having to do with place of residence. Telephone surveys obtain a higher response rate in urban areas and lower in suburban and rural areas (Fowler 2002, 66), which suggests that telephone may be better than face to face for urban surveys, whereas face to face would be better if the researcher is not interested in urban areas.

Krysan et al. (1994) obtained about a 75% response rate among a white sample in the Detroit metropolitan area, regardless of whether they used face-to-face interviews or a mail questionnaire with a prepaid incentive. However, differences were obtained among the African American part of the sample, with an 83% response rate for the face-to-face interviews compared to only 46% for the mail survey, which suggests that mail surveys are less effective for African Americans. The authors admit, however, that the racial differences are confounded with education, income, and urban residence. The mail respondents were more likely to express negative attitudes on racial integration and affirmative action, indicative of a social desirability effect on those questions.

There is concern about interviewing the elderly over the telephone. However, Herzog and Rodgers (1988) found that reinterviewing older adults over the telephone worked well after they had been initially interviewed in a face-to-face mode. The reinterview rate for the elderly was 8% lower than for younger people, and the older group gave significantly more "don't know" responses. Still, response distributions were similar across modes, which suggests that there was little mode effect.

Ellis and Krosnick (1999) have compared the demographic characteristics in several studies that included both telephone and face-to-face interviewing. They found that telephone respondents were younger, more likely to be white and better educated, and had higher incomes, with the largest of these differences being about 5%. Some of these differences are due to people with lower incomes and less education being less likely to have telephones in their home. There also have been some mode comparisons in some National Election Studies. In the 1992 NES, telephone respondents were younger and better educated, had higher incomes, and were higher in social status (Rosenstone, Petrella, and Kinder 1993). Looking at a mixed-mode experiment in the 1998 NES, Wessel, Rahn, and Rudolph (2000) report that phone respondents were more likely to be young, female, minority race, and better educated and to have a high income. Although these studies differ in terms of findings on race, they do provide a warning that phone respondents are likely to be higher in social status as well as younger than respondents in face-to-face surveys. These studies illustrate the importance of comparing sample demographics with known census distributions and weighting when there are important differences.

As to differences between Internet and telephone surveys, Krosnick and Chang (2001) find their 2000 unweighted Harris Interactive Internet survey to be far off the mark on education and gender (see table 12.1a), whereas the Knowledge Networks Internet survey and their telephone survey obtained more similar results. Most, but not all, of the differences diminished with weighting, with the Harris Interactive sample still somewhat off in inclusion of low-education respondents (see table 12.1b). Berrens et al. (2003) report that weighted Internet surveys by Knowledge Networks and Harris Interactive are within 1%–2% of Census Bureau estimates of 23% of the population with college education, whereas the phone sample was much further off, with 41% having college degrees. This suggests that, with proper weighting, Internet surveys can be at least as accurate as other types of surveys, though the volunteer base of the Harris Interactive panel seems to have some demographic distortion.

Question Formats

There are differences when using particular types of questions in different survey modes (Groves 1989, chap. 11). In a face-to-face survey, it is possi-

Table 12.1a Unweighted Demographic Composition of Phone and Internet Surveys

	Census	Telephone	Knowledge Networks	Harris Interactive
Some high school	16.9%	7.0%	6.7%	2.0%
Less than $25,000 income	30.5%	19.0%	14.3%	12.6%
Age 18–24	13.2%	10.0%	7.8%	8.0%
African American	11.9%	9.7%	6.9%	3.6%
Male	48.0%	45.1%	49.2%	60.1%

Source: Adapted from Krosnick and Chang (2001, table 2)

Table 12.1b Weighted Demographic Composition of Phone and Internet Surveys

	Census	Telephone	Knowledge Networks	Harris Interactive
Some high school	16.9%	17.1%	12.3%	7.9%
Less than $25,000 income	30.5%	19.0%	18.0%	24.8%
Age 18–24	13.2%	13.5%	9.8%	14.0%
African American	11.9%	11.9%	10.0%	12.3%
Male	48.0%	46.9%	49.2%	48.2%

Source: Adapted from Krosnick and Chang (2001, table 3)

ble to use "show cards"—such as a card that shows a 0–100 thermometer scale when asking people to rate how much they like social groups or a card that lists several possible answers to a question. Obviously show cards cannot be used for telephone surveys, though some surveys have mailed respondents the appropriate show cards to have by the phone when the interview occurs. The inability to show people lists of possible answers on the phone makes it very awkward to ask people to choose from among a large number of alternatives; people may be more likely to satisfice, such as by choosing the last possibility (a recency effect), since it is hard to keep in mind a long list of alternatives offered by a telephone interviewer.

Branching formats work better than nonbranching formats on the telephone. The branching format on the phone pulls people away from the middle category, whereas many people end up in the middle category for nonbranching questions in face-to-face interviews, but that difference is probably because of the branching; using the exact same format in both modes should produce more similar results.

Finally, differences on 7-point scales in face-to-face versus phone surveys are found to be small. Answers on 100-point scales, such as the thermometer question, are more clustered around 10s (answers such as 10,

20, 30, and so on) in phone interviews than in face-to-face interviews, but in fact there is also considerable clustering in face-to-face interviews on points labeled on the show card (Groves and Kahn 1979).

Answer Quality

Some studies have examined the quality of answers given in different modes, with a focus on whether respondents are more likely to satisfice in some modes than others. Early studies showed that phone surveys lead to more acquiescence and more extremeness in response in phone surveys (Jordan, Marcus, and Reeder 1980) along with less-detailed answers to open-ended questions, particularly among well-educated respondents who are capable of giving wide-ranging answers (Groves and Kahn 1979). These results suggest there will be more satisficing behavior in telephone surveys than in face-to-face surveys, presumably because of the lower engagement of telephone respondents.

Holbrook, Green, and Krosnick (2003) test satisficing levels in three studies in which identical questions were asked in face-to-face and telephone surveys: a 1982 NES experiment, the 1976 University of Michigan Survey Research Center experiment reported by Groves and Kahn (1979), and a 2000 NES experiment. Each of the studies used a lengthy interview schedule, which is more likely to lead to satisficing behavior.

As expected, telephone respondents were found to be more likely to engage in satisficing behavior. For example, after adjusting for demographic differences, 24% of the answers in the telephone part of the 1982 study were "don't knows" compared to 17% in face-to-face interviews. Similarly, 19% of the phone respondents in the 2000 NES study chose a "no opinion" response option, versus only 11% of face-to-face respondents. Holbrook, Green, and Krosnick were able to test acquiescence in only the 1976 and 2000 studies, and they found significantly more "agree" and "yes" responses by face-to-face respondents in both of those studies.

Furthermore, Holbrook, Green, and Krosnick show that the satisficing behavior was greater among less educated respondents. Less educated respondents are also underrepresented in telephone surveys. Thus less educated respondents are doubly disadvantaged in telephone surveys, making it harder for their opinions to be fully represented in such studies.

As to differences between telephone and Internet surveys, Krosnick and Chang (2001) find most satisficing in their phone sample and least satisficing by Harris Interactive respondents, with the Knowledge Networks sample in between.

Results Differences

Researchers have tried to examine differences between survey modes. An extensive comparison based on parallel face-to-face and telephone surveys in the late 1970s found few substantive differences in responses (Groves and Kahn 1979), but several nonsubstantive differences have been emphasized in the literature (Groves and Kahn 1979; Groves 1989, chap. 11). First, there are differences in enjoyment and respondent satisfaction. Most people prefer face-to-face surveys, presumably because of the personal interaction involved (Groves 1979). However, there are also claims that low-motivated respondents may prefer phone surveys since they are quicker. That leads to the second set of differences: pace. Phone interviews are shorter, even when the same questions are asked. Additionally, respondents are more confident in their answers in face-to-face interviews, presumably because of nonverbal approval cues that interviewers give off (Groves 1989, chap. 11). There is also greater intimacy in face-to-face interviews, but there are more social status effects too since respondents are able to see the interviewer's social status.

The differences between results in survey modes are usually confounded with differences in sampling methods. To test for this difference, Aquilino and Wright (1996) conducted an experiment in which two different samples were drawn, one using random-digit dialing with screening questions on the telephone and the other a multistage area probability sampling with in-person screening. The interviewing was done by telephone for both samples, so the only difference was how the sample was originally selected. Although the RDD sample had a lower response rate, the demographic differences between the groups were generally not significant, and answers on most substantive questions (regarding substance use in this survey) were virtually identical. Apparently the differences inherent in the two sampling approaches do not necessarily affect survey results, so differences between phone and in-person interviewing are probably not due to their sampling methods.

Differences in results obtained in different modes have rarely been found, and, as mentioned earlier, those that are obtained are often confounded with other aspects of the survey administration. In an extensive comparison of results in face-to-face and phone surveys, Groves and Kahn (1979, 85–91) found only small differences in the distribution of responses on specific questions. One of the few differences that have been reported is that Gallup phone samples have been about 3% more Republican in partisanship than samples for Gallup's face-to-face surveys (Green, Palmquist, and Schickler 2002, 90). A study of a mixed-mode experiment in the 1998 NES found that telephone respondents were more likely to vote Republican for Congress and were also more likely to be habitual voters and have higher trust (Wessel, Rahn, and Rudolph 2000).

Some differences that might be expected are actually not found. Specifically, there are not large differences in results between face-to-face and telephone surveys for less educated respondents, as some survey researchers hypothesized, nor do the elderly have more problems with the telephone mode (Groves 1989, chap. 11).

De Leeuw and van der Zouwen (1988) conducted a meta-analysis of thirty-one studies in which telephone surveys could be compared with face-to-face interviews. They found only small differences in data quality in terms of such factors as accuracy, social desirability bias, and item non-response. There was a negative correlation between year of publication of the study and effect size, suggesting that the mode difference was diminishing as experience with phone interviewing was increasing. The caution in interpreting these results is that the surveys were fairly early, before CATI methods were fully in place; telephone survey methodology is now more sophisticated than when these mode comparison studies were conducted.

As to mail surveys, Dillman (2000, 226–29) cites evidence that they are less likely to suffer from social desirability effects, question order effects, and acquiescence bias, though de Leeuw's (1992) meta-analysis does not confirm the latter difference. Additionally, as stated in chapters 5 and 6, mail surveys are more likely to show primacy effects, with people choosing the first answer on a list, whereas telephone surveys are more likely to exhibit recency effects, with people choosing the last item. All in all, response rate is a more serious problem for mail surveys than are response effects.

Comparing telephone and Internet surveys by Harris Interactive and Knowledge Networks on the environment, Berrens et al. (2003) find small differences in knowledge about the Kyoto Protocol, but the Internet sample was more liberal and Democratic, which could be because of self-selection to participate in the Internet surveys on this topic. Krosnick and Chang (2001) found their Harris Interactive sample to be more Republican in both partisanship and voting in 2000 than their Knowledge Network and phone samples, presumably because of the volunteer nature of the Harris Interactive sample.

Social Desirability Effects

Social desirability effects occur when respondents may feel that their true answer would not be acceptable to give to the interviewer and substitute a more socially acceptable answer. This behavior is much more likely to occur when there is an actual interviewer, though there can be a smaller social desirability effect in noninterviewer administered surveys if some respondents want to look good to the researcher. The likelihood of mode differences leads to the possibility of switching survey modes as a means of minimizing social desirability effects, since respondents might be expected to edit their answers less when they do not deal directly with an interviewer.

Social desirability effects should be minimal in self-administered questionnaires. In a study in the Detroit metropolitan area testing for such differences between question modes, Krysan (1998) found more answers that could be interpreted as racial prejudice when there was more privacy in responses—a mail questionnaire rather than face-to-face interview, with in-between results when respondents filled out a questionnaire themselves as part of a face-to-face survey. For example, 22% of white respondents in the mail survey indicated they would not vote for a black candidate, versus 14% for respondents who filled out the questionnaire themselves in the face-to-face survey and 11% when they answered questions out loud to an interviewer. More generally, Dillman (2000, 226–29) cites evidence that mail surveys are less likely to suffer from social desirability effects. Also, customers report lower satisfaction with bank and fast food services in mail and interactive voice response surveys than in telephone surveys (Tourangeau, Steiger, and Wilson 2002), which the researchers interpreted as people giving more candid answers when there is no live interviewer. One study found that people give a more positive slant to their health in telephone interviews than mail surveys, suggesting that social desirability effects also occur on health topics (Fowler, Roman, and Di 1998). However, Moon (1998) found that a student sample was more likely to give answers that were not socially desirable when they verbalized their answers into a voice-input computer than when they typed answers into the computer, so computerized self-administered questionnaires may not entirely eliminate social desirability effects.

By extension, social desirability effects might be expected to be greater in face-to-face interviews than on the telephone when respondents do not see the interviewer directly. However, only 70% of those interviewed face to face for the 2000 NES survey said that they voted in the election, compared to 76% of those interviewed by telephone (Finkel and Freedman 2004), which suggests that phone interviews tend to increase claims of performing socially desired behaviors. Similarly, both an early meta-analysis by de Leeuw and van der Zouwen (1988) and the Holbrook, Green, and Krosnick (2003) analysis of three national studies referred to above found significantly greater social desirability effects in telephone interviews. Holbrook, Green, and Krosnick explain this result in terms of the greater rapport with face-to-face interviewing, leading to more confidence that the respondent's confidentiality will be protected, along with less certainty in telephone surveys as to the identity and motives of the interviewers.

Another instance of social desirability effects involves the study of racial differences mentioned at the beginning of this chapter (Schuman, Steeh, and Bobo 1985). White southerners in this study gave more conservative answers on racial questions in the face-to-face interviews than over the phone. Recall, though, that face-to-face interviewers in this study were southern whereas telephone interviewers were northern. The authors interpreted this as more candid answers in face-to-face interviews with southern interviewers, but Groves (1989, 519) points out that giving more conservative answers to southern interviewers and giving more liberal answers to northern interviewers were *both* instances of social desirability effects. Thus there seems to be a complete interaction here between mode effects (with more social desirability problems expected for face-to-face interviews) and conformity with perceived attitudes of interviewers, making it impossible to pull apart the separate effects from this study.

More generally, social desirability effects are likely to differ between countries. Partly as a result, mode effects can differ between countries (Johnson and van de Vijver 2003).

Sensitive Topics

An important instance of social desirability effect involves "sensitive topics" such as one's drug use, sexual behavior, and income, topics that are usually not discussed with strangers. Bradburn, Sudman, and Wansink (2004) prefer to call these "threatening topics." Fowler (1995, 29) usefully points out that the problem is not "sensitive questions" but "sensitive answers"—that whether or not respondents consider a question sensitive will depend on their answer. People clearly differ in what they consider sensitive, but it is always important to ensure confidentiality in order to obtain candid answers on sensitive topics. The lower the cost to the respondent for disclosing something of a private nature, the less social desirability effects should be present; survey modes that provide high anonymity should therefore lead to greater reports of sensitive behavior.

Greater candor would be expected in answering self-administered questionnaires (SAQ) than in responding to surveys using interviewers. In a study of mode differences in reporting sensitive behaviors, including sexual behavior and drug use, Tourangeau and Smith (1996; 1998) find more reporting of sensitive behaviors in SAQ, with the main exception being that men claimed more past sex partners in interviewer-administered surveys. The differences are particularly large on behaviors that are very sensitive, such as far greater reporting of cocaine use in self-administered surveys (see also Aquilino 1994). Tourangeau and Smith find less effect of computerization, but they find a higher rate of reporting with audio computer-assisted self-administered interviews in which respondents take the interview themselves on a computer while wearing headphones through which the questions are read aloud. Audio computer-assisted self-interviews have also been found to lead to more reporting of sexual, contraceptive, and HIV risk-related behavior among older male teenagers (Turner et al. 1998) and more reporting of mental illness symptoms by adults (Epstein, Barker, and Kroutil 2001) than do interviewer-administered surveys. Adolescents are also more likely to admit their smoking tobacco in telephone audio computer-assisted interviews (in which human interviewers recruit respondents but the actual interview is conducted with prerecorded questions that the respondents answer by pressing phone keys) than to telephone interviewers (Currivan et al. 2004; Moskowitz 2004).

Turning to comparisons of telephone and face-to-face surveys, Aquilino and LoSciuto (1990) find small differences for white respondents in alcohol and drug use but more substantial differences for blacks. The authors point out that differences for blacks might be due to coverage error leading to higher socioeconomic status among blacks in the phone sample than among those in the face-to-face survey. Racial differences were negligible in the face-to-face interviewing (in which self-administered answers sheets are used), but fewer blacks than whites reported using these substances in telephone interviews, and that difference was found not to be associated with the race of the interviewer. The differences were minimal on smoking tobacco, which does not raise high social desirability concerns, but larger differences were found on alcohol and drug use.

Groves (1989, chap. 11) reports a difference in reporting income, with phone interviews tending to report higher incomes than face-to-face interviews. This may be due to respondents in face-to-face interviews realizing that the interviewers can assess the reasonableness of their claimed income from their house and furnishings, whereas phone respondents are not constrained by a potential reality check.

Although research suggests that there can be more frank discussion of sensitive topics on the phone, it can be more difficult to establish the legitimacy of asking sensitive questions in that mode. Consider, for example, surveys of sexual behavior. The National Opinion Research Center (NORC) was able to conduct a high-quality face-to-face survey on sexual practices, a topic that had previously been thought to be too sensitive to investigate with a mass survey (Laumann et al. 1994; Miller 1995; see also Michaels and Giami 1999). Calling people on the telephone to ask such questions would be more awkward, since some respondents would suspect that the interview was just a subterfuge for an obscene phone call unless there was a prenotice to the respondent to expect a survey phone call.

Finally, Wright, Aquilino, and Supple (1998) compare pencil-and-paper SAQ with computer-assisted self-administered interviewing (CASI) on smoking, alcohol, and drug use. There is an expectation that people will be more willing to give sensitive information with computer-assisted self-interviewing. However, some respondents might be concerned about the security and/or confidentiality of their responses on the computer. Using the 1995 National Household Survey of Drug Abuse study of individuals aged 12–34 years old from urban and suburban areas, they obtained few significant main effects. Adolescents were more sensitive to the mode than older respondents. In particular, teenagers reported significantly higher levels of alcohol use, illicit drug use, and psychological distress in CASI than pencil-and-paper SAQ. At the same time, this study obtained some evidence of less candid answers with computers. People with higher levels of mistrust were less likely to report substance use in the computer mode, and lower-income people were more distrustful and had less experience with computers, differences that may disappear as a broader range of people have experience with computer use.

All in all, these studies show that there are mode differences on sensitive topics, even though there is limited evidence on mode differences more generally. It should be added that what is a sensitive question will vary between countries as well as between cultures. People in many other countries would not find sexual topics as sensitive a topic as do people in the United States, for example, while people in some other countries would consider it unusual—and dangerous—to discuss their political attitudes with interviewers.

Mixed Modes

Dual-mode (also known as *mixed-mode*) surveys have advantages because they allow comparisons to determine the extent of mode effects. They are particularly useful in overcoming noncoverage problems, since the noncoverage involved in one mode is likely to be different from the noncoverage in the other mode. However, dual-mode surveys also impose extra costs, particularly when they require two separate administrative operations. The effects of dual-mode surveys are likely to differ from one question to another.

Mixed-mode surveys are often used in panel surveys, with a more economical mode used in follow-up waves than was used to generate the original sample. For example, face-to-face interviews in a first wave can obtain phone numbers that are then used to contact people in later waves. Similarly, self-administered questionnaires in a doctor's office can obtain home addresses that can be used for follow-up mail surveys. However, changes in results between waves can be due to differences in administration rather than to change in the attitudes or behavior of interest. That is particularly the case when question formats used in the original mode are not as effective in the later mode.

Sometimes multiple modes are used in a single data collection on sensitive topics, such as asking people to fill out a self-administered questionnaire after a face-to-face survey. This approach can help minimize social desirability effects. Also, as indicated in chapter 8, some surveys switch modes in refusal conversion so as to increase response rates. For example, nonrespondents can be sent a very short questionnaire saying that these few questions will let the researchers know if people who did not answer differ from those who did. Or interviewers in face-to-face surveys could leave mail-back self-administered questionnaires on the doors of people not at home. People without e-mail addresses can be mailed surveys or phoned. Finally, some mixed-mode surveys give people a choice as to whether to complete an Internet survey or whether to use a more conventional mode, like phone or mail.

The problem with using multiple modes is that different interviews may not be comparable if they were obtained in different ways. This effect can be measured if assignment to mode is done randomly but is confounded with respondent differences in studies that allow respondents to choose which mode they prefer to use. Minimally, it can be useful to compare results obtained in the different modes to see if there are systematic differences before combining them in a survey report.

Dillman (2000, 232–40) describes a *unimode design* strategy for writing survey questions so they can be used for mixed-mode surveys. This entails recognizing from the start that multiple administration modes may be used and therefore writing the question in such a way that it can be used across modes. This may require reducing the number of answer categories to a small number that can be used effectively regardless of the mode. The limit of this logic may be seen with strongly agree/strongly disagree scales. Dillman (238) explicitly suggests avoiding branching formats for such scales in mixed-mode surveys because the skip patterns are too complicated for self-administered questionnaires. Krosnick and Berent (1993), however, find that branching formats increase the reliability of scales. Thus unimode design is likely to result in using suboptimal question formats.

The difficulty in designing identical questions across modes can be seen in handling "don't know" situations. Interviewer-administered surveys rarely include "don't know" alternatives, though there is often a box for the interviewer to check when the respondent answers "don't know." This intermediate strategy allows for "don't know" responses without encouraging them. That strategy cannot be employed in self-administered surveys. Internet surveys do not have such an intermediate strategy—either they encourage such responses by providing "don't know" as a response option, or they disallow it by not permitting that response. The handling of "don't knows" is inevitably different in these two modes, but my preference would be to make it easy to skip the Internet question so as to make the situations as similar as possible.

Survey Modes and Survey Constraints

There are important cost and time differences that affect the choice of survey mode. There are large cost differences between face-to-face and telephone surveys. Obviously face-to-face surveys have substantial travel costs, whereas telephone surveys require setting up a telephone calling facility. The cost of long-distance telephone calls has gone down so much that the cost of conducting a national study from a single phone bank is not much higher than the cost of conducting a local survey.

Some survey organizations employ additional techniques to cut costs. Instead of equipping a central phone facility, some organizations have interviewers call from home to a switch that chooses a phone number to call, while accessing the survey through a broadband connection to the Web. This saves the money required to furnish a central facility while still making it possible to monitor the phone calls. Some operations are actually contracting the interviewing out to other countries, particularly India, where interviewer hourly wages are lower.

The most detailed time and cost figures are from the Groves and Kahn (1979) report of a comparable national face-to-face and telephone survey. There was a considerable difference in time per interview, with face-to-face interviews occurring at the rate of 1 half-hour interview per 8.7 hours of interviewer time (including travel time) versus 1 per 3.3 in phone surveys. There were also large differences in the amount of time required for administrative coordination—716 hours for the face-to-face interviews versus just 211 for the phone survey. These time differences translate into substantially lower costs for telephone interviewing.

There are few estimates of the overall cost differences. A one-city study (Tuchfarber and Klecka 1976) found that phone interviewing cost only a quarter of face-to-face. The phone survey in Groves and Kahn's (1979) national study cost half as much as their face-to-face interviewing, and it would probably be a smaller fraction today because of lower long-distance telephone charges. Groves et al. (2004, 161) estimate the costs of phone surveys as one-fifth or one-tenth that of national face-to-face surveys. As to telephone versus mail, they cite figures of telephone costing 1.2 to 1.7 times as much as mail surveys. However, it can nearly be 5 or 10 times as costly to do a phone survey, since a mail survey can, when necessary, be sent out on a shoestring budget, whereas that is not possible for a large phone survey.

Berrens et al. (2003) give some comparative cost figures for Internet and phone surveys. Their 18-minute Knowledge Networks survey cost $60,000 for 2,000 completed surveys, and $72,000 for 6,000 with Harris Interactive, versus $50,000 for 1,700 completions for their phone survey. These cost differences seem minimal, especially with Harris Interactive providing so many more interviews for its higher cost. However, the real cost in the Internet surveys is for the initial programming; the marginal cost of extra Internet respondents is exceedingly low.

There are also cost differences between other approaches, though there are not published data on comparative costs. Self-administered questionnaires, such as to university classes, are very inexpensive. Mail surveys are somewhat more expensive, because they require postage, return postage, and, ideally, multiple mailings. Also, there can be considerable clerical costs involved in well-executed mail surveys, including keeping track of the questionnaires as they come in so that reminders can be sent at appropriate times. Putting a static survey up on the Internet is inexpensive. Creating an interactive survey, whether for telephone interviewing

(CATI), personal interviewing (CAPI), interactive voice response (IVR), voice-recognition entry (VRE), or an Internet survey is more expensive because of the programming (and debugging) costs involved as well as the need to purchase quality software.

Another point to emphasize is that face-to-face interviews can be longer than those in other modes. There are expectations that telephone interviews will be limited in length, with most being 5 to 15 minutes. The Gallup Organization, for example, tries to keep its phone interviews shorter than 12 minutes, and most survey organizations try to discourage phone interviews that would last longer than 30 minutes. The greater cost of transporting interviewers to face-to-face interviews means that those interviews have to be long to be cost effective; it would not be worth sending interviewers out to households for only 5- or 10-minute interviews. The norm of courtesy in face-to-face interactions in people's homes permits interviews in that setting to be as long as 30 to 60 minutes.

An actual example of the choice involved between these modes comes from the National Election Studies (NES). These surveys were started in the 1950s, when face-to-face interviewing prevailed, and NES did not switch to the phone mode when telephone interviewing became common. Costs for national face-to-face interviewing have skyrocketed in the meantime. The National Science Foundation (NSF) signaled in 2000 that it was imposing cost limitations in the future. The plan was to move to phone interviewing for the election surveys by 2004, but NES successfully argued to NSF that it needed to use both modes in 2000 so that it could compare results. After all, NES had accumulated half a century's worth of data using one survey mode, and an abrupt change would make it impossible to tell whether changes in time series were due to political developments or the mode change. Comparison of data collected by the two modes in 2000 was designed to estimate how much the mode change would affect comparisons of pre-2000 and post-2000 data. Large differences were found between the two modes in the 2000 data for several questions. It became clear that data collected through the telephone would not be very comparable to the half-century of face-to-face data. NSF relented and decided to fund face-to-face interviews again for 2004 and presumably beyond, with cost saving instead obtained through not funding a midterm election survey in 2002. This exemplifies the larger dilemmas involved in mode choice: the survey modes that provide the best data are often the most expensive, and changing modes to reduce costs can risk loss of continuity in longitudinal work.

13

Comparability Effects
THE LIMITS OF EQUIVALENCE

Comparability issues arise when different surveys are compared. Surveys taken by different organizations, in different countries, and/or at different times are often compared, but the differences that are found are not necessarily meaningful.

The first section of this chapter focuses on the meaning of equivalence between surveys, after which the second section briefly examines three related comparability issues: "house effects," cross-cultural differences, and over-time differences. *House effects* refers to the differences that different survey organizations sometimes obtain even when they ask the same question to the same population. Cross-cultural differences are the factors that preclude total equivalence when the same question is asked in multiple countries. Over-time differences relate to changes in the meaning of survey questions over time.

The sources of differences between poll results can be determined only if there is full disclosure of the details of survey administration. Therefore, the third section of this chapter considers disclosure rules.

The Theory of Equivalence

It is only natural to compare surveys asking similar questions, whether the surveys are taken by different polling organizations, taken in different countries, or taken at different points in time. But comparisons work only when the surveys are equivalent, which leads to the problem of what constitutes equivalence in surveys. This problem is sometimes evident during election campaigns, when different polls are giving discrepant vote predictions. Slight differences between surveys can easily be explained away in terms of sampling error, but how should significant differences be interpreted?

Surveys should give similar results if they use the same questions and same procedures to interview samples of the same population at the same point in time. If the same questions and same procedures are used to interview samples of the same population at different points, then meaningful time trends can be observed. If the same questions and same procedures are used to interview samples of different populations, then the two populations can be meaningfully compared.

The equivalence problem is that the comparisons are rarely as precise as implied in the previous paragraph. How do you know whether the procedures used by two different survey organizations are exactly identical? How do you ask the same question in two different languages in cross-cultural research? How can you tell whether the meaning of a question has changed when you are comparing surveys taken many years apart? Comparisons require equivalence, but equivalence is difficult to establish.

Each of the factors reviewed in the earlier chapters of this book can lead to equivalence problems. Different survey organizations use different interviewing styles, use different strategies for filtering questions, give interviewers different instructions on probing "don't know" answers, use different numbers of callbacks, obtain their samples from different frames, and take different numbers of interviews per sampled cluster, each of which can affect the results. These problems are multiplied further in cross-national research, since it is difficult to ascertain whether a question really has the same meaning in two different cultures.

Although there are not simple solutions to these problems, a few basic dimensions for equivalence can easily be listed. First, equivalence assumes using similar survey procedures, including, among other things, the same mode, interviewing style, interviewer instructions, and sampling procedure. Second, equivalence requires using the same questions, in terms of both wording and meaning. Third, equivalence presupposes the same context, such as the same question order.

Unfortunately, the equivalence problem is less developed than other topics discussed in this book. There are no theoretical statements of the equivalence problems in surveys, and few studies have dealt explicitly with equivalence issues.

Equivalence Problems

This section examines equivalence in three types of survey settings: comparisons across results from different survey organizations, comparisons across distinct populations, and comparisons over time.

House Effects

There used to be only a handful of preelection surveys, but nowadays hundreds of preelection surveys are conducted for U.S. presidential elections. Typically they give similar results, but sometimes there are differences, at which point newspapers publish articles on why the results differ (see box on page 302). Sometimes these differences can be explained easily, such as differences in sampling frames between all adults and likely voters, but other times the articles refer more vaguely to differences between the ways different organizations conduct their surveys.

The term *house effects* points out that surveys on the same topic conducted by different survey organizations can give different results. Of course, sampling error implies that identical results should never be expected, but differences are sometimes well beyond what can be explained by sampling error. Part of the reason is that different researchers word questions on the same topic somewhat differently, but the concept of house effects is that the results differ more than would be explained by wording differences. As such, it can be seen as a residual effect that remains after one has controlled for differences in types of questions asked. House effects are included here under the rubric of survey administration because the differences are often due to the directions, training, and supervision given to interviewers, which are invisible outside of the survey organization. Usually, the term is used generally to refer to systematic differences between results obtained by different survey organizations that cannot be explained by differences in the timing of the surveys, question wording, or question order.

The term *house effects* is sometimes used in a stronger sense, to imply that particular survey organizations give biased results. This would be very serious if some pollsters had partisan biases. In the 1960s, for example, the Harris Poll, founded by Louis Harris, who had been John F. Kennedy's pollster in the 1960 campaign, was thought to favor liberal Democratic positions, whereas the Gallup Poll was thought to favor establishment Republican positions.[1]

There is little evidence, however, to prove such bias in poll results. For example, studies that have analyzed the effects of differences in wording of the party identification question across survey organizations find that the differences can be explained in terms of such routine factors as question wording. Borrelli, Lockerbie, and Niemi (1987) explain most of the differences in the Democratic-Republican partisanship gap as measured by different polling organizations between 1980 and 1984 in terms of their sampling frame (if they only sampled voters), their question wording (whether they ask about the person's partisanship "today" or "generally"), and their timing (how close to the election they were taken). Their multiple regression equation accounted for 78% of the variance, providing little room for house effects.[2] Kiewiet and Rivers's (1985, 73–75) analysis of the 1984 presidential election found some significant house effects. However, Gelman and King's (1993, 424) study of 1988 rejected the hypothesis that different polling organizations obtained systematically different results, except that there were significant differences between organizations in the proportion of the electorate that was undecided. In an analysis of trial heat polls prior to the 1992 presidential election, Lau (1994) finds that the length of time the survey was in the field, whether it polled only on weekdays, and whether it looked only at likely voters affected error in support for the candidates, whereas house differences per se mattered little in his multivariate equations. Many Web sites tracked election trial heats in the 2004 presidential election, and some analysts claimed that the differences were statistically significant. The poll with the most pro-Bush results (Fox News) and the one with the most pro-Kerry results (Zogby) are both associated with conservative pollsters, with Zogby's results due to his weighting on the basis of partisanship to compensate for fewer Democrats in 2004 polls than in recent presidential elections. Thus, even when systematic house differences are found, they could be due to differences in survey administration rather than to intentional partisan bias.

House effects can exist, though, even if they do not involve biases. Two early studies by Tom Smith found limited house effects, mainly having to do with differences in "don't knows." His first study (Smith 1978) examined 33 instances in the 1970s when the General Social Survey and other survey organizations asked identical questions in about the same time frame. Significant differences in proportions were found on 10 of these questions. Part of this was due to Roper's obtaining more "don't know" responses than the GSS. Smith ascribed one difference between GSS and NES results to a difference in the order of response alternatives, since people tend to answer the first named alternative in face-to-face interviews, whereas he ascribed another difference to a historical effect of a real-world event that occurred between the two surveys. In a later study, Smith (1982) compared 1980 GSS and NES surveys that purposely included some identical questions in order to look for house effects. Significant differences in marginal distributions were found on 16 of 17 items, partly due to more "don't know" responses in the NES because it used "no opinion" filters on previous questions. His analysis suggests that additional reasons for distribution differences include real change between the periods of interviewing and context effects due to differences in preceding questions. Smith concludes that the results are similar enough to show that the surveys are producing reliable and reproducible results, though differences in "don't knows" should always be examined when surveys obtained by different survey organizations are compared. Thus, differences in encouraging or discouraging "don't know" responses could be among the most important house effects.

There can also be house effects in reports of preelection polls. Some organizations report results only for people who usually vote, while others report results for the whole sample. Election pollsters differ in whether they treat soft supporters of a candidate as undecided, how long they are in the field, and whether they interview on weekends. Some report weighted results while others do not, and the bases of weighting differ between survey organizations. In particular, a few weight on the basis of party identification, which requires assuming prior knowledge of its distribution. Preelection polls often diverge considerably from one another, partly because of these differences in details, and the polling organizations generally do not disclose these details, so these can essentially be considered house effects.[3]

Some studies that combine surveys taken from different houses explicitly control for house effects. For example, Erikson and Wlezien's (1999) study of polls taken by 31 different polling organizations during the 1996 presidential election adjusted for house differences by regressing the results of all of the polls on 30 dummy variables representing the organizations plus dummy variables for each of the survey dates, and then used the regression coefficients for the survey dates as the house-adjusted poll results for their dependent variable.

The usual focus of house effects has to do with different average tendencies in results, but another aspect is that some survey organizations obtain more variable results than others. This problem has been acknowledged in the literature, but no solutions have been advanced. Another focus can be on how similar correlations of variables are between different survey organizations. Smith (1982) found that seven-point scales in the

Poll Position All Depends upon Who's Asking Whom and When

By David Jackson
The Dallas Morning News
September 18, 2004

WASHINGTON—(KRT)—President Bush has a double-digit lead over John Kerry.
Unless it's a dead heat.

It all depends on which poll you believe.

Surveys released this week range from a 13-point Bush edge to a 1-point Kerry comeback, reflecting the fact that polls can be as volatile as the political races they try to reflect.

"When there's a lot of information in the environment—a lot of controversial information . . . we often find polls that end up with quite conflicting results," said David W. Moore, senior editor with Gallup.

Republicans love the new Gallup poll, which gives Bush a lead of 55 percent to 42 percent among likely voters. The lead shrinks to eight among registered voters, 52–44 percent. The poll has an error margin of plus or minus 4 percentage points.

Democrats prefer a Harris poll giving Kerry a 1-point lead, 48–47.

And sometimes the variances are within one survey. One wave of questions in the Pew Research Center poll, posed Sept. 11–14, showed a 46–46 tie among registered voters. But questions taken the three previous days gave Bush a 12-point lead, 52–40. Both surveys had error margins of 3 percent.

Some analysts attributed the differences to a fast-paced race: Violence in Iraq, uncertainty over the economy, Republican claims of Kerry flip-flopping, Kerry accusations that Bush is misleading the nation, questions about Bush's National Guard service, and disputes over whether documents used to report on that service were real.

The ways polls are conducted are another factor, analysts said.

There are differing definitions of likely voters, and the demographic and political makeup of the sampled voters. Some polls are believed to lean Republican, others Democratic.

The rise of cell phones has made it harder for pollsters to track down respondents, analysts said, and a rising number of people don't want to participate at all.

"There are a lot of things that could explain these discrepancies," said Karlyn Bowman, who analyzes polls for the Washington-based American Enterprise Institute.

Analysts and campaign aides said the best thing to do is average several polls at the same time. Kerry aides said Bush may enjoy a lead of 2 points or so nationally, but the convention bounce that had given him a commanding lead has dissipated.

"We are looking at a race that is very tight, both nationally and in the battleground states," said Kerry strategist Joe Lockhart.

Said Bush campaign spokesman Reed Dickens: "At this point, we would much rather have our candidate's message, record, and poll position than theirs."

GSS were intercorrelated more than in the NES, but the differences between correlations of three-point spending scales were more random. The correlations between demographics and attitude scales were similar in the two studies, with differences not being statistically significant. He concluded that the GSS and NES results were essentially similar, but it is not clear how different the correlations in the studies would have to be for one to decide that two studies are not truly comparable.

Cross-Cultural Differences

Comparability issues directly affect cross-cultural survey research. This can be seen as an extension of house effects, since surveys taken in different countries are inevitably conducted by different survey organizations. Even when these different organizations are under the umbrella of the same international company, such as Gallup International, there are invariably differences in how different branches conduct surveys. Sampling may be performed differently, and interviewers may be trained differently. Phone surveys may work in some countries, whereas they cannot be used in other countries because people are not used to divulging information to strangers on the telephone. The extent of nonresponse can vary between the countries, and it is particularly difficult to achieve comparable wording across countries. Thus, while cross-national survey research is an exciting field, special caution is appropriate when results from surveys taken in different countries are compared.

The initial approach to cross-cultural research was simply taking a survey conducted in one country and moving it to another country, as when election studies became popular and spread from the United States to some European nations. A more interesting—and challenging—approach is developing new surveys to be conducted in parallel form in several nations, such as the increasingly popular Eurobarometer survey, the new European Social Survey, and the collaborative International Social Survey Program. It should also be noted that the problems involved in cross-cultural surveys also apply to conducting a single survey in a culturally diverse nation, such as a country with two dominant languages. Indeed, a larger question is the extent to which survey results apply to all subcultures in the United States; some results may actually be culture specific. Surveys of the United States usually omit non-English-speaking respondents, except for an increasing use of Spanish-language versions in California—though the questions may not mean the same thing to this subpopulation.

A large problem for cross-cultural surveys is developing the questionnaire. The first step is deciding to what extent the same questions should be used for each nation (known as *etic* mode) versus using country-specific questions (*emic* mode; Przeworski and Teune 1970). The most common choice is "ask the same question" (ASQ), with translation when need be (Harkness 2003, 35). This requires translators, followed by reviewers to check the translations and adjudicators to make decisions when reviewers disagree with the original translators. One translation process is known as *back translation*—questions are translated from language A to language B, and then the language B version is translated back to language A to check whether that yields the intended question. Back translation is often not sufficient, however, particularly because it treats idioms too literally, so the intended meaning of the question is lost in translation. A better process is *decentering*—using iterative translation to develop questions in each language simultaneously, with paraphrasing and translating back and forth (Werner and Campbell 1970). Decentering is commonly used in developing new surveys that are to be used simultaneously in several nations. Another translation problem is that different languages have different grammatical characteristics. For example, most nouns in English are gender neutral, whereas nouns in many other languages have a gender. Direct translation of questions from English to those other languages would have to make choices as to how to treat gender-specific nouns.

The difficulty in developing equivalent measures across cultural groups actually has several dimensions (van de Vijver 2003). One problem is whether the construct is the same across groups. Terms can have different meanings in different countries. A common example is the phrase *social security*, which refers in the United States to a specific pension program for the elderly but refers in Europe to social welfare programs very generally. A second problem is whether the construct's structure is the same. For example, personality dimensions may differ across cultures. Third, response scales may not have universal meanings. For example, the feeling thermometer question was developed in the United States to ask people how warm or cold they feel to particular social groups on a 0 to 100 degree thermometer scale. However, it is an empirical question whether that thermometer scale means the same thing to respondents in countries using the Celsius thermometer scale as it does in the United States, where the Fahrenheit scale is used. One solution would be to measure and standardize the strength of labels that are used (Smith 1997). Fourth, scalar scores may not mean the same thing across cultures, as in the debate over whether IQ scores are culturally dependent.

A further issue in cross-cultural research is that results based on the United States and other Western societies may not generalize. For example,

Schwarz (2003) argues that East Asian cultures foster interdependent perspectives, so that respondents there would be more knowledgeable about how people fit in and about others' behavior as well as one's own and would be more sensitive to the conversational context. There is a Western bias to survey research, and simply applying the Western mode of surveys to non-Western societies may not produce meaningful results. Treating surveys taken in poor countries or countries with repressive governments as equivalent to surveys in Western nations would be unrealistic.

The problems involved in cross-cultural research are reviewed in an edited volume by Harkness et al. (2003), as well as a working book on demographic variables (Hoffmeyer-Zlotnik and Wolf 2003).

Over-Time Differences

A third comparability issue involves comparisons of survey results over time, for both short and long periods of time. Tracking polls during election campaigns exemplify the problems involved in comparisons over short periods of time. There is a natural tendency to interpret even small changes as meaningful, though the differences are often within the range of sampling error. If polls taken after a presidential speech on a topic differ from those taken before the speech, the media inevitably interpret the change as due to the speech, without recognizing how fleeting that change can be.

Compounding this problem, the media often compare results at different times in surveys conducted by different sampling organizations, interacting house effects and over-time differences. A careful analysis often shows that the polls taken at different times by different organizations asked somewhat different questions, so the apparent change is partially due to the question wording differences. Rarely is it possible to parse out the extent to which the changes are due to coverage differences, sampling differences, differences in types of interviewers hired, differences in training of interviewers, number of callbacks, and so on. The best rule is to believe that real change has occurred only when polls taken by different organizations all move in the same direction.

There are further problems that can occur when one is looking for trends over lengthy periods of time. Even when the question wording is kept constant, the meaning of terms in the question can change over time. For example, surveys show that the American public was more politically conservative in the 1990s than in the 1960s. Before we accept that result completely, however, it is important to consider whether the terms *liberal* and *conservative* meant the same thing in those two decades. As American politics changed, these terms had fewer economic connotations and more social issue meanings, so simple comparisons of poll results across the decades can be misleading. Furthermore, the procedures for taking surveys have changed over time, and this makes it awkward to compare surveys taken decades apart.

Bishop (2005, chap. 5) makes a similar point regarding how dramatic events can change the meanings of standard survey questions. For example, he argues that the terrorist attacks of September 11, 2001, changed the meaning of the presidential approval question, so that the issue became whether people approved of his handling of the immediate crisis. The event transformed a normally vague question into a very specific one.

Another problem is that changes in behavioral items over time can be due to changes in reporting rather than changes in actual behavior. For example, increases in rape found in crime surveys over the years may be due to increased willingness of victims to report their victimization rather than to an actual increase in rapes.

Differences in survey results over time can also be due to changes in the composition of the population. In the United States, for example, the aging of the post–World War II baby boom, combined with higher life expectancy, will lead to greater frequencies of attitudes and behaviors associated with older people. There are actually three interrelated possible effects related to age: (1) cohort effects that occur when people born in one era have different attitudes from the attitudes of those born in another era, (2) aging effects that are uniform across all age cohorts (such as if all age groups become more conservative at the same rate), and (3) period effects as cultural changes affect all age cohorts simultaneously. These effects are interrelated, such that any change that can be explained by one could also be explained by the other two together, though the usual approach is to apply a parsimony rule and use the simplest explanation. Harding and Jencks (2003) provide a nice example with their careful analysis of liberalized attitudes toward premarital sex from the 1960s through the 1990s.

As longitudinal studies become more common, there are more reports on over-time stability and change in attitudes and behavior. However, there are still few systematic studies of over-time differences in surveys that are sensitive to the potential problem of differences' being artifactual. Stimson (1999, appendix) has developed a useful computer routine for testing whether survey questions on the same topic taken at irregular points in time measure the same underlying concept and then fitting a smoothed curve to the data to emphasize the main trends rather than the effects of sampling error (see also Wcalc at http://www.unc.edu/~jstimson/resource.html).

Disclosure Rules

The possibilities of house effects, cross-cultural differences, and over-time differences make it important that reports on surveys disclose the details of their procedures so that readers can see exactly how the survey was conducted. House effects can be analyzed systematically only when exact procedures are disclosed. Recognizing cross-cultural differences is possible only when the questionnaires are made available in every language that was used in the survey. Over-time comparisons are valid only when it is possible to tell how stable the survey administration was over the time period.

The AAPOR Code of Professional Ethics and Practices includes standards for minimal disclosure of how a survey is conducted (see box). The information to be made public includes the survey's sponsor, the organization that conducted the survey, the exact wording of questions, the population being studied, the sampling frame, the sampling procedure, the number of cases, the completion rate, screening procedures, sampling error estimates, weighting information, and the location and dates of interviewing.

In a rare instance, AAPOR made a public finding that a major pollster,

---

AAPOR Code of Professional Ethics and Practices

III. Standard for Minimal Disclosure
Good professional practice imposes the obligation upon all public opinion researchers to include, in any report of research results, or to make available when that report is released, certain essential information about how the research was conducted. At a minimum, the following items should be disclosed:
1. Who sponsored the survey, and who conducted it.
2. The exact wording of questions asked, including the text of any preceding instruction or explanation to the interviewer or respondents that might reasonably be expected to affect the response.
3. A definition of the population under study, and a description of the sampling frame used to identify this population.
4. A description of the sample selection procedure, giving a clear indication of the method by which the respondents were selected by the researcher, or whether the respondents were entirely self-selected.
5. Size of samples and, if applicable, completion rates and information on eligibility criteria and screening procedures.
6. A discussion of the precision of the findings, including, if appropriate, estimates of sampling error, and a description of any weighting or estimating procedures used.
7. Which results are based on parts of the sample, rather than on the total sample.
8. Method, location, and dates of data collection.

Source: American Association for Public Opinion Research, http://www.aapor.org

Frank Luntz of Luntz Research Corporation, violated its Code of Professional Ethics and Practices. Luntz had refused to make public his methods in researching public attitudes about the Republican House candidates' 1994 "Contract with America"; he claimed at least 60% of the public supported each plank of that legislative agenda (Traugott and Powers 2000).

Different disclosure rules may be appropriate for different survey situations. For example, the National Council of Public Polls (NCPP) was formed, at least in part, because media organizations believed that their audience was not interested in the full set of information required by AAPOR disclosure rules (Miller, Merkle, and Wang 1991). The NCCP disclosure rules are more limited, though they still emphasize the need to disclose in press releases such information as the poll's sponsorships, dates of interviewing, interviewing mode, population sampled, size of sample, wording of reported questions, and percentages used as a basis for conclusions (see http://www.ncpp.org).

Comparability Effects and Survey Constraints

There is little possibility of minimizing house effects that arise when one is comparing surveys that were not planned to be comparable. Comparisons of such polls are inherently questionable.

Conducting cross-cultural research raises more concerns. Using questions that work in one culture without checking their cross-national validity would raise ethical issues of research imperialism, as occurred in the early days of survey research when questions that were successful in the United States were exported to other nations without first ascertaining their validity there. Nowadays it is more common to have cross-national research teams for such surveys, providing a check on whether the questions are appropriate in each nation. Yet developing separate questionnaires with equivalent questions for different nations is costly and takes considerable time.

Longitudinal comparisons raise more concerns than house effects but less than cross-cultural research. Designing a continuing survey requires continual attention to whether questions maintain their meaning over time. It is too easy for researchers simply to continue using the same questions without noticing that circumstances and/or changes in language use have changed the meaning of key concepts. Analysts comparing surveys taken years apart have a special responsibility to consider how comparable the survey questions really are.

PART V

Total Survey Error

14

Ethics in Surveys
RESPECTING RESPONDENTS' RIGHTS

Treatments of the survey process do not always include discussions of survey ethics, but ethics is an important topic that should be considered in all types of research. Many ethics rules for research with human subjects have been developed over the past half-century, and these rules directly affect surveys. These are not sources of survey error in the same sense as the topics of chapters 4 through 11, but they constitute restrictions on surveys that should be kept in mind. After reviewing the basic ethical standards, this chapter discusses professional codes of conduct and governmental regulations on human subjects research. (Related topics include interviewer falsification in chapter 4 and disclosure rules in chapter 13).

Ethical Standards

Ethical rules for survey research are relatively new developments. They are based mainly on a few simple principles: the importance of obtaining informed consent from participants, the need to ensure confidentiality of responses, and taking account of the power relationships in surveys.[1]

Informed Consent

The most basic ethics rule for human subjects research is that the participants must give informed consent to participate in the research. This rule encompasses several aspects: that the participation be voluntary and that they should be informed of possible risks.

The first protections for human subjects were drafted in the aftermath of World War II. Nazi doctors in German concentration camps conducted inhumane research on detainees. As part of the war crime trials, the victorious Allies wrote the Nuremberg Code, which mainly emphasized the need for participation in research to be voluntary. This requirement leads to restricting research on prisoners and others who cannot give meaningful consent. Students in classes are also captive audiences, so this rule implies that they should not be required to participate in research, except to the extent that research is part of the actual instruction. Consequently, when students are offered extra credit for participation in studies, the instructor is often required to make an alternative extra-credit assignment available to students who do not want to participate in the research projects.

As a result of several controversial research studies that included deception of subjects and especially as a result of concern about the safety of biomedical research, a blue-ribbon commission issued the Belmont Report (National Commission for the Protection of Human Subjects of Biomedical and Behavioral Research 1979). This report proposed three principles for human subjects research: beneficence (minimizing possible harm), justice (that the participants benefit from the research), and respect for persons in the form of obtaining their knowing consent to participate.

If consent to participate is to be meaningful, research participants must first be told about possible risks of the research. One potential risk is that asking or answering a question could cause some respondents mental distress. Stress is most likely to arise in surveys in in-depth interviews in which respondents are asked to remember traumatic episodes, such as in interviews of military veterans about a war or of New Yorkers about the attacks of September 11, 2001. The usual response is that no one has ever suffered lasting psychological harm from an interview. The risks involved in participating in a survey are no greater than these people would encounter in everyday life. Asking questions can cause momentary upset, but most people find it therapeutic to talk about their problems. If people are embarrassed by an answer, they can refuse to answer or can lie, and interviewers are trained to accept answers as legitimate even if they are suspicious of them. Indeed, the benefits of the survey usually outweigh the risks. Respondents generally report they enjoy surveys, and the potential benefits to society from surveys on sensitive topics (such as HIV-related behavior) are considerable. Some research suggests that people express less embarrassment about sensitive questions in an interview if they are told first that it will include some questions about the topics, such as alcohol and sex (Singer 1978), but, in fact, there is not enough research about this to be confident of the result.

When there is a chance of risk to respondents, AAPOR recommends telling people that the questions can cause discomfort and that they can skip questions that they do not want to answer, with a debriefing at the end of the interview to see if any matters were upsetting and to see if respondents have further questions.

The potential for risk is greatest when respondents admit to illegal or stigmatizing behavior that could later be made public. As described in the next section, it is vital in this situation that identifier information, such as the person's name, address, and/or phone number, be separated from the data after the interview has been verified and that the identifier information be stored securely. Also, the researcher should make certain that respondents cannot be identified through analysis of the data; thus sometimes identities are concealed by altering identifying demographic information.

The usual rule is that people must give written consent that they agree to participate in research studies. Written consent is, of course, impractical in telephone surveys, though respondents must still be given some basic information and at least implicitly consent to the interview. Some interviewing operations tell the respondent something like "This interview is voluntary and confidential; if I ask you a question that you do not want to answer, just let me know, and we will go on to the next question." Respondents can consent in Internet surveys by clicking an "accept" button after being informed of risks and benefits, or implicitly by choosing to participate in the survey. It is also often necessary to make sure that the person is not a minor, as by asking, "Are you at least eighteen years of age?" unless, of course, the survey is an approved study of minors. Survey researchers generally fear that stressing the voluntary nature of participation will have the effect of decreasing response rates, but it is still important to make sure that participation is meaningfully voluntary.

Confidentiality

Another basic ethical rule for survey research is that the confidentiality of the respondent should be protected. Interviewers generally have to promise confidentiality in order to get the interview, so maintaining confidentiality is partially a matter of keeping this promise. Confidentiality is particularly vital when the survey obtains sensitive information.[2] Researchers are sometimes concerned that confidentiality assurances would decrease response rates. However, a meta-analysis (Singer,

Von Thurn, and Miller 1995) has found that response rates are higher on studies involving sensitive topics when strong assurances of confidentiality are given.

Some surveys are totally anonymous, without any material that identifies the respondent. For example, there would be no way to know who filled out which questionnaire when a large lecture class fills out self-administered questionnaires. However, potential identifying material is usually obtained in interviewer-assisted surveys, such as the respondents' addresses in face-to-face surveys, their phone numbers in telephone surveys, and their e-mail addresses in e-mail surveys, in which case it is important to maintain confidentiality.

The best procedure to achieve confidentiality is to have as few identifiers as possible on interviews and in computer files. When possible, the researcher should keep the identifiers in a separate limited-access file, such as by giving each interview a unique number with the decoding scheme existing only in that separate file. If the respondent's name was obtained, it should be deleted as soon as possible. Similarly, addresses and phone numbers should be deleted, or maintained with restricted access if reinterviews are planned. Similarly, it is best to keep geographic identifiers broad enough that the exact location of the interview cannot be determined. Questionnaires with identifying material on them should be kept in a locked location and should be destroyed when they are no longer needed.

The confidentiality problem extends to the release of public use data. Sometimes enough information is obtained about respondents to identify them. For example, if a survey of teachers includes the state and city in which each school is located, the age, gender, and race of its principal, and the number of students per grade, a teacher given access to the data might be able to recognize the school. In this situation, it is incumbent on the researcher to modify enough of the nonessential identifying material to preclude identification, such as by altering the age of the principal. The identification problem also occurs in business surveys. Businesses are often asked to provide information that they would want to keep private for competitive reasons, so it is important that data are not reported in such a way as to breach promised confidentiality, even in tabulations of results.

Disclosure analysis involves looking for "indirect identifiers" in a data set that could pose a risk for later reidentification of respondents, so at least the respondent would have deniability if identified. This would be particularly important in dealing with subgroups, as, for example, if a survey file in the

U.S. happened to include one Buddhist doctor in a small town who has an income of more than $300,000. Researchers often "blur" such data so that reidentification cannot be achieved with certainty. Several methods have been developed to minimize the chances of identifying respondents. Biemer and Lyberg (2003, 105) list five methods for ensuring confidentiality: anonymization by removing identifiers, suppression by deleting a variable value that permits identification (especially when so few cases have a particular value that identification of cases is possible), collapsing categories ("coarsening the data") along with "top coding" so all high values are given the same value, adding noise (a random value, sometimes known as *perturbation*) to the value of a variable, and encryption. Additionally, some organizations swap some data values from one case to another, so that the frequencies on variables are unchanged. These studies should publish the percentages of items and cases involved in such data swapping, so it is known how much uncertainty this adds to subsequent data analysis. The Census Bureau handles the disclosure problem by releasing only samples from its data collections. It is common in some establishment surveys to delete data values in published tables if it is clear that the results in the cell in question must have come from only one or two firms.

Special procedures should be taken to protect the confidentiality of surveys on topics that potentially raise legal issues. Prosecutors can seek access to survey data when they believe that it is relevant to a criminal case, and courts are likely to order that access be given. Other participants in legal proceedings sometimes seek access to survey records. For example, the lawyer for an Ohio State University assistant professor who was denied tenure requested a university survey on sexual climate. Our survey operation (Polimetrics, at the time) realized that the dean of her college was instantly identifiable, given that there were codes in the data for each respondent's college and administrative position, so we declined to provide the data in a usable form. Fortunately, the assistant professor's lawyer accepted our response rather than testing the applicability of the state's open record law in court.

An unusual confidentiality issue arises when one is using information on the Internet. For example, would it be appropriate to analyze people's statements about public policy issues that they have posted on a message board or in a chat room? Search engines often can locate such postings, making it possible to sample from among them. The people posting such messages did not consent to participate in the research, yet they posted their views in a public site. One standard is to get the approval of the site administrator before using such data, but there is no universally accepted standard. A useful discussion of such problems appears in a special issue of *Ethics and Information Technology* (2002, vol. 4, issue 3).

For elite interviewing, the distinctions are between answering "on the record," "not for attribution," "on background," and "off the record" (Goldstein 2002). "Not for attribution" means that answers and quotes can be used, so long as the respondent is not identified as the source. "On background" means that answers can be used only to inform the investigator's further research, while "off the record" means that the answers cannot be used in any way.

Additionally, it is common for survey organizations to require all of their employees to sign confidentiality agreements, promising that they will not tell people information they obtain as part of their employment. This is particularly important in small communities where interviewers are likely to recognize the names of people who are in their sample; it would be inappropriate for the interviewers to gossip about how those respondents answered the questions.

Power Relationships

Another ethical issue involves the proper power relationships in surveys. George Gallup (American Institute of Public Opinion 1939, 6) originally envisioned surveys as empowering people, providing a means by which leaders would know "the true state of public opinion" and could take that into account in their actions. Nowadays we are more realistic both about the possibilities of determining the true state of public opinion and about whether leaders would always follow that opinion, but the notion of surveys as empowering remains potent.

However, some (e.g., Mishler 1986) would argue that the standardized interview puts respondents into a dependent situation, essentially stripping them of power. Forcing answers into narrow categories prevents respondents from giving their own stories in their own voices (see Gubrium and Holstein 2002, chap. 1). This is not simply a matter of the use of closed-versus open-ended questions, since coding open-ended material into categories for systematic analysis would provoke similar criticisms of depersonalizing the respondents (see Briggs 1986). One solution would be unstructured in-depth interviewing combined with qualitative analysis, instead of large-scale standardized surveys.

Feminist perspectives raise further questions as to both what the answers of women respondents mean (are they speaking as women?) and the control relationships when the interviewer and respondent are of opposite genders. Comparable issues arise in interviewing members of minorities, even when the interviewer is part of that minority group. This is a difficult topic to resolve, but minimally it suggests that considerable sensitivity is in order when writing questions, so as to think through how those questions would affect different types of respondents.

It has also been argued that surveys involve control by the researcher of the interviewer, carefully limiting how the interviewer can behave when taking the interview. Another control issue occurs in cross-cultural research when researchers from one country determine the questions to be used in other countries. In recent years it has become more common to have cross-national research teams jointly prepare the questions, which lessens the problem of cultural imperialism in research.

An additional argument is that polling has transformed the nature of public opinion. There have always been many ways of expressing public opinion, such as petitioning government, discussion in salons and coffee-houses, and writing letters to newspapers. These forms often involved deliberation to achieve consensus as to the common good (Herbst 1993). By contrast, modern public opinion polling focuses on individual views, with less opportunity for sharing of information in ways that have people consider both sides of an argument. The deliberative poll is one attempt to shift the balance back (see chapter 3), but its applicability is quite limited. Giving voice to the views of common people through polling may em-power them, but perhaps in a manner that changes the nature of demo-cratic dialogue for the worse.

Ethical issues, by their very nature, are inherently difficult to resolve, but the power relationships are particularly unsolvable. Some would see these issues as illustrating how social research extends social inequalities, whereas others would still view surveys as giving the public a chance to voice its opinions and affect public actions.

Professional Conduct Codes

When ethical standards became commonplace in scientific research, pro-fessional organizations developed codes of ethical conduct. Like other professions, survey research organizations have instituted their own codes of professional conduct. The most important of these is the American Association for Public Opinion Research code, which is reprinted in the appendix to this chapter. It speaks of the importance of confidentiality and the need to avoid harming respondents, but it also addresses other con-cerns including the importance of appropriate methods, analysis, and interpretations.

Additionally, AAPOR has condemned particular polling practices, most notably push polling. Push polls occur when election campaigns claim to be doing a poll to see how people would react if they were told damaging information about the candidate's opponent. Push polls are really just smear campaigns disguised as surveys, and they serve only to increase public distrust of surveys in general.

AAPOR also has been critical of phone-in and click-in polls. It issued a press release during the 2000 campaign, disapproving of television network claims that their click-in polls and focus groups provided accurate representations of public opinion on the presidential debates. It also condemned the misuse of respondent information when an organization that had claimed to conduct a survey gave respondents' identity and answers to the Wisconsin Republican Party for its 1988 get-out-the-vote drive.

Two other survey organizations have professional conduct codes. The Council of American Survey Research Organizations (CASRO), which focuses on commercial survey companies (see http://www.casro.org/), has provided guidelines for many aspects of the survey operation. Its most innovative standards, shown in the box on the opposite page, are for Internet research, applying the principle of respondent privacy to this new survey mode. The National Council of Public Polls (NCPP) guidelines also suggest that Internet-based surveys be designed to be representative, provide evidence of being demographically representative, not be click-in polls, and prevent people from participating multiple times.

Governmental Regulations

Dissatisfaction with protections for human subjects in some scientific fields has resulted in the federal government's requiring that all organizations that receive federal funds—which includes virtually all U.S. universities—have institutional review boards (IRBs) to review all proposals for human subjects research. The legal basis for this is the National Research Act of 1974, which led to the writing of the Federal Regulations for the Protection of Human Subjects of Research. Researchers must submit detailed descriptions of their proposed research procedures to their local IRB, which must grant approval *before* the research is conducted. Universities will not accept dissertations on human subjects that do not have IRB approval, and human subjects research that is carried out without IRB preapproval could be found (by the IRB) to constitute "scientific misconduct."

CASRO Standards for Internet Research

The unique characteristics of Internet research require specific notice that the principle of respondent privacy applies to this new technology and data collection methodology. The general principle of this section of the Code is that survey research organizations will not use unsolicited emails to recruit respondents for surveys.

1. Research organizations are required to verify that individuals contacted for research by email have a reasonable expectation that they will receive email contact for research. Such agreement can be assumed when ALL of the following conditions exist:

    a. A substantive pre-existing relationship exists between the individuals contacted and the research organization, the client or the list owners contracting the research (the latter being so identified);

    b. Individuals have a reasonable expectation, based on the pre-existing relationship, that they may be contacted for research;

    c. Individuals are offered the choice to be removed from future email contact in each invitation; and,

    d. The invitation list excludes all individuals who have previously taken the appropriate and timely steps to request the list owner to remove them.

2. Research organizations are prohibited from using any subterfuge in obtaining email addresses of potential respondents, such as collecting email addresses from public domains, using technologies or techniques to collect email addresses without individuals' awareness, and collecting email addresses under the guise of some other activity.

3. Research organizations are prohibited from using false or misleading return email addresses when recruiting respondents over the Internet.

4. When receiving email lists from clients or list owners, research organizations are required to have the client or list provider verify that individuals listed have a reasonable expectation that they will receive email contact, as defined in (1) above.

Source: Council of American Survey Research Organizations, http://www.casro.org

Many IRBs are more familiar with the risks involved in medical research than in survey research, and unfortunately they sometimes try to apply the same rules to surveys as they would to medical research where there is potential risk of death. As a result, the AAPOR Web site, http://www.aapor.org, gives several practical suggestions for how to respond to some concerns that IRBs often raise about surveys.

Certain categories of research are considered exempt, though exemption requires a formal written exemption. Generally the chair (or a designee) of the IRB must approve the exemption request, so that there is some written evidence that an exemption has been granted. Categories that are normally exempt include (1) research on education, (2) many surveys, (3) research on politicians, (4) use of existing data, (5) evaluation research, and (6) food research. However, exemption does not apply to research on children and other special populations (such as prisoners) nor to research that involves deception and/or stress to participants. Also, exemption does not apply if disclosure of participants' responses could place them at risk in legal terms, financially, or in terms of their reputation.

Much survey research is able to pass through this web of regulation fairly easily, so long as the research is on adults, does not involve deceiving respondents, and would not create stress in the respondents. Deception is usually not necessary in surveys. If deception is employed, an IRB will require that respondents be told about the deception in a debriefing at the end of the interview. The IRB will also insist that the confidentiality of respondents be guaranteed, unless they consent to have their identity revealed.

IRBs are concerned with guaranteeing informed consent. They are used to requiring written consent forms for medical research studies. Such forms are rarely used in survey research, however, since people can simply refuse to participate in surveys and can skip particular questions they do not want to answer. Federal regulations (CFR 46.117c) indicate that IRBs can waive requirements for signed consent forms when the research poses only minimal risk of harm to subjects and involves no procedures for which written consent is normally required outside of the research context, but local IRBs are responsible for interpreting these regulations. An IRB is most likely to require written consent forms when sensitive information is being obtained, when there are limits to the confidentiality being guaranteed to respondents, when there are reasons to believe that participation in the survey is not entirely voluntary, and/or when the researcher plans to combine the survey answers with other information about the respondent, such as health records. The federal regulations do permit waiver of signed consent if the proposed research cannot be practicably carried out without the waiver, as in random-digit-dialing telephone samples in which respondents' addresses are not known in advance to the researcher—though some survey researchers have had to argue with local IRBs that wanted to require written consent forms for telephone surveys.

Generally IRBs also require that certain types of information be disclosed to the respondent at the beginning of the interview. This includes the name of the company doing the interviewing, the sponsor, and a very brief description of the survey (e.g., "We are talking to people tonight about their feelings on current affairs"). The respondent is also to be told whether or not the answers are confidential, be assured that participation in the survey is voluntary, and be reminded that questions can be skipped if he or she desires.

Researchers in many fields feel that the IRB process has become overly onerous (Shea 2000). Local IRBs have wide discretion, and they vary widely in their interpretation of the rules. It is common for an IRB to require changes in research plans before granting approval and to require any later changes in procedures to be cleared with it. Because the IRB wants assurances that subjects will not be unduly stressed in the study, it may insist on seeing the questions to be asked of respondents, and it may expect to see any changes made to those questions. The IRB process can take several weeks, partly because IRBs often meet once or twice a month and partly because of the time involved to get approval after required changes are made. This can lead to an unfortunate incentive for researchers to not be fully candid with their IRBs, which leads to a spiral effect with IRBs becoming more suspicious of the information that researchers give them.

In addition to federal government restrictions that have led to IRBs, there are some federal laws and regulations regarding commercial phone calls to residences. These laws are usually intended to limit telemarketing, but some are written so broadly that they could affect surveys. In particular, the 1991 Telephone Consumer Protection Act and the 1995 Telemarketing and Consumer Fraud and Abuse Prevention Act are relevant federal regulations, though phone calls for research purposes are exempt from regulation. Also, the federal do-not-call list specifically exempts legitimate survey research. To complicate matters further, many state governments have enacted laws that affect telemarketing directly and may affect surveys indirectly, with the specifics naturally varying between states.

Ethics and Survey Constraints

Maintaining ethics in research is an important responsibility but is often considered an inconvenience by researchers. Going through an institutional review board can take time, which is often in short supply when one is facing a deadline to get a survey in the field. Ethical constraints can be seen as an additional constraint on surveys, forcing investigators to think through what information they really require.

Another ethical consideration involved in face-to-face interviewing is protecting the interviewer. Interviewers should not be sent to high-crime areas and should not be expected to go to potentially dangerous areas, especially at night. Interviewers who are afraid of going to an area because of safety issues should not be forced to do so. This can lower the response rate and increase the time required to complete the interviewing, but it is a necessary constraint.

When designing surveys, researchers always consider their budgetary constraints and how much time they have to obtain results. Ethical considerations should be routinely added to this list of items to consider in designing studies.

Appendix: AAPOR Code of Professional Ethics

Code of Professional Ethics and Practices

We, the members of the American Association for Public Opinion Research, subscribe to the principles expressed in the following code. Our goals are to support sound and ethical practice in the conduct of public opinion research and in the use of such research for policy and decision-making in the public and private sectors, as well as to improve public understanding of opinion research methods and the proper use of opinion research results.

We pledge ourselves to maintain high standards of scientific competence and integrity in conducting, analyzing, and reporting our work in our relations with survey respondents, with our clients, with those who eventually use the research for decision-making purposes, and with the general public. We further pledge ourselves to reject all tasks or assignments that would require activities inconsistent with the principles of this code.

The Code

I. Principles of Professional Practice in the Conduct of Our Work

A. We shall exercise due care in developing research designs and survey instruments, and in collecting, processing, and analyzing data, taking all reasonable steps to assure the reliability and validity of results.

1. We shall recommend and employ only those tools and methods of analysis which, in our professional judgment, are well suited to the research problem at hand.
2. We shall not select research tools and methods of analysis because of their capacity to yield misleading conclusions.
3. We shall not knowingly make interpretations of research results, nor shall we be tacitly permit interpretations that are inconsistent with the data available.

4. We shall not knowingly imply that interpretations should
    be accorded greater confidence than the data actually
    warrant.

B. We shall describe our methods and findings accurately and in
appropriate detail in all research reports, adhering to the standards for
minimal disclosure specified in Section III.

C. If any of our work becomes the subject of a formal investigation of
an alleged violation of this Code, undertaken with the approval of the
AAPOR Executive Council, we shall provide additional information on the
survey in such detail that a fellow survey practitioner would be able to
conduct a professional evaluation of the survey.

II. Principles of Professional Responsibility in Our Dealings with People

A. THE PUBLIC:

1. If we become aware of the appearance in public of serious
    distortions of our research, we shall publicly disclose what is
    required to correct these distortions, including, as appropriate, a
    statement to the public media, legislative body, regulatory agency,
    or other appropriate group, in or before which the distorted
    findings were presented.

B. CLIENTS OR SPONSORS:

1. When undertaking work for a private client, we shall hold
    confidential all proprietary information obtained about the
    client and about the conduct and findings of the research
    undertaken for the client, except when the dissemination of
    the information is expressly authorized by the client, or when
    disclosure becomes necessary under terms of Section I-C or II-A of
    this Code.

2. We shall be mindful of the limitations of our techniques and
    capabilities and shall accept only those research assignments
    which we can reasonably expect to accomplish within these
    limitations.

C. THE PROFESSION:

1. We recognize our responsibility to contribute to the science of
    public opinion research and to disseminate as freely as possible the
    ideas and findings which emerge from our research.

2. We shall not cite our membership in the Association as evidence of
    professional competence, since the Association does not so certify
    any persons or organizations.

D. THE RESPONDENT:

1. We shall strive to avoid the use of practices or methods that may harm, humiliate, or seriously mislead survey respondents.

2. Unless the respondent waives confidentiality for specified uses, we shall hold as privileged and confidential all information that might identify a respondent with his or her responses. We shall also not disclose or use the names of respondents for non-research purposes unless the respondents grant us permission to do so.

III. Standards for Minimal Disclosure

This section is reprinted in chapter 13.

15

Survey Errors
COPING WITH SURVEY CONSTRAINTS

The total survey error approach emphasizes the several possible sources of survey error, along with constraints that affect the minimization of those errors and various effects that are inherent in surveys. Taken together, these different elements provide an overarching perspective on survey research, a paradigm for a new science. Yet it is a science that will keep evolving, and technology change will directly affect how surveys are conducted in the future, just as it has in the past.

The Theory of Survey Error

This book has emphasized survey errors, but that should not be taken as a warning against conducting survey research. Instead, the explicit focus on sources of error should be seen as an advantage of the survey research literature. All research is susceptible to a variety of types of error, but rarely are the different sources of error diagnosed and dissected as completely as they have been in the survey field. It would be unrealistic to require a hermetically sealed error-free environment for the conduct of human research. What is necessary instead is an awareness of the potentials for error along with an understanding of how to minimize each source of error, as has been the focus of this book.

Types of Error

To review very briefly: measurement error occurs in all research, including surveys. Behavior of interviewers is one source of measurement error in face-to-face and telephone surveys. Measurement error due to interviewers can be minimized through interviewer training and supervision, but surveys obtained by interviewers will always contain some amount of interviewer-related error.

Regardless of whether interviewers are used, the answers that respondents give are not necessarily error free. Measurement error due to respondents can be minimized by care in wording questions and in questionnaire construction. However, it is important to recognize that there is no perfect question wording or perfect question order, and survey results must always be taken with a grain of salt since they depend on how the question was worded and since questions can never be asked free of context.

Nonresponse error is endemic to surveys. Item nonresponse is a common problem, as some people answer "don't know" and some skip questions. Analyzing only cases with complete data on the variables of interest can lead to fallacious results. Statisticians have developed several procedures for imputing missing data, which is to say assigning meaningful scores for the missing data, but these procedures are not yet widely used in some disciplines.

Unit nonresponse is also an important issue. It is rarely possible to contact every designated respondent, and some will refuse to cooperate. Attempts to minimize unit nonresponse usually involve repeated callbacks to contact the designated respondents and the offer of incentives to increase cooperation rates. Since these techniques will never be fully successful, it is always necessary to consider whether unit nonresponse is likely to bias the survey results.

Coverage error is less visible than most other types of survey-related error. It is important that the sample be drawn from a sampling frame that resembles the population of interest as much as possible. Mismatches between the sampling frame and the target population can throw off estimates of population parameters. Using multiple sampling frames can sometimes alleviate this problem, though that creates a need to combine the data collected from the different frames.

Sampling error is inevitable whenever a sample is used to represent a larger population. However, sampling error should be seen as only the tip of the proverbial iceberg. Probability samples avoid the potential bias of nonprobability samples while allowing the sampling error to be computed. Sampling error can be decreased when researchers increase the planned sample size, but the total survey error approach reminds one that minimizing sampling error is of negligible usefulness if other sources of error are ignored.

As explained in previous chapters, different types of error lead to different statistical problems. Researchers often hope that survey error is random so that their statistics will not be biased. However, even random measurement error and random nonresponse error decrease the reliability of variables, which has the effect of attenuating correlations and regression coefficients. Systematic error has more serious effects. Systematic coverage, nonresponse, and/or measurement error lead to biased means and other statistics. Additionally, there is correlated error when the same interviewer conducts several interviews, when the same coder codes open-ended questions from multiple interviews, and when multiple cases are selected from the same cluster. Correlated error leads to a loss of efficiency and larger standard errors, which makes it harder to achieve statistical significance.

In summary, then, the existence of error in surveys should not be viewed as debilitating. Error is inevitable in all research, but what is important is recognizing its existence and properly taking it into account when interpreting results. There is no reason to believe there is more error in surveys than in experiments or other forms of research; rather, the survey field has become more comprehensive in considering the various types of error than have most other research fields.

At the same time, survey research nicely illustrates the practical limitations on research. There are elegant theories underlying survey sampling and several other topics discussed in this book, but the best survey researchers are the ones who have figured out how to make effective compromises between theory and real-world limitations. The Mitofsky-Waksberg method for telephone sampling is one example, as is the focus on satisficing as a way to understand respondent behavior.

Survey-Related Effects

In addition to these several types of errors, the survey research literature speaks of several "effects." Usually these are situations where there is not considered to be a "correct" standard, just inevitable implications of decisions that must be made in research. For example, it is common to speak of interviewer effects, including interviewer style and probing style effects as well as race-of-interviewer and gender-of-interviewer effects and possibly interviewer expectancy effects. There can be question-wording effects, including response effects such as primacy and recency, as well as filter effects. There are context and framing effects in questionnaires in addition to several possible question order effects (assimilation, contrast, carryover, subtractive, consistency effects for adjacent questions and more generally salience, accessibility, rapport, and fatigue effects). There can be day-of-the-week and time-of-the-year effects in doing interviews. There can be effects due to the location of the interview, including third-party presence effects. Panel studies are susceptible to interviewer familiarity, conditioning, and reactive effects. There are sampling design effects. There are questionnaire design effects in written questionnaires and display effects in Internet surveys. There are possible mode effects, as well as concern about differences in acquiescence, social desirability, and anonymity effects between different modes. There can be house effects and other comparability effects in cross-national or over-time studies.

The research literature attempts to measure these various effects so that survey results can be better understood, even if these effects are not "errors" that can be eliminated. A creative approach for dealing with some of these survey-related effects is to use a randomization strategy, such as different question versions or orders, so as to measure the effects and to randomize them out. Randomization increases the cost of programming the questionnaire, however, and increases the chances of programming errors.

Survey Errors and Survey Constraints

Conducting surveys involves a trade-off between seeking to minimize errors and recognizing practical constraints. As has been pointed out repeatedly in the preceding chapters, there are inevitable constraints involving cost, time, and ethics that have to be taken into account in a realistic manner.

The cost side of the equation is always important in research. Quadrupling the sample size can halve sampling error, but that may not be cost effective because the other sources of error can be greater than the sampling error. Sometimes it is better to spend money on more callbacks or on refusal conversions rather than on a larger sample. Similarly, spending more money on pretests and interviewer training can be cost effective.

The trade-off between survey errors and costs is controversial, especially with regard to response rate. High response rates are increasingly difficult to obtain. The response rate can always be improved by spending more money, whether on callbacks, refusal conversion, or incentives for respondents. Some survey organizations interpret the total survey error approach as implying that response rate is not all that important since there are always other sources of error, an interpretation that provides a convenient excuse when a high response rate is not obtained. Yet substantive researchers realize that their work will receive more credence with a higher response rate. Referees are more likely to recommend publication of articles in academic journals when a high response rate is reported, and clients of commercial surveys are more likely to accept survey results when the response rate is high. My own preference is toward working hard to obtain a reasonably high response rate, though I agree that it is important that clients understand that a very high response rate is not necessarily cost effective.

Time considerations are also important in surveys. Some surveys have to be completed on a rigid schedule, especially daily tracking polls. When clients require daily reports on survey results, computer-assisted data collection is essential. Time considerations even affect the type of questions asked: self-coded questions are virtually mandatory when speedy results are required, whereas open-ended questions can be used for academic surveys in which the client does not need to obtain the data immediately. Multiple callbacks and refusal conversions are also possible only when time pressures are not excessive, which means that it can be hard to minimize error when time is of the essence.

Ethics can be seen as a further constraint on survey research. Both researchers and survey organizations should think through the ethical implications of their surveys, as well as dealing, when necessary, with institutional review boards. There are valid reasons for ethical guidelines, and media publicity on violations of ethical guidelines makes it more difficult for all researchers to conduct surveys. Survey ethics involves respecting respondents and possible respondents and their rights, recognizing the needs of interviewers, and reporting survey results honestly. High-quality survey research can be conducted with full honoring of ethical guidelines without increased cost, and any claims to the contrary should be regarded with suspicion.

The choice of survey mode is usually dependent on cost, time, and other feasibility constraints. As seen in chapter 12, it is difficult to examine mode differences by themselves, since differences in results between modes are confounded with numerous other differences between surveys conducted in the different modes. Still, repeated experiments by multiple survey houses with comparable surveys using different modes would be most useful, since more data would permit better separation between mode and other effects.

The Future of Survey Research

Survey research is a field that is constantly evolving. Procedures for sampling households and conducting face-to-face interviews were perfected in the 1960s, only to be largely supplanted by telephone interviewing within two decades. Similarly, procedures for sampling telephone numbers and conducting phone interviews were refined in the 1980s, but technological changes since then have posed severe challenges to phone interviewing. Call screening, telephone numbers assigned to computer modems and fax machines, cellular phones, and phoning over the Internet all make it harder to get a good phone sample. The increased difficulty of obtaining telephone samples may lead to some revival of face-to-face interviewing. Meanwhile, Internet surveys provide a possible means for data collection that could supersede phone surveys, though we still need to learn more about how best to conduct Internet surveys.

There are several lessons to draw. Personal communication technology will continue to change, and survey research must adapt to those changes. Some established survey modes will become less feasible, while new modes will be developed. What is important is to keep in mind the principles underlying quality research and to adapt them for new technologies. Surveys will remain an important research tool, though procedures for conducting them are likely to keep evolving.

The Total Survey Error Approach

One mark of a field's becoming scientific is the development of a paradigm for its study. The early survey research field was at best quasi-scientific. Even after probabilistic sampling procedures were developed, the understanding of interviewing was mainly from accumulated lore plus a small dose of the social psychology of the time.

The development of the total survey error approach represents an important paradigm shift. All the elements of the survey process have been brought together in a unified framework, with consideration of the several different types of effects that arise in surveys. Furthermore, considering survey costs and other constraints as balancing off the need to minimize errors makes this into a comprehensive paradigm.

This new science of survey research is inherently interdisciplinary. The theories brought together in the survey realm range from sampling and missing-data imputation in statistics to compliance, interviewing, and survey response in social psychology. As the survey research field matures with the development of certification and degree programs at universities, the top survey organizations are the ones that are interdisciplinary in the best sense of that term. The survey experiment in computer-assisted data collection—when random parts of the sample are asked different question versions or questions in different orders—is an important case of borrowing from experimental disciplines, importing the power of experiments into the survey.

Yet it is appropriate to end this book with a pair of cautions. First, even our best knowledge of many aspects of the survey process is still tentative. As stressed in several of the above chapters, our knowledge is often based on just a few studies. There are some important studies of different types of question wording, different survey modes, and so on, but it would be a mistake to regard those tests as definitive until they are shown to be robust across different survey organizations and topics.

The second caution is that the total survey error paradigm itself may eventually be displaced by a new approach. The total survey error approach is still relatively new, but further work may suggest insights that lead to even better approaches. This is an important aspect of progress in science: developing a new paradigm while recognizing that no paradigm is sacrosanct. The total survey error approach has, for the first time, provided a comprehensive scientific approach to the survey research process, which may, in turn, produce insights that lead to the development of other new ways of understanding the survey process.

Indeed, there are already signs of cracks in the total survey error approach. Several survey researchers instead emphasize their own perspectives, such as the total design method and the role of satisficing in the survey response. Others would still emphasize the importance of maximizing response rate, rather than viewing this as a trade-off with other ways to minimize total survey error. In the end, the question is whether viewing the different aspects from an integrated perspective provides purchase on increasing our understanding of the survey process. Thus we should expect continued work outside the total survey error paradigm at the same time that the theories underlying this paradigm are developed further.

Appendix

Meta-analysis

Several of the articles mentioned in this book (including Knäuper 1999; Tourangeau et al. 1989; Singer, Von Thurn, and Miller 1988; de Leeuw and van der Zouwen 1988; de Leeuw 1992; and Church 1993) employ meta-analysis techniques. Meta-analysis can be viewed as a quantitative literature review, seeking to combine separate studies to look for effects that hold across them. This has become a common procedure in many disciplines. For example, medical research is often based on studies of very small numbers of individuals, so pooling across studies becomes a useful strategy. An effect that is too small to be significant in separate studies can be found to be significant in a meta-analysis. Meta-analysis has been applied to several survey research topics, such as the effects of incentives and other specific procedures on response rates.

There are several forms of meta-analysis, including multiple regression analysis, combined significance tests, pooling correlations, and averaging effect sizes. The most basic approach employs regression analysis, with each separate study being a case in the analysis. Consider a meta-analysis of the determinants of response rates. The dependent variable would be the response rate observed in each study. The independent variables would correspond to whether particular procedures were used in the surveys, such as coding each survey as to whether an advance letter was employed, whether prepaid incentives were used, and so on. The regression coefficients would then show the effect of each of these procedures on the response rate.

However, regression analysis makes a series of assumptions that can be problematic (Fox, Crask, and Kim 1988). The full details of a study are often not included in an article, so there are invariably missing data about study features. Some variables are often coded subjectively, such as coding the salience of the survey topic to the respondent. Third, each study is usually counted equally, though it is possible to use weighted least squares to count large surveys more heavily than smaller ones. High correlation among the predictors (multicollinearity) can increase the standard errors of the regression coefficients and therefore affect the determination of statistical significance. Finally, there must be many more studies than study characteristics for the regression procedure to succeed.

An alternative meta-analysis approach is to accumulate evidence across separate studies to determine the overall significance level for rejecting the null hypothesis. For example, Fox, Crask, and Kim (1988) compare studies that report using a particular approach (the "treatment" group) to studies that do not use it (the "control" group), taking into account the variances of the individual study effects. Significance can be judged through a combined test, such as Fisher's combined test: $\chi^2 = -2 \times \Sigma\ln(p_i)$, where p_i is the significance of the term in the i-th study and the degrees of freedom for the chi-square test is twice the number of studies being combined.[1] Wolf (1986, 18) comments favorably on this procedure, viewing it as conservative, even though it can give misleading results when studies with strong but opposite findings are combined. In any event, significance by itself does not indicate whether the effect is large or small.

Hunter and Schmidt (2004, chap. 3) give a procedure for pooling correlations from separate studies, including controls for error in measurement if some reliability information is known. They also develop procedures for testing for moderator variables that cause differences in correlations between other variables, which can occur when corrected variance figures are smaller in subsets than in the full set of data. However, examining all possible combinations of studies for moderator effects is vulnerable to chance effects if the number of studies is not very large.

Finally, effect size can be judged through taking the average of the standardized effects in each study. The standardized effect size, d, is the difference between the means for treatment ($\bar{X}_t$) and control ($\bar{X}_c$) groups, divided by its standard deviation s (based on either the control group standard deviation or a pooled standard deviation for the study). So $d = (\bar{X}_t - \bar{X}_c)/s$. The average effect size is usually a weighted average, with each study i

weighted by the inverse of its sampling variance: so $\overline{d} = \Sigma\ w_i d_i / \Sigma\ w_i$, where $w_i = 1/se_i^2$; the weight effectively gives more emphasis to studies with larger numbers of cases. The standard deviation of the effect size can be used to establish its confidence interval. However, the d value is attenuated by measurement error in the dependent variable, and differences in that variable's reliability across studies can spuriously affect its value. Alternatively, Lentner and Bishop (1986) develop a minimum variance unbiased estimate of the effect size, weighting the individual effects with weights that sum to 1 and are inversely proportional to their variances. Lipsey and Wilson (2001, chap. 3) usefully summarize how to compute effect sizes for proportions, means, mean gains, mean differences, proportion differences, odds-ratios, and Pearson correlations. Different studies often report effects in different ways, so that one must convert them into a common effect-size metric in order to conduct a meta-analysis.

As above, moderator variables would exist if the effect sizes were different for some subsets of studies than for others. This can be tested by dividing the studies into subsets or by correlating the effect size with a feature of the study (Hunter and Schmidt 2004, 293). Studies that do not mention how they handled a possible survey feature (such as advance letters for surveys) are simply excluded from the analysis of the effect of that feature. This procedure can be seen as providing aggregate estimates of effect sizes, but at a bivariate level rather than fully multivariate. Looking at interaction effects is one way to move toward a more multivariate analysis.

An important question in meta-analysis is which studies to include. Narayan and Krosnick (1996) limited their meta-analysis to studies in which the effect of interest was at least marginally significant for the full sample, arguing that there is no need to look for third-variable correlates of effects if there is no effect in the first place. They claim that including insignificant studies would have watered down effects because zero-magnitude effects would have occurred in all subgroups, but it is still possible that significant effects would have been found for some subgroups. That might have been appropriate in the Narayan and Krosnick study, since they were comparing average effect sizes for different third-variable groups (education in their case), but in general using only studies that find significant effects would seem to bias the meta-analysis toward finding effects.

Yet publication bias always limits which studies can be included in a meta-analysis. Rosenthal (1979) describes this as the "file drawer problem." Journals generally publish only significant results, so meta-analysis of published journal articles focuses on effects that have been found significant. Studies that do not obtain significance are rarely published and end up instead in the researcher's file drawers. Since 5% of results will be significant by chance alone at the 0.05 level, a meta-analysis of only significant findings can be misleading. Thus meta-analysis can be biased toward finding significant effects. Rosenthal provides a technique for calculating how many unpublished studies with no effects there would have to be for a reported significant effect to be nonsignificant.

Finally, not every published study is of equal quality. Most meta-analyses count each study equally, without taking study quality into account. Indeed, survey reports rarely include enough detail to tell how good the data are. However, including weak studies in the meta-analysis can distort the results.

Further sources on meta-analysis include Wolf (1986), Rosenthal (1991), Hunter and Schmidt (2004), Hedges and Olkin (1985), and Lipsey and Wilson (2001).

Notes

Chapter One

1. In their 2004 book, Groves et al. call this new science "survey methodology"—the study of survey methods. I use the more traditional term "survey research," but we are both clearly describing the same scientific field.

2. Straw polls were used in local elections in the United States as early as 1787 (Jensen 1980) and for presidential elections by 1824 (Smith 1990a), but they were not conducted extensively until the 1896 election. Early straw polls have been studied by Herbst (1993).

3. This right-triangle representation also demonstrates the statistical fact that the total error is the square root of the sum of the squared sampling error and the squared nonsampling error.

4. Another alternative approach is to focus on "survey quality," as in Biemer and Lyberg's study (2003). That perspective recognizes total survey error while placing it in a broader context of quality improvement by the survey organization throughout the survey process.

Chapter Two

1. The bias of a statistic (such as a mean or a regression coefficient) is the difference between the value of that statistic as observed in the survey and the true value of that statistic for the population of interest.

2. The observed variance of a statistic reflects the distribution of its values around its observed mean value, not around its true value for the population of interest.

Chapter Three

1. Mail surveys and Internet surveys are actually self-administered questionnaires, but it is useful to distinguish them here because they pose different problems in terms of formatting and achieving high response rates.

2. A special form of panel study is the *cohort study*, in which a single age cohort, such as high school seniors, are followed over the years through fresh interviews of

original respondents. A *retrospective panel study* asks people about a period in the past as well as the current period, but it depends on recall, and it necessarily misses people who could have been interviewed in the earlier period but cannot be in the current period.

3. This can be prevented by placing cookies on the person's computer or by the survey organization's not allowing repeated interviews from the same computer (based on the IP number of its Internet connection). Web surveys that invite particular people to participate usually control access by issuing individual identification numbers or passwords to selected respondents.

Chapter Four

1. See the insightful discussion of survey research as seen by interviewers in Converse and Schuman (1974). This book nicely describes the tension between interviewers as technicians applying standard techniques and interviewers as human beings who build warm relationships with respondents.

2. By contrast, elite interviews (such as of government officials or business leaders) are generally "unstructured," with the interviewer, for example, asking simply, "Tell me about . . ."; or they may be "semistructured," with the interviewer having a set of topics to raise in the questioning but having flexibility with the question order.

3. In a previous laboratory experiment using different respondents for the two types of interviews, Schober and Conrad (1997) had found 60% greater accuracy with conversational interviewing when the mapping between concepts and the question wording was complicated.

4. Rho is the between-interviewer variance on a question divided by the sum of the between-interviewer and within-interviewer variance on that question.

5. Hox, de Leeuw, and Kreft (1991) confirm the finding of interviewer effects associated with the number of comments on open-ended questions, finding a high rho of 0.13 on the number of open-ended comments varying by interviewer.

6. There also can be special issues involved in obtaining interviews from people of the opposite gender, homosexuals, and older people (Gubrium and Holstein 2002, chaps. 10–13).

7. One study (Mathiowetz and Cannell 1980) found 0.85 intercoder agreement on question wording, defining, and feedback but a lower 0.75 agreement level on pacing and clarity.

Chapter Five

1. The emphasis on the interview as both a conversation and a cognitive task is due to Sudman, Bradburn, and Schwarz (1996).

2. Groves (1989) adds a fifth component, prior to Tourangeau's four: the encoding of information into memory that occurs before the interview. The encoding process will be discussed as part of the retrieval stage.

3. Studies often model recall of past activities with an exponential decay function, such that the probability of recall is ae^{-bt}, where t is the time since the event and a and b are constants that can be estimated from recall data. In this type of decay function, there is a considerable loss at first, after which recall eventually steadies out. The value of b controls how sharp is the rate of decay, which differs between different types of activities.

4. Bradburn and Sudman (1979) asked respondents how uneasy various survey topics would make "most people." Sexual behavior, drugs use, and getting drunk ranked

highest on the list, with income, gambling, and drinking next, while few people viewed education, occupation, social or sports activities in this light.

5. Because even that wording still led to an overreport of the vote, the 2000 NES survey, after giving the first sentence, asked people to choose among the following four statements: "I did not vote (in the election this November)," "I thought about voting this time—but didn't," "I usually vote, but didn't this time," "I am sure I voted."

6. Belli et al. (1999) show that overreporting turnout can be reduced by pointing out to respondents the possibility of memory confusion and encouraging them to think carefully, which suggests that the problem with turnout questions may be recall mistakes rather than social desirability.

7. One-twelfth of a sample should have been born in January, so if the randomized method asks that question with a probability of one-third in a sample of 900 people, 25 people would be expected to answer "yes" out of the 300 expected to be asked their month of birth. If 85 people actually answered "yes," that would translate to an estimated rate of 10% answering "yes" to the sensitive question $((85 - 25)/(900 - 300))$.

8. The two-route theories are similar to Petty and Cacioppo's (1986) distinction between central and peripheral routes to persuasion.

9. An important class of hypothetical questions that is considered valid appears in surveys on contingent valuation (Freeman 2003). Contingent valuation is a technique for estimating the economic benefits of services that are not traded in traditional markets, particularly as regards willingness to pay for improvements in public goods and as regards valuation of damages. This approach is widely used in cost-benefit analysis of government projects, especially in the environmental area, and has been accepted in court cases. As an example, people might be asked if they would be willing to pay $50 for cleanup of an oil spill, after they are given a full explanation of the problem. This is usually asked as a yes-or-no question (called *referendum format*), with follow-up questions on reasons for the answer to ensure that it was not just due to the hypothetical nature of the payment. Contingent valuation questions are considered to be truth revealing if three conditions are met: decisive implementation (the probability of the project's being implemented depends on the response), compelled payment (if the project passes, payment will be collected from the respondent), and decoupled payment (the individual's response does not change the payment if the project passes).

10. Another relevant distinction is between the cognitive and affective components of attitudes, where *cognitive* refers to thoughts about the attitude object and *affective* refers to emotions toward that object; separate questions would be needed to measure these components. Additionally, some attitudes may be multidimensional, in which case multiple questions would be needed to explore them fully. Finally, some topics may invoke *nonseparable preferences* (Lacy 2001a; 2001b)—two attitudes at once, with people's preferences on one dependent on the status quo state of the other. For example, which political party people would like to have control the Congress may depend on which party controls the White House. Survey questions in such cases should be phrased as conditional on what is the status quo.

11. There is also a greater tendency for telephone respondents to choose the items near the beginning of the list than ones in the middle of the list, since those at the beginning may enter long-term memory.

12. The validity of a measure can be assessed in several ways, such as correlating it with a true measure of the variable (a *gold standard criterion*) and correlating it with other variables that it theoretically is related to (*construct validity*).

Chapter Six

1. Lacy (2001a) instead would explain this behavior in terms of preferences for the two being nonseparable. He finds question order effects to be significant when preferences are nonseparable.

2. In addition to directional effects, question order can have "opinionational" effects, where one question affects whether or not a person is willing to give their opinion on a later question (Benton and Daly 1991).

3. Also the standard errors of questions asked of a half-sample are 41% higher than if they were asked of the full sample, which makes it harder to achieve statistical significance.

Chapter Seven

1. Strictly speaking, missing data can be missing at random but nonignorable, but that situation is rare (see Allison 2002, 5; King et al. 2001, n. 6).

2. An example would be if variables A and B were perfectly positively correlated for cases with data on both variables but have different correlations with a third variable. For instance, say the values for 4 people on variables A, B, and C are as follows: person 1: 1, 1, 1; person 2: 2, missing, 2; person 3: missing, 0, 3; person 4: 3, 3, missing. Then $r_{AB} = 1.0$, so A and B should have equal correlations with any other variable, but r_{AC} and r_{BC} are unequal ($r_{AC} = 1.0$, but $r_{BC} = -1.0$), which is clearly inconsistent.

3. Likewise, an additional dummy variable can be used for the missing category of a categorical variable, but Allison (2002, 11) again indicates that leads to biased estimates.

4. This procedure may seem unusual in that it depends on the order of cases in the data set. The hot-deck procedure was developed, and named, back when data were punched on computer cards, and this procedure avoided the expense involved in going through the deck of data cards more than once. That expense problem now is only of historical interest, and multiple-pass variants with random selection of donors have been devised.

5. Another variant replaces missing values with the item mean, corrected for the "ability" of the respondent as determined by the respondent's score on observed items compared to the mean scores on those items. Consider a three-item index in which different people have missing data on each of the items. If a person was one-third a standard-deviation unit below the mean on the two items she responded to, for example, she would be assigned a value one-third a standard-deviation unit below the mean of the item that she did not answer. Huisman's (2000) simulation exercise found "corrected item mean substitution" to be superior to several other deterministic procedures for estimating the actual missing values, though it overestimates the index's reliability.

6. The random errors should have a mean of zero, with a variance equal to the residual variance from the regression using the cases without missing data. Alternatively, a randomly selected regression residual can be used as the error term.

7. Other approaches have been proposed for dealing with nonignorable missing data. A *Bayesian* approach involves augmenting the data from the cases with full data in the survey with known information from other sources, such as combining the relationship between age and income in a survey with the known relationship between them in Census Bureau data to obtain estimates for missing data on income in the sample. Alternatively, *maximum likelihood estimation* can be used if there is a model of the joint distribution of all variables with missing data; Allison (2002, chap. 4) considers

this the preferred method when dealing with a linear model where there is a large amount of missing data. Allison (2002) also reviews *pattern-mixture models*, which assume that whether the data are missing affects the distribution of the variable; while admitting that causal logic rarely favors this approach, Allison argues that the causal direction is not the issue and that such models are easy to apply.

Chapter Eight

1. Fowler (2002, 45) provides a useful table showing, for a given response rate, the range of possible percentages when the sample percentage is 50%. For example, if the response rate is 90%, an obtained 50% rate can have occurred if the actual rate was between 45% and 55%.

2. If q is the proportion of people not successfully recontacted and w is the wave number, the proportion of people recontacted at wave w would be $(1 - q)^{w-1}$ if the effect is cumulative.

3. The Knowledge Networks cooperation rate was slightly higher, 31%, calculated as a percentage of contacted households that participated in the original telephone interview, with omission of the 11% of eligible sampled telephone numbers in which the household refused to participate in the original Knowledge Networks phone interview. The CSR cooperation rate was 51% when households that refused to participate in the respondent selection procedure were dropped.

4. The Tailored Design Method differs from Dillman's (1978) earlier Total Design Method in using a prenotice letter, including a stamped return envelope, and, most important, in the use of financial incentives.

5. However, Couper, Traugott, and Lamias (2001) find that a graphic progress indicator that increases download times can increase the time it takes to complete a survey.

6. There are actually several alternative conventions for what the weighted number of cases should be. Instead of the population size as here (100), it can be the actual number of interviews (here 73), or it can be a weighting that adjusts for intentional disproportionate sampling (see chapter 10). To weight to the actual number of interviews, the weights above should be multiplied by (73/100) to yield weights of 1.22 for specialists and 0.82 for family doctors.

7. Brehm actually first estimated the probability of participation, generated residuals, and then used the residual as a predictor in the substantive equation, but the maximum likelihood estimation approach described above is more efficient, now that advances in computer power make it computationally feasible.

8. Groves and Couper (1998, 310–19) develop a set of model-based weights based on their best models for contact and cooperation and then apply them to several variables in two face-to-face government-sponsored surveys. The weights turn out to be similar to traditional postsurvey adjustment weights, both of which have small effects for the variables they examine in these studies. See also the adjustments used by Brehm (2000), which show important differences on estimates of effects of some variables in the 1988–1990–1992 Senate Election Studies because of sample truncation.

9. As to substitutions, AAPOR specifies that all replaced cases should be accounted for in the final disposition codes. "For example, if a household refuses, no one is reached at an initial substitute household, and an interview is completed at a second substitute household, then the total number of cases would increase by two and the three cases would be listed as one refusal, one no one at residence, and one interview." The same procedure is used for substitutions within households. Additionally, substitutions and proxies should be clearly identified and reported.

10. CASRO takes an intermediate approach by including a fraction of this category as failures, where the fraction is $e = I/(I + SO)$, where SO represents "screen outs"; this has the effect of assuming that the proportion of eligible cases among those with unknown eligibility status is the same as the proportion among those with known eligibility status. In addition to cases that failed the screening criteria, it includes under screen outs the situations of language problems (unless the population of interest includes that language), nonworking phone numbers, no housing units at that location, and fax/data line.

11. The other formulas are $RR2 = 1/(1 + 2 + 3)$, $RR4 = 1/(1 + 2 + 3e)$, and $RR5 = I/(1 + 2)$.

12. The other formula is $CON2 = (1 + R + O)/(1 + 2 + 3e)$.

13. The other formulas are $COOP2 = 1/(1 + R + O)$, $COOP3 = I/(1 + R)$.

14. Using its estimation of eligibility for cases of unknown eligibility, CASRO uses the in-between formula: $REF2 = R/(1 + 2 + 3e)$.

Chapter Nine

1. Lessler and Kalsbeek (1992, 44) develop a very precise definition of a sampling frame as the "materials, procedures, and devices that identify, distinguish, and allow access to elements of the target population, . . . composed of a finite set of units to which probability sampling is applied."

2. Iannacchione, Staab, and Redden (2003) suggest a procedure for using the U.S. Postal Service's computerized list of delivery point addresses as a sampling frame for metropolitan areas, showing that this frame missed only 2% of households in the Dallas metropolitan area.

3. The face-to-face National Health Interview Survey asked people from 1963 to 1986 to provide their phone number (Thornberry and Massey 1988). They found a telephone coverage rate in the low 80% range in the early 1960s. It hit 85% by 1967, the high 80% range in the late 1960s, and then 90% since the early 1970s.

4. Note there is not a coverage problem for household samples if a person has a cellular phone in addition to a regular landline phone in the house.

5. Specifically, the instruction given to interviewers is to attempt to conduct an interview at every dwelling unit between the one selected in the sample (say 1045 E. Huron) and the next known dwelling unit on the map (say 1049 E. Huron), if any unit is found in that "half-open interval." That is, the interview would not go to the next known unit (1049) but would try any unknown unit (such as all the units in a new small apartment building at 1047).

6. Note that what is relevant is the number of phone lines; having multiple phones on the same line does not constitute a multiplicity problem.

Chapter Ten

1. The formulas given here are for samples, rather than for the full population. The $n - 1$ denominator is used instead of n so that the estimator of the sample variance is unbiased.

2. If the sampling fraction is large, this should be multiplied by the finite population correction factor, as shown above.

3. Squire (1988) uses a 1937 Gallup survey that asked about participation in the *Literary Digest* poll to show that nonresponse was also a major source of the debacle, with Roosevelt having more than a two-to-one majority among those who did not return the straw vote ballot. See also Cahalan (1989).

4. See http://www.ajc.org/InTheMedia/Publications.asp?did=1030&pid=2041.

5. Couper (2000) lists five types of probability-based Web surveys, including intercept surveys, list-based surveys, providing a Web option for mixed-mode surveys, recruiting panels of Internet users, and recruiting panels of the full population. These types of Web surveys are discussed in appropriate locations throughout this book.

6. The weights discussed here are "sampling weights," which are separate from the poststratification weights that deal with differential response rates for different demographic groups (see chapter 8).

7. The weighted number of cases with oversampling is often adjusted to the total number of interviews. Say that 950 interviews are taken with non–Asian Americans and double sampling leads to 100 interviews with Asian Americans. The procedure described in the text would weight non–Asian Americans at 1.0 and Asian Americans to 0.5, so that the weighted total number of cases would be 1,000. Alternatively, the total weighted number of cases could be set to the total number of interviews, 1,050, by using a weight of $(950/1,000) \times (1,050/950) = 1.05$ for non–Asian Americans and $(50/1,000) \times (1,050/100) = 0.525$ for Asian Americans.

8. See also the discussion of design effects in analyzing interviewer effects in chapter 4.

9. This deft combines the defts for the face-to-face and the telephone parts of the 2000 NES sample and is mainly due to weighting considerations.

10. See http://www.surveysampling.com.

11. A change in the distribution of phone numbers has been proposed so that they are allotted to telephone companies in groups of 1,000 instead of 10,000. That would presumably lead to modification of procedures for sampling phone numbers.

12. The Mitofsky-Waksberg method requires considerable knowledge of how telephone numbers are assigned. This is available in the United States but not necessarily in other countries. For example, as Moon (1999, 200–202) points out, random-digit dialing is impractical in Britain because there are 50 times as many potential phone numbers there as households and too little is known about the telephone number assignment process for the Mitofsky-Waksberg system to be workable.

13. Information about WesVar can be obtained at http://www.westat.com/wesvar/. Information about SUDAAN is available at http://www.rti.org/sudaan/.

14. Technically, the analysis can proceed as if the data came from a simple random sample if the dependent variable is not related to the sample selection probabilities, in which case the sample design is considered "ignorable" and "noninformative."

Chapter Eleven

1. The deff due to coders is $1 + \rho(\bar{m} - 1)(1 - r)$, where ρ is the intraclass correlation for coders, $\bar{m}$ is the average number of cases coded by an individual coder, and r is the reliability of the code for a variable.

Chapter Thirteen

1. Evidence has surfaced that George Gallup indeed gave the Eisenhower White House early indications of the results of his polls (Eisinger 2003, 114–15).

2. MacKuen, Erikson, and Stimson (1992) report that Gallup's phone surveys found 2.9% more Republican identifiers than the CBS–*New York Times* phone surveys during the 1985–91 period, though the two series moved roughly in tandem through most of that period. In line with Borrelli, Lockerbie, and Niemi (1987), though, this difference could be due to differences in question wording, since Gallup focuses on party

affiliation "as of today" whereas CBS–*New York Times* asks about partisanship "generally speaking."

3. Voss, Gelman, and King (1995) provide one of the few reports on the details of survey administration for different survey organizations, in this instance media polls of the 1988 presidential election.

Chapter Fourteen

1. There are other ethics issues that can affect some survey situations, such as conflicts of interest when a person working for one organization steers a survey contract to a survey firm in which she or he has a financial interest.

2. Rules for confidentiality, as well as for data privacy more generally, are summarized at the American Statistical Association's Web site: http://www.amstat.org/comm/cmtepc/.

Appendix

1. A related approach is to convert the p-values into z-scores through a standard normal transformation, sum the z-scores, divide that sum by the square root of the number of studies, and then transform that back to a p-value that can be used for significance testing.

References

AAPOR (American Association for Public Opinion Research). 2001. "Annual Membership Meeting." *Public Opinion Quarterly* 65:470–78.

———. 2004. *Standard Definitions: Final Dispositions of Case Codes and Outcome Rates for Surveys.* 3rd ed. Lenexa, KS: AAPOR.

Abramson, Paul R., and Charles W. Ostrom. 1994. "Question Wording and Partisanship: Change and Continuity in Party Loyalties during the 1992 Election Campaign." *Public Opinion Quarterly* 58:21–48.

Achen, Christopher H. 1975. "Mass Political Attitudes and the Survey Response." *American Political Science Review* 69:1218–31.

Alba, J. W., and L. Hasher. 1983. "Is Memory Schematic?" *Psychological Bulletin* 93:203–31.

Allison, Paul D. 2002. *Missing Data.* Thousand Oaks, CA: Sage.

Allsop, Dee, and Herbert F. Weisberg. 1988. "Measuring Change in Party Identification in an Election Campaign." *American Journal of Political Science* 32:996–1017.

Alvarez, R. Michael, and John Brehm. 2002. *Hard Choices, Easy Answers: Values, Information, and Public Opinion.* Princeton, NJ: Princeton University Press.

Alvarez, R. Michael, Robert P. Sherman, and Carla VanBeselaere. 2003. "Subject Acquisition for Web-Based Surveys." *Political Analysis* 11:23–43.

Alwin, Duane F. 1997. "Feeling Thermometers versus 7-Point Scales: Which Are Better?" *Sociological Methods and Research* 25:318–40.

Alwin, Duane F., and Jon A. Krosnick. 1985. "The Measurement of Values in Surveys: A Comparison of Ratings and Rankings." *Public Opinion Quarterly* 49:535–52.

American Institute of Public Opinion. 1939. *The New Science of Public Opinion Measurement.* New York: American Institute of Public Opinion.

American Statistical Association. 2003. "Interviewer Falsification in Survey Research." http://www.amstat.org/sections/srms/falsification.pdf.

Andersen, Ronald, Judith Kasper, Martin R. Frankel, and associates. 1979. *Total Survey Error.* San Francisco: Jossey-Bass.

Anderson, Barbara A., Brian D. Silver, and Paul R. Abramson. 1988a. "The Effects of the Race of the Interviewer on Measures of Electoral Participation by Blacks in SRC National Election Studies." *Public Opinion Quarterly* 52:53–83.

———. 1988b. "The Effects of Race of the Interviewer on Race-Related Attitudes of Black Respondents in SRC/CPS National Election Studies." *Public Opinion Quarterly* 52:289–324.

Aquilino, William S. 1993. "Effects of Spouse Presence during the Interview on Survey Responses concerning Marriage." *Public Opinion Quarterly* 57:358–76.

———. 1994. "Interview Mode Effects in Surveys of Drug and Alcohol Use: A Field Experiment." *Public Opinion Quarterly* 58:210–40.

Aquilino, William S., and Leonard A. LoSciuto. 1990. "Effects of Interviewer Mode on Self-Reported Drug Use." *Public Opinion Quarterly* 54:362–95.

Aquilino, William S., and Debra Wright. 1996. "Substance Use Estimates from RDD and Area Probability Samples: Impact of Differential Screening Methods and Unit Nonresponse." *Public Opinion Quarterly* 60:563–73.

Asher, Herbert. 1974. "Some Consequences of Measurement Error in Survey Data." *American Journal of Political Science* 18:469–85.

Ayidiya, Stephen A., and McKee J. McClendon. 1990. "Response Effects in Mail Surveys." *Public Opinion Quarterly* 54:229–47.

Bachman, Jerald G., and Patrick M. O'Malley. 1981. "When Four Months Equal a Year: Inconsistencies in Student Reports of Drug Use." *Public Opinion Quarterly* 45:536–48.

Bailar, Barbara A. 1989. "Information Needs, Surveys, and Measurement Errors." In *Panel Surveys*, ed. Daniel Kasprzyk et al. New York: Wiley.

Bailey, Liberty Hyde. 1906. "Why Do the Boys Leave the Farm?" *Century Magazine* 72:410–16.

Bartels, Larry M. 2000. "Panel Effects in the American National Election Studies." *Political Analysis* 8:1–20.

Bassili, John N. 1996. "The How and Why of Response Latency Measures in Telephone Surveys." In *Answering Questions*, ed. Norbert Schwarz and Seymour Sudman. San Francisco: Jossey-Bass.

Bassili, John N., and Joseph F. Fletcher. 1991. "Response Time Measurement in Survey Research: A Method for CATI and a New Look at Attitudes." *Public Opinion Quarterly* 55:331–46.

Bassili, John N., and B. Stacey Scott. 1996. "Response Latency as a Signal to Question Problems in Survey Research." *Public Opinion Quarterly* 60:390–99.

Beatty, P. 2004. "The Dynamics of Cognitive Interviewing." In *Methods for Testing and Evaluating Survey Questionnaires*, ed. Stanley Presser et al. New York: Wiley.

Belinfante, A. 2000. *Telephone Subscribership in the United States*. Washington, DC: Industry Analysis Division, Common Carrier Bureau, Federal Communications Commission.

Belli, Robert F., William L. Shay, and Frank P. Stafford. 2001. "Event History Calendars and Question List Surveys." *Public Opinion Quarterly* 65:45–74.

Belli, Robert F., Michael W. Traugott, Margaret Young, and Katherine A. McGonagle. 1999. "Reducing Vote Overreporting in Surveys: Social Desirability, Memory Failure, and Source Monitoring." *Public Opinion Quarterly* 63:90–108.

Belson, William A. 1981. *The Design and Understanding of Survey Questions*. Aldershot, UK: Gower.

———. 1986. *Validity in Survey Research*. Aldershot, UK: Gower.

Bem, D., and H. McConnell. 1974. "Testing the Self-Perception Explanation of Dissonance Phenomena: On the Salience of Premanipulation Attitudes." *Journal of Personality and Social Psychology* 14:23–31.

Benson, Lawrence E. 1941. "Studies in Secret-Ballot Technique." *Public Opinion Quarterly* 5:79–82.

Benton, J. E., and J. L. Daly. 1991. "A Question Order Effect in a Local Government Survey." *Public Opinion Quarterly* 55:640–42.

Berinsky, Adam J. 1999. "The Two Faces of Public Opinion." *American Journal of Public Opinion* 43:1209–30.

———. 2002. "Political Context and the Survey Response." *Journal of Politics* 64:567–84.

———. 2004. *Silent Voices*. Princeton, NJ: Princeton University Press.

Bernstein, Robert, Anita Chadha, and Robert Montjoy. 2001. "Overreporting Voting." *Public Opinion Quarterly* 65:22–44.

Berrens, Robert P., Alok K. Bohara, Hank Jenkins-Smith, Carol Silva, and David L. Weimer. 2003. "The Advent of Internet Surveys for Political Research: A Comparison of Telephone and Internet Surveys." *Political Analysis* 11:1–22.

Best, Samuel J., and Brian S. Krueger. 2004. *Internet Data Collection*. Thousand Oaks, CA: Sage.

Bethlehem, Jelke, and Frank van de Pol. 1998. "The Future of Data Editing." In *Computer Assisted Survey Information Collection*, ed. Mick P. Couper et al. New York: Wiley.

Biemer, Paul P., et al., eds. 1991. *Measurement Errors in Surveys*. New York: Wiley.

Biemer, Paul P., and Lars E. Lyberg. 2003. *Introduction to Survey Quality*. New York: Wiley.

Billiet, Jacques, and Geert Loosveldt. 1988. "Improvement of the Quality of Responses to Factual Survey Questions by Interviewer Training." *Public Opinion Quarterly* 52:190–211.

Bingham, W., and B. Moore. 1924. *How to Interview*. New York: Harper and Row.

Bishop, George F. 1987. "Experiments with Middle Response Alternatives in Survey Questions." *Public Opinion Quarterly* 51:220–32.

———. 1990. "Issue Involvement and Response Effects in Surveys." *Public Opinion Quarterly* 54:209–18.

———. 2005. *The Illusion of Public Opinion*. Lanham, MD: Rowman and Littlefield.

Bishop, George F., and Bonnie S. Fisher. 1995. " 'Secret Ballots' and Self-Reports in an Exit-Poll Experiment." *Public Opinion Quarterly* 59:568–88.

Bishop, George F., Hans-J. Hippler, Norbert Schwarz, and Fritz Strack. 1988. "A Comparison of Response Effects in Self Administered and Telephone Surveys." In *Telephone Survey Methodology.*, ed. Robert Groves et al. New York: Wiley.

Bishop, George F., Robert Oldendick, and Alfred Tuchfarber. 1980. "Pseudo-opinions on Public Affairs." *Public Opinion Quarterly* 44:198–209.

———. 1984. "What Must My Interest in Politics Be If I Just Told You 'I Don't Know?'" *Public Opinion Quarterly* 48:510–19.

———. 1986. "Opinions on Fictitious Issues: The Pressure to Answer Survey Questions." *Public Opinion Quarterly* 50:240–50.

Blair, Edward A., and S. Burton. 1987. "Cognitive Processes Used by Survey Respondents to Answer Behavioral Frequency Questions." *Journal of Consumer Research* 14:280–88.

Blair, Johnny, Geeta Menon, and Barbara Bickart. 1991. "Measurement Effects in Self vs. Proxy Responses to Survey Questions: An Information-Processing Perspective." In *Measurement Errors in Surveys*, ed. Paul Biemer et al. New York: Wiley.

Blais, André, Neil Nevitte, Elisabeth Gidengil, and Richard Nadeau. 2000. "Do People Have Feelings about Leaders about Whom They Say They Know Nothing?" *Public Opinion Quarterly* 64:452–63.

Blau, Peter M. 1964. *Exchange and Power in Social Life*. New York: Wiley.

Booth, Charles. 1889. *Life and Labour of the People*. London: Macmillan.

Borrelli, Stephen, Brad Lockerbie, and Richard G. Niemi. 1987. "Why the Democratic-Republican Partisanship Gap Varies from Poll to Poll." *Public Opinion Quarterly* 51:115–19.

Bowers, William L. 1974. *The Country Life Movement in America, 1900–1920*. Port Washington, NY: Kennikat.

Box-Steffensmeier, Janet M., Gary C. Jacobson, and J. Tobin Grant. 2000. "Question Wording and the House Vote Choice." *Public Opinion Quarterly* 64:257–70.

Bradburn, Norman M., and Carrie Miles. 1979. "Vague Quantifiers." *Public Opinion Quarterly* 43:92–101.

Bradburn, Norman M., and Seymour Sudman. 1979. *Improving Interview Method and Questionnaire Design*. San Francisco: Jossey-Bass.

Bradburn, Norman M., Seymour Sudman, and Brian Wansink. 2004. *Asking Questions*. San Francisco: Jossey-Bass.

Brehm, John. 1993. *The Phantom Respondents*. Ann Arbor: University of Michigan Press.

———. 2000. "Alternative Corrections for Sample Truncation." *Political Analysis* 8:183–99.

Brick, J. Michael, and Graham Kalton. 1996. "Handling Missing Data in Survey Research." *Statistical Methods in Medical Research* 5:215–38.

Brick, J. Michael, Jill Montaquila, and Fritz Scheuren. 2002. "Estimating Residency Rates for Undetermined Telephone Numbers." *Public Opinion Quarterly* 66: 18–39.

Brick, J. Michael, Joseph Waksberg, and Scott Keeter. 1996. "Using Data on Interruptions in Telephone Service as Coverage Adjustments." *Survey Methodology* 22:185–97.

Brick, J. Michael, Joseph Waksberg, Dale Kulp, and Amy Starer. 1995. "Bias in List-Assisted Telephone Samples." *Public Opinion Quarterly* 59:218–35.

Briggs, Charles L. 1986. *Learning How to Ask: A Sociolinguistic Appraisal of the Role of the Interview in Social Science Research*. Cambridge: Cambridge University Press.

Bryant, B. E. 1975. "Respondent Selection in a Time of Changing Household Composition." *Journal of Marketing Research* 12:129–35.

Bulmer, Martin, Kevin Bales, and Kathryn Kish Sklar, eds. 1991. *The Social Survey in Historical Perspective, 1880–1940*. Cambridge: Cambridge University Press.

Burgess, Richard D. 1989. "Major Issues and Implications of Tracing Survey Respondents." In *Panel Surveys*, ed. Daniel Kasprzyk et al. New York: Wiley.

Burton, Scot, and Edward A. Blair. 1991. "Task Conditions, Response Formulation Processes, and Response Accuracy for Behavioral Frequency Questions in Surveys." *Public Opinion Quarterly* 55:50–79.

Business Market Research Association. 2002. "No Means No! The Harsh Reality of Respondent Rejection." *MRBusiness*, no. 6, 6 March. http://www.bmra.org.uk/mrbusiness/content.asp?ezine=73&article=155.

Cahalan, Don. 1989. "The *Digest* Poll Rides Again!" *Public Opinion Quarterly* 53: 129–33.

Campbell, Angus, Philip E. Converse, Warren E. Miller, and Donald E. Stokes. 1960. *The American Voter*. New York: Wiley.

Cannell, Charles F., S. A. Lawson, and D. L. Hausser. 1975. *A Technique for Evaluating Interviewer Performance*. Ann Arbor: Institute for Social Research, University of Michigan.

Cannell, Charles F., K. Marquis, and A. Laurent. 1977. "A Summary of Studies." *Vital and Health Statistics*, series 2, 69. Washington, DC: Government Printing Office.

Cannell, Charles F., Peter V. Miller, and Lois Oksenberg. 1981. "Research on Interviewing Techniques." In *Sociological Methodology, 1981*, ed. S. Leinhardt. San Francisco: Jossey-Bass.

Cannell, Charles F., Lois Oksenberg, and Jean M. Converse. 1977. *Experiments in Interviewing Techniques: Field Experiments in Health Reporting, 1971–1977*. Hyattsville, MD: National Center for Health Services Research (NCHSR).

Cantor, David. 1989. "Substantive Implications of Selected Operational Longitudinal Design Features: The National Crime Study as a Case Study." In *Panel Surveys*, ed. Daniel Kasprzyk et al. New York: Wiley.

Casady, R., and James M. Lepkowski. 1993. "Stratified Telephone Survey Designs." *Survey Methodology* 19:103–13.

Catania, Joseph A., Diane Binson, Jesse Canchola, Lance M. Pollack, Walter Hauck, and Thomas J. Coates. 1996. "Effects of Interviewer Gender, Interviewer Choice, and Item Wording on Responses to Questions concerning Sexual Behavior." *Public Opinion Quarterly* 60:345–375.

Church, Allan H. 1993. "Estimating the Effect of Incentives on Mail Survey Response Rates: A Meta-analysis." *Public Opinion Quarterly* 57:62–79.

Cialdini, Robert B. 1988. *Influence: Science and Practice*. Glenview, IL: Scott Foresman.

Clausen, Aage. 1968. "Response Validity: Vote Report." *Public Opinion Quarterly* 41:56–64.

Clymer, Adam. 2001. "The Unbearable Lightness of Public Opinion Polls." *New York Times*, July 22, sec. 4, p. 3.

Coale, Ansley J., and Frederick F. Stephan. 1962. "The Case of the Indians and the Teen-Age Widows." *Journal of the American Statistical Association* 57:338–47.

Conrad, Frederick, and Michael Schober. 2000. "Clarifying Question Meaning in a Household Telephone Survey." *Public Opinion Quarterly* 64:1–28.

Converse, Jean M. 1987. *Survey Research in the United States: Roots and Emergence, 1890–1960*. Berkeley: University of California Press.

Converse, Jean M., and Stanley Presser. 1986. *Survey Questions: Handcrafting the Standardized Questionnaire*. Beverly Hills, CA: Sage.

Converse, Jean M., and Howard Schuman. 1974. *Conversations at Random: Survey Research as Interviewers See It*. New York: Wiley.

Converse, Philip E. 1964. "The Nature of Belief Systems in Mass Publics." In *Ideology and Discontent.*, ed. David Apter. New York: Free Press.

Cook, C., F. Heath, and R. Thompson. 2000. "A Meta-analysis of Response Rates in Web- or Internet-Based Surveys." *Educational and Psychological Measurement* 60:821–36.

Cooper, S. L. 1964. "Random Selection by Telephone: A New and Improved Method." *Journal of Marketing Research* 1:45–48.

Couper, Mick P. 1997. "Survey Introductions and Data Quality." *Public Opinion Quarterly* 61:317–38.

———. 2000. "Web Surveys: A Review of Issues and Approaches." *Public Opinion Quarterly* 64:464–94.

Couper, Mick P., et al., eds. 1998. *Computer Assisted Survey Information Collection.* New York: Wiley.

Couper, Mick P., and Robert M. Groves. 2002. "Introductory Interactions in Telephone Surveys and Nonresponse." In *Standardization and Tacit Knowledge: Interaction and Practice in the Survey Interview*, ed. Douglas Maynard et al. New York: Wiley.

Couper, Mick P., and William Nicholls. 1998. "The History and Development of Computer Assisted Survey Information Collection Methods." In *Computer Assisted Survey Information Collection*, ed. Mick P. Couper et al. New York: Wiley.

Couper, Mick P., and Benjamin Rowe. 1996. "Evaluation of a Computer-Assisted Self-Interview Component in a Computer-Assisted Personal Interview Survey." *Public Opinion Quarterly* 60:89–105.

Couper, Mick P., Michael W. Traugott, and Mark J. Lamias. 2001. "Web Survey Design and Administration." *Public Opinion Quarterly* 65:230–53.

Crespi, Irving. 1988. *Pre-election Polling: Sources of Accuracy and Error.* New York: Russell Sage.

Currivan, Douglas B., et al. 2004. "Does Telephone Audio Computer-Assisted Self-Interviewing Improve the Accuracy of Prevalence Estimates of Youth Smoking?" *Public Opinion Quarterly* 68:542–64.

Curtin, Richard, Stanley Presser, and Eleanor Singer. 2000. "The Effects of Response Rate Changes on the Index of Consumer Sentiment." *Public Opinion Quarterly* 64:413–28.

Davern, Michael, Todd H. Rockwood, Randy Sherrod, and Stephen Campbell. 2003. "Prepaid Monetary Incentives and Data Quality in Face-to-Face Interviews." *Public Opinion Quarterly* 67:139–47.

Davis, Darren W. 1997. "Nonrandom Measurement Error and Race of Interviewer Effects among African Americans." *Public Opinion Quarterly* 61:183–207.

de Leeuw, Edith D. 1992. *Data Quality in Mail, Telephone, and Face-to-Face Surveys.* Amsterdam: TT Publications.

de Leeuw, Edith D., and Johannes van der Zouwen. 1988. "Data Quality in Telephone and Face to Face Surveys: A Comparative Meta-analysis." In *Telephone Survey Methodology*, ed. Robert M. Groves et al. New York: Wiley.

Delli Carpini, Michael X., and Scott Keeter. 1996. *What Americans Know about Politics and Why It Matters.* New Haven, CT: Yale University Press.

DeMaio, T., and A. Landreth. 2004. "Do Different Cognitive Interview Methods Produce Different Results?" In *Methods for Testing and Evaluating Survey Questionnaires*, ed. Stanley Presser et al. New York: Wiley.

Dewald, William G., Jerry G. Thursby, and Richard G. Anderson. 1986. "Replication in Empirical Economics: The *Journal of Money, Credit, and Banking* Project." *American Economic Review* 76:587–603.

Dijkstra, Wil, and Johannes H. Smit. 2002. "Persuading Reluctant Recipients in Telephone Surveys." In *Survey Nonresponse*, ed. Robert M. Groves et al. New York: Wiley.

Dillman, Don A. 1978. *Mail and Telephone Surveys: The Total Design Method.* New York: Wiley.

———. 2000. *Mail and Internet Surveys: The Tailored Design Method.* 2nd ed. New York: Wiley.

Dillman, Don A., and Dennis K. Bowker. 2001. "The Web Questionnaire Challenge to Survey Methodologists." In *Dimensions of Internet Science*, ed. Ulf-Dietrich Reips and Michael Bosnjak. Lengerich, Germany.: Pabst Science.

DuBois, W. E. B. 1899. *The Philadelphia Negro: A Social Study.* Philadelphia: Ginn.

Eisenhower, Donna, Nancy A. Mathiowetz, and David Morganstein. 1991. "Recall Error: Sources and Bias Reduction Techniques." In *Measurement Errors in Surveys*, ed. Paul Biemer et al. New York: Wiley.

Eisinger, Robert M. 2003. *The Evolution of Presidential Polling.* Cambridge: Cambridge University Press.

Ellis, Charles, and Jon A. Krosnick. 1999. "Comparing Telephone and Face-to-Face Surveys in Terms of Sample Representativeness: A Meta-analysis of Demographic Characteristics." NES Technical Report 59. http://www.umich.edu/~nes.

Epstein, Joan Faith, Peggy Barker, and Larry Kroutil. 2001. "Mode Effects in Self-Reported Mental Health Data." *Public Opinion Quarterly* 65:529–49.

Erikson, Robert S., and Christopher Wlezien. 1999. "Presidential Polls as a Time Series." *Public Opinion Quarterly* 63:163–77.

Ferber, Robert. 1956. "The Effect of Respondent Ignorance on Survey Results." *Journal of the American Statistical Association* 51:576–86.

Finkel, Steven E., and Paul Freedman. 2004. "The Half-Hearted Rise: Voter Turnout in the 2000 Election." In *Models of Voting in Presidential Elections*, ed. Herbert F. Weisberg and Clyde Wilcox. Stanford, CA: Stanford University Press.

Finkel, Steven E., Thomas M. Guterbock, and Marian J. Borg. 1991. "Race-of-Interviewer Effects in a Preelection Poll: Virginia 1989." *Public Opinion Quarterly* 55:313–30.

Fiske, Susan T., and Shelley E. Taylor. 1991. *Social Cognition.* 2nd ed. New York: McGraw-Hill.

Foddy, W. 1995. "Probing: A Dangerous Practice in Social Surveys?" *Quality and Quantity* 29:73–86.

Forsyth, Barbara H., and Judith T. Lessler. 1991. "Cognitive Laboratory Methods." In *Measurement Error in Surveys*, ed. Paul Biemer et al. New York: Wiley.

Fowler, Floyd J., Jr. 1991. "Reducing Interviewer-Related Error through Interviewer Training, Supervision, and Other Methods." In *Measurement Error in Surveys*, ed. Paul Biemer et al. New York: Wiley.

———. 1992. "How Unclear Terms Affect Survey Data." *Public Opinion Quarterly* 56:218–31.

———. 1995. *Improving Survey Questions: Design and Evaluation.* Thousand Oaks, CA: Sage.

———. 2002. *Survey Research Methods.* 3rd ed. Thousand Oaks, CA: Sage.

———. 2004. "Getting beyond Pretesting and Cognitive Interviews." In *Methods for Testing and Evaluating Survey Questionnaires*, ed. Stanley Presser et al. New York: Wiley.

Fowler, Floyd J., Jr., and Charles F. Cannell. 1996. "Using Behavioral Coding to Identify Cognitive Problems with Survey Questions." In *Answering Questions*, ed. Norbert Schwarz and Seymour Sudman. San Francisco: Jossey-Bass.

Fowler, Floyd J., Jr., and Thomas W. Mangione. 1990. *Standardized Survey Interviewing: Minimizing Interviewer-Related Error.* Newbury Park, CA: Sage.

Fowler, Floyd J., Jr., Anthony Roman, and Zhu Di. 1998. "Mode Effects in a Survey of Medicare Prostate Surgery Patients." *Public Opinion Quarterly* 62:29–46.

Fox, Richard J., Melvin R. Crask, and Jonghoon Kim. 1988. "Mail Survey Response Rate: A Meta-analysis of Selected Techniques for Inducing Response." *Public Opinion Quarterly* 52:467–91.

Freeman, A. Myrick, III. 2003. *The Measurement of Environmental and Resource Values.* 2nd ed. Washington, DC: Resources for the Future.

Fuller, Wayne. 1991. "Regression Estimation in the Presence of Measurement Error." In *Measurement Error in Surveys*, ed. Paul P. Biemer et al. New York: Wiley.

Furse, David H., and David W. Stewart. 1982. "Monetary Incentives versus Promised Contribution to Charity: New Evidence on Mail Survey Response." *Journal of Marketing Research* 19:375–80.

Gaskell, George D., Daniel B. Wright, and Colin A. O'Muircheartaigh. 2000. "Telescoping of Landmark Events." *Public Opinion Quarterly* 64:77–89.

Geer, John G. 1988. "What Do Open-Ended Questions Measure?" *Public Opinion Quarterly* 52:365–71.

———. 1991. "Do Open-Ended Questions Measure 'Salient' Issues?" *Public Opinion Quarterly* 55:360–70.

Gelman, Andrew, and Gary King. 1993. "Why Are American Presidential Election Polls So Variable When Votes Are So Predictable?" *British Journal of Political Science* 23:409–51.

Gelman, Andrew, and Thomas C. Little. 1998. "Improving on Probability Weighting for Household Size." *Public Opinion Quarterly* 62:398–404.

Gilljam, M., and Donald Granberg. 1993. "Should We Take Don't Know for an Answer?" *Public Opinion Quarterly* 57:348–57.

Goldstein, Kenneth M. 2002. "Getting in the Door: Sampling and Completing Elite Interviews." *P.S.: Political Science and Politics* 35:669–72.

Goldstein, Kenneth M., and M. Kent Jennings. 2002. "The Effect of Advanced Letters on Cooperation in a List Assisted Telephone Survey." *Public Opinion Quarterly* 66:608–17.

Gosnell, Harold F. 1927. *Getting Out the Vote.* Chicago: University of Chicago Press.

Granquist, Leopold. 1995. "Improving the Traditional Editing Process." In *Business Survey Methods*, ed. Brenda G. Cox et al. New York: Wiley.

Green, Donald, Bradley Palmquist, and Eric Schickler. 2002. *Partisan Hearts and Minds: Political Parties and the Social Identities of Voters.* New Haven, CT: Yale University Press.

Grice, H. Paul. 1975. "Logic and Conversation." In *Syntax and Semantics: Speech Acts*, ed. P. Cole and J. L. Morgan, 3:41–58. New York: Academic Press.

Groves, Robert M. 1979. "Actors and Questions in Telephone and Personal Interview Surveys." *Public Opinion Quarterly* 43:190–205.

———. 1989. *Survey Errors and Survey Costs.* 1989. New York: Wiley.

———. 1990. "Theories and Methods of Telephone Surveys." *Annual Review of Sociology* 16:221–40.

Groves, Robert M., et al., eds. 1988. *Telephone Survey Methodology.* New York: Wiley.

———, eds. 2002. *Survey Nonresponse.* New York: Wiley.

———. 2004. *Survey Methodology.* New York: Wiley.

Groves, Robert M., Robert B. Cialdini, and Mick P. Couper. 1992. "Understanding the Decision to Participate in a Survey." *Public Opinion Quarterly* 56:475–95.

Groves, Robert M., and Mick P. Couper. 1998. *Nonresponse in Household Interview Surveys.* New York: Wiley.

Groves, Robert M., and N. H. Fultz. 1985. "Gender Effects among Telephone Interviewers in a Survey of Economic Attitudes." *Sociological Methods and Research* 14:31–52.

Groves, Robert M., and Robert L. Kahn. 1979. *Surveys by Telephone: A National Comparison with Personal Interviews.* New York: Academic.

Groves, Robert M., and Lars Lyberg. 1988. "An Overview of Nonresponse Issues in Telephone Surveys." In *Telephone Survey Methodology,* ed. Robert Groves et al. New York: Wiley.

Groves, Robert M., and K. McGonagle. 2001. "A Theory-Guided Interviewer Training Protocol regarding Survey Participation." *Journal of Official Statistics* 17:249–65.

Groves, Robert M., Stanley Presser, and Sarah Dipko. 2004. "The Role of Topic Interest in Survey Participation Decisions." *Public Opinion Quarterly* 68:2–32.

Groves, Robert M., Eleanor Singer, and Amy Corning. 2000. "Leverage-Saliency Theory of Survey Participation." *Public Opinion Quarterly* 64:299–308.

Groves, Robert M., and Robert D. Tortora. 1998. "Integrating CASIC into Existing Designs and Organizations: A Survey of the Field." In *Computer Assisted Survey Information Collection,* ed. Mick P. Couper et al. New York: Wiley.

Gubrium, Jaber F., and James A. Holstein, eds. 2002. *Handbook of Interview Research: Context and Method.* Thousand Oaks, CA: Sage.

Hagan, D., and C. Collier. 1983. "Must Respondent Selection Procedures for Telephone Surveys Be Invasive?" *Public Opinion Quarterly* 47:547–56.

Hansen, Morris, W. Hurwitz, and M. Bershad. 1961. "Measurement Errors in Censuses and Surveys." *Bulletin of the International Statistical Institute* 38:359–74.

Hansen, Morris, W. Hurwitz, and W. Madow. 1953. *Sample Survey Methods and Theory.* New York: Wiley.

Harding, David J., and Christopher Jencks. 2003. "Changing Attitudes toward Premarital Sex: Cohort, Period, and Aging Effects." *Public Opinion Quarterly* 67:211–26.

Harkness, Janet A. 2003. "Questionnaire Translation." In *Cross-Cultural Survey Methods,* ed. Janet A. Harkness, Fons J. R. van de Vijver, and Peter Ph. Mohler, 35–56. New York: Wiley.

Harkness, Janet A., Fons J. R. van de Vijver, and Peter Ph. Mohler, eds. 2003. *Cross-Cultural Survey Methods.* New York: Wiley.

Hastie, Reid, and Bernadette Park. 1986. "The Relationship between Memory and Judgment Depends Whether the Task Is Memory-Based or On-Line." *Psychological Review* 93:258–68.

Hatchett, Shirley, and Howard Schuman. 1975–76. "White Respondents and Race-of-Interviewer Effects." *Public Opinion Quarterly* 39:523–28.

Heckman, James J. 1979. "Sample Selection Bias as a Specification Error." *Econometrica* 47:153–61.

Hedges, L. V., and I. Olkin. 1985. *Statistical Methods for Meta-analysis.* Boston: Academic.

Heeringa, Steven G., et al. 1994. *1990 SRC National Sample.* Ann Arbor: University of Michigan, Institute for Social Research.

Henderson, L., and D. Allen. 1981. "NLS Data Entry Quality Control: The Fourth Followup Survey." Washington, DC: National Center for Education Statistics, Office of Educational Research and Improvement.

Henry, Gary T. 1990. *Practical Sampling.* Newbury Park, CA: Sage.

Herbst, Susan. 1993. *Numbered Voices: How Opinion Polling Has Shaped American Politics.* Chicago: University of Chicago Press.

Herzog, A. Regula, and Willard L. Rodgers. 1988. "Interviewing Older Adults: Mode Comparison Using Data from a Face-to-Face Survey and a Telephone Resurvey." *Public Opinion Quarterly* 52:84–99.

Hess, Irene. 1985. *Sampling for Social Research Surveys, 1947–1980.* Ann Arbor: Institute for Social Research.

Hite, Shere. 1976. *The Hite Report.* New York: Dell.

Hoch, S. J. 1987. "Perceived Consensus and Predictive Accuracy: The Pros and Cons of Projection." *Journal of Personality and Social Psychology* 53:221–34.

Hoffmeyer-Zlotnik, Jürgen P., and Christof Wolf. 2003. *Advances in Cross-National Comparison: A European Working Book for Demographic and Socio-economic Variables.* Dordrecht, Netherlands: Kluwer Academic/Plenum.

Holbrook, Allyson L., Melanie C. Green, and Jon A. Krosnick. 2003. "Telephone vs. Face-to-Face Interviewing of National Probability Samples with Long Questionnaires: Comparisons of Respondent Satisficing and Social Desirability Response Bias." *Public Opinion Quarterly* 67:79–125.

Houtkoop-Steenstra, Hanneke. 2000. *Interaction and the Standardized Survey Interview: The Living Questionnaire.* Cambridge: Cambridge University Press.

Houtkoop-Steenstra, Hanneke, and Huub van den Bergh. 2002. "Effects of Introductions in Large-Scale Telephone Survey Interviews." In *Standardization and Tacit Knowledge,* ed. Douglas W. Maynard et al. New York: Wiley.

Hox, Joop, Edith de Leeuw, and Ita Kreft. 1991. "The Effect of Interviewer and Respondent Characteristics on the Quality of Survey Data: A Multilevel Model." In *Measurement Errors in Surveys,* ed. Paul Biemer et al. New York: Wiley.

Hubbard, Raymond, and Eldon L. Little. 1988. "Promised Contributions to Charity and Mail Survey Responses: Replication with Extension." *Public Opinion Quarterly* 52:223–30.

Huddy, Leonie, et al. 1997. "The Effect of Interviewer Gender on the Survey Response." *Political Behavior* 19:197–220.

Hughes, A., et al. 2002. "Impact of Interviewer Experience on Respondent Reports of Substance Use." In *Redesigning an Ongoing National Household Survey,* ed. J. Gfroerer et al., 161–84. Washington, DC: Substance Abuse and Mental Health Services Administration.

Huisman, Mark. 2000. "Imputation of Missing Item Responses: Some Simple Techniques." *Quality and Quantity* 34:331–51.

Hunter, John E., and Frank L. Schmidt. 2004. *Methods of Meta-analysis.* 2nd ed. Thousand Oaks, CA: Sage.

Hyman, Herbert H. 1954. *Interviewing in Social Research.* Chicago: University of Chicago Press.

———. 1955. *Survey Design and Analysis.* Glencoe, IL: Free Press.

Hyman, Herbert H., and Paul B. Sheatsley. 1950. "The Current State of American Public Opinion." In *The Teaching of Contemporary Public Affairs,* ed. J. C. Payne, 11–34. New York: National Education Association.

Iannacchione, Vincent G., Jennifer M. Staab, and David T. Redden. 2003. "Evaluating the Use of Residential Mailing Lists in a Metropolitan Household Survey." *Public Opinion Quarterly* 67:202–10.

Jabine, T., K. King, and R. Petroni. 1990. *Quality Profile for the Survey of Income and Program Participation (SIPP).* Washington, DC: U.S. Bureau of the Census.

Jabine, T., M. Straf, Judith M. Tanur, and Roger Tourangeau, eds. 1984. *Cognitive Aspects of Survey Methodology: Building a Bridge between Disciplines.* Washington, DC: National Academy Press.

James, Jeannine, and Richard Bolstein. 1990. "The Effect of Monetary Incentives and Follow-Up Mailings on the Response Rate and Response Quality in Mail Surveys." *Public Opinion Quarterly* 54:346–61.

Jensen, Richard. 1980. "Democracy by Numbers." *Public Opinion* 3, no. 1A (February–March): 53–59.

Johnson, Robert A., Dean R. Gerstein, and Kenneth A. Rasinski. 1998. "Adjusting Survey Estimates for Response Bias: An Application to Trends in Alcohol and Marijuana Use." *Public Opinion Quarterly* 62:354–77.

Johnson, Timothy P., and Fons J. R. van de Vijver. 2003. "Social Desirability in Cross-Cultural Research." In *Cross-Cultural Survey Methods*, ed. Janet A. Harkness, Fons J. R. van de Vijver, and Peter Ph. Mohler, 195–204. New York: Wiley.

Johnston, Richard, and Henry E. Brady. 2002. "The Rolling Cross-Section Design." *Electoral Studies* 21:283–95.

Jordan, L. A., A. C. Marcus, and L. G. Reeder, 1980. "Response Styles in Telephone and Household Interviewing: A Field Experiment." *Public Opinion Quarterly* 44:210–22.

Jowell, Roger, Barry Hedges, Peter Lynn, Graham Farrant, and Anthony Heath. 1993. "The 1992 British Election: The Failure of the Polls." *Public Opinion Quarterly* 57:238–63.

Judd, Charles M., R. A. Drake, J. W. Downing, and Jon A. Krosnick. 1991. "Some Dynamic Properties of Attitude Structures: Context-Induced Response Facilitation and Polarization." *Journal of Personality and Social Psychology* 60:193–202.

Juster, F., and J. Smith. 1997. "Improving the Quality of Economic Data." *Journal of the American Statistical Association* 92:1268–78.

Kaldenberg, Dennis O., Harold F. Koenig, and Boris W. Becker. 1994. "Mail Survey Response Rate Patterns in a Population of the Elderly: Does Response Deteriorate with Age?" *Public Opinion Quarterly* 58:68–76.

Kalton, Graham. 1983. *Introduction to Survey Sampling*. Beverly Hills, CA: Sage.

Kalton, Graham, and Michael Brick. 2000. "Weighting in Household Panel Surveys." In *Researching Social and Economic Change: The Uses of Household Panel Studies*, ed. David Rose. London: Routledge.

Kalton, Graham, and Constance F. Citro. 2000. "Panel Surveys: Adding the Fourth Dimension." In *Researching Social and Economic Change: The Uses of Household Panel Studies*, ed. David Rose. London: Routledge.

Kalton, Graham, Daniel K. Kasprzyk, and D. B. McMillen. 1989. "Nonsampling Errors in Panel Surveys." In *Panel Surveys*, ed. Daniel Kasprzyk et al. New York: Wiley.

Kane, Emily W., and Laura J. Macaulay. 1993. "Interviewer Gender and Gender Attitudes." *Public Opinion Quarterly* 57:1–28.

Kaplowitz, Michael D., Timothy D. Hadlock, and Ralph Levine. 2004. "A Comparison of Web and Mail Survey Response Rates." *Public Opinion Quarterly* 68:94–101.

Katosh, John P., and Michael W. Traugott. 1981. "The Consequences of Validated and Self-Reported Voting Measures." *Public Opinion Quarterly* 45:519–35.

Keeter, Scott. 1995. "Estimating Telephone Noncoverage Bias with a Telephone Survey." *Public Opinion Quarterly* 59:196–217.

Keeter, Scott, Carolyn Miller, Andrew Kohut, Robert M. Groves, and Stanley Presser. 2000. "Consequences of Reducing Nonresponse in a National Telephone Survey." *Public Opinion Quarterly* 64:125–48.

Kelman, Herbert C. 1953. "Compliance, Identification, and Internalization: Three Processes of Attitude Change." *Human Relations* 6:185–214.

Kessel, John H., and Herbert F. Weisberg. 1999. "Comparing Models of the Vote: The Answers Depend on the Questions." In *Reelection 1996: How Americans Voted*, ed.

Herbert F. Weisberg and Janet M. Box-Steffensmeier, 88–98. Chappaqua, NY: Chatham House.

Kiewiet, D. Roderick, and Douglas Rivers. 1985. "The Economic Basis of Reagan's Appeal." In *The New Direction in American Politics*, ed. John Chubb and Paul Peterson, 66–78. Washington, DC: Brookings Institution.

King, Gary, et al. 2001. "Analyzing Incomplete Political Science Data: An Alternative Algorithm for Multiple Imputation." *American Political Science Review* 95:49–69.

Kinsey, Alfred, Wardell Pomeroy, and Clyde Martin. 1948. *Sexual Behavior in the Human Male*. Philadelphia: W. B. Saunders.

Kish, Leslie. 1965. *Survey Sampling*. New York: Wiley.

Knäuper, Bärbel. 1998. "Filter Questions and Question Interpretation: Presuppositions at Work." *Public Opinion Quarterly* 62:70–78.

———. 1999. "The Impact of Age and Education on Response Order Effects in Attitude Measurement." *Public Opinion Quarterly* 63:347–70.

Knight, Kathleen, and Robert S. Erikson. 1997. "Ideology in the 1990s." In *Understanding Public Opinion*, ed. Barbara Norrander and Clyde Wilcox, 89–110. Washington, DC: CQ Press.

Koch, Nadine S., and Jolly A. Emrey. 2001. "The Internet and Opinion Measurement: Surveying Marginalized Populations." *Social Science Quarterly* 82:131–38.

Konner, Joan. 2003. "The Case for Caution: This System Is Dangerously Flawed." *Public Opinion Quarterly* 67:5–18.

Krosnick, Jon A. 1991. "Response Strategies for Coping with the Cognitive Demands of Attitude Measures in Surveys." *Applied Cognitive Psychology* 5:213–236.

———. 1999. "Survey Research." *Annual Review of Psychology* 50:537–67.

———. 2002. "The Causes of No-Opinion Responses to Attitude Measures in Surveys: They Are Really What They Appear to Be." In *Survey Nonresponse*, ed. Robert M. Groves et al. New York: Wiley.

Krosnick, Jon A., et al. 2002. "The Impact of 'No Opinion' Response Opinions on Data Quality: Non-attitude Reduction or an Invitation to Satisfice?" *Public Opinion Quarterly* 66:371–403.

Krosnick, Jon A., and Duane F. Alwin. 1987. "An Evaluation of a Cognitive Theory of Response-Order Effects in Survey Measurement." *Public Opinion Quarterly* 51:201–19.

———. 1988. "A Test of the Form-Resistant Correlation Hypothesis: Ratings, Rankings, and the Measurement of Values." *Public Opinion Quarterly* 52:526–38.

Krosnick, Jon A., and Matthew Berent. 1993. "Comparisons of Party Identification and Policy Preferences: The Impact of Survey Question Format." *American Journal of Political Science* 37:941–64.

Krosnick, Jon A., and LinChiat Chang. 2001. "A Comparison of the Random Digit Dialing Telephone Survey Methodology with Internet Survey Methodology as Implemented by Knowledge Networks and Harris Interactive." Paper presented at the annual AAPOR conference, Montreal.

Krosnick, Jon A., Matthew Courser, Kenneth Mulligan, and LinChiat Chang. 2001. "Exploring the Causes of Vote Choice in the 2000 Presidential Election." Paper presented at the annual meeting of the American Political Science Association, San Francisco.

Krosnick, Jon A., and Leandre R. Fabrigar. 1997. "Designing Rating Scales for Effective Measurement in Surveys." In *Survey Measurement and Process Quality*, ed. Lars Lyberg et al. New York: Wiley.

Krysan, Maria. 1998. "Privacy and the Expression of White Racial Attitudes." *Public Opinion Quarterly* 62:506–44.

Krysan, Maria, Howard Schuman, Lesli Jo Scott, and Paul Beatty. 1994. "Response Rates and Response Content in Mail versus Face-to-Face Surveys." *Public Opinion Quarterly* 58:381–99.

Lacy, Dean. 2001a. "A Theory of Nonseparable Preferences in Survey Responses." *American Journal of Political Science* 45:239–58.

———. 2001b. "Nonseparable Preferences, Measurement Error, and Unstable Survey Responses." *Political Analysis* 9:95–115.

Ladd, Everett Carll. 1994. "The Holocaust Poll Error: A Modern Cautionary Tale." *Public Perspective* 5 (July/August): 3–5.

Lambro, Donald. 2001. "Pollsters Dismiss Survey Giving Low Marks to Bush." *Washington Times*, June 26, A4.

Lau, Richard R. 1994. "An Analysis of the Accuracy of 'Trial Heat' Polls during the 1992 Presidential Election." *Public Opinion Quarterly* 58:2–20.

Laumann, Edward O., John H. Gagnon, Robert T. Michael, and Stuart Michaels. 1994. *The Social Organization of Sexuality: Sexual Practices in the United States.* Chicago: University of Chicago Press.

Lavrakas, Paul J. 1993. *Telephone Survey Methods: Sampling, Selection and Supervision.* 2nd ed. Newbury Park, CA: Sage.

Lavrakas, Paul J., S. Bauman, and Daniel Merkle. 1993. "The 'Last Birthday' Selection Method and Within-Unit Coverage Problems." Paper presented at the annual meeting of the American Association for Public Opinion Research, St. Charles, IL.

Leal, David, and Frederick Hess. 1999. "Survey Bias on the Front Porch: Are All Subjects Interviewed Equally?" *American Politics Quarterly* 27:468–87.

Lentner, Marvin, and Thomas Bishop. 1986. *Experimental Design and Analysis.* Blacksburg, VA: Valley Book.

Lepkowski, James M. 1989. "Treatment of Wave Nonresponse in Panel Surveys." In *Panel Surveys*, ed. Daniel Kasprzyk et al. New York: Wiley.

Lepkowski, James M., and Mick P. Couper. 2002. "Nonresponse in the Second Wave of Longitudinal Household Surveys." In *Survey Nonresponse*, ed. Robert M. Groves et al. New York: Wiley.

Lepkowski, James M., Sally Ann Sadosky, and Paul S. Weiss. 1998. "Mode, Behavior, and Data Recording Error." In *Computer Assisted Survey Information Collection*, ed. Mick P. Couper et al. New York: Wiley.

Lessler, Judith T., and William D. Kalsbeek. 1992. *Nonsampling Errors in Surveys.* New York: Wiley.

Lewis, Michael. 2000. "The Two-Bucks-a-Minute Democracy." *New York Times Magazine*, November 5. http://query.nytimes.com/gst/abstract.html?res=F10 A17FF3E540C768 CDDA80994D8404482&incamp=archive:search (accessed January 18, 2005).

Lichtenstein, S., et al. 1978. "Judged Frequency of Lethal Events." *Journal of Experimental Psychology: Human Learning and Memory* 4:551–78.

Lillard, Lee, et al. 1986. "What Do We Really Know about Wages? The Importance of Nonreporting and Census Imputation." *Journal of Political Economy* 94: 490–506.

Lin, I-Fen, and Nora Cate Schaeffer. 1995. "Using Survey Participants to Estimate the Impact of Nonparticipation." *Public Opinion Quarterly* 59:236–58.

Link, Michael W., and Robert W. Oldendick. 1999. "Call Screening: Is It Really a Problem for Survey Research?" *Public Opinion Quarterly* 63:577–89.

Lipsey, M. W., and D. B. Wilson. 2001. *Practical Meta-analysis*. Thousand Oaks, CA: Sage.

Little, Roderick. 2003. "Bayesian Methods for Unit and Item Nonresponse." In *Analysis of Survey Data*, ed. R. L. Chambers and C. J. Skinner. New York: Wiley.

Little, Roderick, and Donald Rubin. 2002. *Statistical Analysis with Missing Data*. 2nd ed. New York: Wiley.

Lockerbie, Brad, and Stephen A. Borrelli. 1990. "Question Wording and Public Support for Contra Aid, 1983–1986." *Public Opinion Quarterly* 54:195–208.

Lodge, Milton, Kathleen McGraw, and Patrick Stroh. 1989. "An Impression-Driven Model of Candidate Evaluation." *American Political Science Review* 83:399–419.

Loftus, Elizabeth F., and D. C. Fathi. 1985. "Retrieving Multiple Autobiographical Memories." *Social Cognition* 3:280–95.

Loftus, Elizabeth F., Mark R. Klinger, Kyle D. Smith, and Judith Fiedler. 1990. "A Tale of Two Questions: Benefits of Asking More Than One Question." *Public Opinion Quarterly* 54:330–45.

Loftus, Elizabeth F., and W. Marburger. 1983. "Since the Eruption of Mt. St. Helens Did Anyone Beat You Up? Improving the Accuracy of Retrospective Reports with Landmark Events." *Memory and Cognition* 11:114–20.

Losch, Mary E., Aaron Maitland, Gene Lutz, Peter Mariolis, and Steven C. Gleason. 2002. "The Effect of Time of Year of Data Collection on Sample Efficiency." *Public Opinion Quarterly* 66:594–607.

Luevano, Patricia. 1994. "Response Rates in the National Election Studies, 1948–1992." NES Technical Report 44. http://www.umich.edu/~nes.

Lyberg, Lars, et al., eds. 1997. *Survey Measurement and Process Quality*. New York: Wiley.

Lynn, Peter, et al. 2002. "The Effects of Extended Interviewer Efforts on Nonresponse Bias." In *Survey Nonresponse*, ed. Robert M. Groves et al. New York: Wiley.

MacKuen, Michael B., Robert S. Erikson, and James A. Stimson. 1992. "Question Wording and Macropartisanship." *American Political Science Review* 86:475–81.

Mann, Thomas E., and Raymond E. Wolfinger. 1980. "Candidates and Parties in Congressional Elections." *American Political Science Review* 74:617–32.

Marquis, K. H., Charles F. Cannell, and A. Laurent. 1972. "Reporting Health Events in Household Interviews: Effects of Reinforcement, Question Length, and Reinterviews." *Vital and Health Statistics*, series 2, 45. Washington, DC: U.S. Government Printing Office.

Martin, Elizabeth A. 1999. "Who Knows Who Lives Here? Within-Household Disagreement as a Source of Survey Coverage Error." *Public Opinion Quarterly* 63:220–236.

Martin, Elizabeth A., and Anne E. Polivka. 1995. "Diagnostics for Redesigning Survey Questionnaires: Measuring Work in the Current Population Survey." *Public Opinion Quarterly* 59:547–67.

Marton, Krisztina. 2004. "Effects of Questionnaire Fieldwork Characteristics on Call Outcome Rates and Data Quality in a Monthly CATI Survey." Ph.D. diss., Ohio State University.

Mason, Robert, Virginia Lesser, and Michael W. Traugott. 2002. "Effect of Item Nonresponse on Nonresponse Error and Inference." In *Survey Nonresponse*, ed. Robert M. Groves et al. New York: Wiley.

Mathiowetz, Nancy A. 1992. "Errors in Reports of Occupation." *Public Opinion Quarterly* 56:352–55.

———. 1998. "Respondent Expressions of Uncertainty: Data Source for Imputation." *Public Opinion Quarterly* 62:47–56.

Mathiowetz, Nancy A., and Charles F. Cannell. 1980. "Coding Interviewer Behavior as a Method of Evaluating Performance." *Proceedings of the Section on Survey Research Methods, American Statistical Association,* 525–28.

Mathiowetz, Nancy A., and Robert M. Groves. 1985. "The Effects of Respondent Rules on Health Survey Reports." *American Journal of Public Health* 75:639–44.

Maynard, Douglas W., and Nora Cate Schaeffer. 2002a. "Standardization and Its Discontents." In *Standardization and Tacit Knowledge: Interaction and Practice in the Survey Interview,* ed. Douglas Maynard et al. New York: Wiley.

———. 2002b. "Opening and Closing the Gate: The Work of Optimism in Recruiting Survey Respondents." In *Standardization and Tacit Knowledge: Interaction and Practice in the Survey Interview,* ed. Douglas Maynard et al. New York: Wiley.

Maynard, Douglas W., Hanneke Houtkoop-Steenstra, Nora Cate Schaeffer, and Johannes van der Zouwen, eds. 2002. *Standardization and Tacit Knowledge: Interaction and Practice in the Survey Interview.* New York: Wiley.

McCarty, John A., and L. J. Shrum. 2000. "The Measurement of Personal Values in Survey Research: A Test of Alternative Rating Procedures." *Public Opinion Quarterly* 64:271–98.

McDonald, Michael P., and Samuel L. Popkin. 2001. "The Myth of the Vanishing Voter." *American Political Science Review* 95:963–74.

McGuire, William J. 1960. "A Syllogistic Analysis of Cognitive Relationships." In *Attitude Organization and Change,* ed. Morris Rosenberg et al. New Haven, CT: Yale University Press.

Means, B., and Elizabeth F. Loftus. 1991. "When Personal History Repeats Itself: Decomposing Memories for Recurring Events." *Applied Cognitive Psychology* 5:297–318.

Menard, Scott. 2002. *Longitudinal Research.* 2nd ed. Thousand Oaks, CA: Sage.

Mensch, Barbara S., and Denise B. Kandel. 1988. "Underreporting of Substance Use in a National Longitudinal Youth Cohort." *Public Opinion Quarterly* 52:100–24.

Merkle, Daniel M. 1996. "The National Issues Convention Deliberative Poll." *Public Opinion Quarterly* 60:588–619.

Merkle, Daniel M., S. L. Bauman, and Paul Lavrakas. 1991. "Nonresponse Bias: Refusal Conversions and Call-Backs in RDD Telephone Surveys." Paper presented at the Midwest Association for Public Opinion Research conference, Chicago.

Merkle, Daniel M., and Murray Edelman. 2000. "A Review of the 1996 Voter News Service Exit Polls from a Total Survey Error Perspective." In *Election Polls, the News Media, and Democracy,* ed. Paul Lavrakas and Michael Traugott. New York: Chatham House.

———. 2002. "Nonresponse in Exit Polls." In *Survey Nonresponse,* ed. Robert M. Groves et al. New York: Wiley.

Michaels, Stuart, and Alain Giami. 1999. "Sexual Acts and Sexual Relationships: Asking about Sex in Surveys." *Public Opinion Quarterly* 63:401–20.

Miller, J. 1984. "A New Survey Technique for Studying Deviant Behavior." Ph.D. diss., George Washington University.

Miller, Joanne M., and David A. M. Peterson. 2004. "Theoretical and Empirical Implications of Attitude Strength." *American Journal of Political Science* 66:847–67.

Miller, Peter V. 1991. "Which Side Are You On? The 1990 Nicaraguan Poll Debacle." *Public Opinion Quarterly* 55:281–302.

———. 1995. "They Said It Couldn't Be Done: The National Health and Social Life Survey." *Public Opinion Quarterly* 59:404–19.

Miller, Peter V., Daniel M. Merkle, and Paul Wang. 1991. "Journalism with Footnotes: Reporting the 'Technical Details' of Polls." In *Polling and Presidential Election Coverage*, ed. Paul Lavrakas and Jack Holley. Newbury Park, CA: Sage.

Miller, Thomas W., and Kurian J. Panjikaran. 2001. *Studies in Comparability: The Propensity Scoring Approach*. Madison, WI: A. C. Nielsen Center for Marketing Research, University of Wisconsin.

Mishler, Eliot G. 1986. *Research Interviewing: Context and Narrative*. Cambridge, MA: Harvard University Press.

Mitofsky, Warren J. 1991. "A Short History of Exit Polls." In *Polling and Presidential Election Coverage*, ed. Paul Lavrakas and Jack Holley. Newbury Park, CA: Sage.

Mitofsky, Warren J., and Murray Edelman. 1995. "A Review of the 1992 VRS Exit Polls." In *Presidential Polls and the News Media*, ed. Paul Lavrakas, Michael Traugott, and Peter Miller. Boulder, CO: Westview.

Mondak, Jeffery J. 2001. "Developing Valid Knowledge Scales." *American Journal of Political Science* 45:224–38.

Moon, Nick. 1999. *Opinion Polls: History, Theory, and Practice*. Manchester, UK: Manchester University Press.

Moon, Youngme. 1998. "Impression Management in Computer-Based Interviews." *Public Opinion Quarterly* 62:610–22.

Mooney, Chris. 2003. "John Zogby's Creative Polls." *American Prospect* 14 (2):29–33.

Moore, David W. 1995. *The Superpollsters: How They Measure and Manipulate Public Opinion in America*. 2nd ed. New York: Four Walls Eight Windows.

———. 2002. "Measuring New Types of Question-Order Effects: Additive and Subtractive." *Public Opinion Quarterly* 66:80–91.

Moore, Jeffrey C., Linda L. Stinson, and Edward J. Welniak Jr. 1999. "Income Reporting in Surveys: Cognitive Issues and Measurement Errors." In *Cognition and Survey Research*, ed. Monroe Sirken et al. New York: Wiley.

Morton-Williams, Jean. 1993. *Interviewer Approaches*. Aldershot, UK: Dartmouth.

Moskowitz, Joel M. 2004. "Assessment of Cigarette Smoking and Smoking Susceptibility among Youth." *Public Opinion Quarterly* 68:565–87.

Mosteller, F., et al. 1948. "The Pre-election Polls of 1948: Report of the Committee on Analysis of Pre-election Polls and Forecasts." *Public Opinion Quarterly* 12:595–622.

Mulligan, Kenneth, J. Tobin Grant, Stephen T. Mockabee, and Joseph Quin Monson. 2003. "Response Latency Methodology for Survey Research: Measurement and Modeling Strategies." *Political Analysis* 11:289–301.

Murray, D., et al. 1987. "The Validity of Smoking Self-Reports by Adolescents: A Reexamination of the Bogus Pipeline Procedure." *Addictive Behaviors* 12:7–15.

Nadeau, Richard, Richard G. Niemi, and Jeffrey Levine. 1993. "Innumeracy about Minority Populations." *Public Opinion Quarterly* 57:332–47.

Narayan, Sowmya and Jon A. Krosnick. 1996. "Education Moderates Some Response Effects in Attitude Measurement." *Public Opinion Quarterly* 60:58–88.

National Commission for the Protection of Human Subjects of Biomedical and Behavioral Research. 1979. *The Belmont Report: Ethical Principles and Guidelines for the Protection of Human Subjects of Research*. Washington, DC: Government Printing Office.

National Election Studies (NES) Ad Hoc Committee. 1999. "The State of Scientific Knowledge on the Advantages and Disadvantages of Telephone vs. Face-to-Face Interviewing." NES Technical Report 55. http://www.umich.edu/~nes.

Nelson, Thomas E., and Donald R. Kinder. 1996. "Issue Frames and Group-Centrism in American Public Opinion." *Journal of Politics* 58:1055–78.

Neter, John, and Joseph Waksberg. 1964. "A Study of Response Errors in Expenditures Data from Household Surveys." *Journal of the American Statistical Association* 59:18–55.

Nicholls, William L., II., Reginald P. Baker, and Jean Martin. 1997. "The Effect of New Data Collection Technologies on Survey Data Quality." In *Survey Measurement and Process Quality*, ed. Lars Lyberg et al. New York: Wiley.

Niemi, Richard G. 1974. *How Family Members Perceive Each Other*. New Haven, CT: Yale University Press.

Nisbett, Richard E., and Timothy D. Wilson. 1977. "Telling More Than We Can Know: Verbal Reports on Mental Processes." *Psychological Review* 84:231–59.

Norwood, Janet L., and Judith M. Tanur. 1994. "Measuring Unemployment in the Nineties." *Public Opinion Quarterly* 58:277–94.

Oksenberg, Lois. 1981. *Analysis of Monitored Telephone Interviews*. Report to the U.S. Bureau of the Census for JSA 80–23. Ann Arbor: Survey Research Center, University of Michigan.

Oksenberg, Lois, and Charles Cannell. 1988. "Effects of Interviewer Vocal Characteristics on Nonresponse." In *Telephone Survey Methodology*, ed. Robert Groves et al., 257–69. New York: Wiley.

Oldendick, Robert O., and Michael W. Link. 1994. "The Answering Machine Generation: Who Are They and What Problem Do They Pose for Survey Research?" *Public Opinion Quarterly* 58:264–273.

O'Muircheartaigh, Colm. 1989. "Sources of Nonsampling Error: Discussion." In *Panel Surveys*, ed. Daniel Kasprzyk et al. New York: Wiley.

O'Muircheartaigh, Colm A., George D. Gaskell, and Daniel B. Wright. 1993. "Intensifiers in Behavioral Frequency Questions." *Public Opinion Quarterly* 57:552–65.

O'Rourke, D., and J. Blair. 1983. "Improving Random Respondent Selection in Telephone Surveys." *Journal of Marketing Research* 20:428–32.

Parten, Mildred. 1950. *Surveys, Polls, and Samples*. New York: Harper.

Payne, Stanley. 1951. *The Art of Asking Questions*. Princeton, NJ: Princeton University Press.

Petty, Richard E., and John T. Cacioppo. 1986. *Communication and Persuasion: Central and Peripheral Routes to Attitude Change*. New York: Springer-Verlag.

Petty, Richard E., and Jon A. Krosnick, eds. 1995. *Attitude Strength*. Mahwah, NJ: Erlbaum.

Petty, Richard E., Greg A. Rennier, and John T. Cacioppo. 1987. "Assertion versus Interrogation Format in Opinion Surveys: Questions Enhance Thoughtful Responding." *Public Opinion Quarterly* 51:481–94.

Pew Research Center for the People and the Press. 2004. "Polls Face Growing Resistance, but Still Representative." http://people-press.org/reports/display.php3?ReportID=211.

Piazza, Thomas. 1993. "Meeting the Challenge of Answering Machines." *Public Opinion Quarterly* 57:219–31.

Piekarski, Linda, Gwen Kaplan, and Jessica Prestengaard. 1999. "Telephony and Telephone Sampling: The Dynamics of Change." Paper presented at the annual AAPOR conference, St. Petersburg, FL.

Pierzchala, M. 1990. "A Review of the State of the Art in Automated Data Editing and Imputation." *Journal of Official Statistics* 6:355–77.

Pollner, Melvin, and Richard E. Adams. 1997. "The Effect of Spouse Presence on Appraisals of Emotional Support and Household Strain." *Public Opinion Quarterly* 61:615–26.

Presser, Stanley, et al., eds. 2004. *Methods for Testing and Evaluating Survey Questionnaires*. New York: Wiley.

Presser, Stanley, and Johnny Blair. 1994. "Survey Pretesting: Do Different Methods Produce Different Results?" In *Sociological Methodology 1994*, ed. P. Marsden, 73–104. San Francisco: Jossey-Bass.

Presser, Stanley, and Howard Schuman. 1980. "The Measurement of a Middle Position in Attitude Surveys." *Public Opinion Quarterly* 44:70–85.

Presser, Stanley, and Shanyang Zhao. 1992. "Attributes of Questions and Interviewers as Correlates of Interviewing Performance." *Public Opinion Quarterly* 56:236–40.

Przeworski, Adam, and Henry Teune. 1970. *The Logic of Comparative Social Inquiry*. New York: Wiley.

Rasinski, Kenneth A. 1989. "The Effect of Question Wording on Public Support for Government Spending." *Public Opinion Quarterly* 53:388–94.

Rasinski, Kenneth A., David Mingay, and Norman M. Bradburn. 1994. "Do Respondents Really 'Mark All That Apply' on Self-Administered Questions?" *Public Opinion Quarterly* 58:400–08.

Residents of Hull House. 1895. *Hull-House Maps and Papers: A Presentation of Nationalities and Wages in a Congested District of Chicago*. New York: Crowell.

Rhodes, P. J. 1994. "Race-of-Interviewer Effects." *Sociology* 28:547–58.

Rice, Stuart A. 1928. *Quantitative Methods in Politics*. New York: Alfred A. Knopf.

Rizzo, Louis, J. Michael Brick, and Inho Park. 2004. "A Minimally Intrusive Method for Sampling Persons in Random Digit Dial Surveys." *Public Opinion Quarterly* 68:267–74.

Robertson, D. H., and D. N. Bellenger. 1978. "A New Method of Increasing Mail Survey Responses: Contributions to Charity." *Journal of Marketing Research* 15:632–33.

Roese, N. J., and Jamieson, D. W. 1993. "Twenty Years of Bogus Pipeline Research: A Critical Review and Meta-analysis." *Psychological Bulletin* 114:365–75.

Rose, David. 2000a. "Household Panel Studies: An Overview." In *Researching Social and Economic Change: The Uses of Household Panel Studies*, ed. David Rose. London: Routledge.

———, ed. 2000b. *Researching Social and Economic Change: The Uses of Household Panel Studies*. London: Routledge.

Rosenstone, Steven J., Margaret Petrella, and Donald R. Kinder. 1993. "Excessive Reliance on Telephone Interviews and Short-Form Questionnaires in the 1992 National Election Study: Assessing the Consequences for Data Quality." NES Technical Report 43. http://www.umich.edu/~nes.

Rosenthal, Robert. 1979. "The 'File Drawer Problem' and Tolerance for Null Results." *Psychological Bulletin* 86:638–41.

———. 1991. *Meta-analytic Procedures for Social Research*. Beverly Hills, CA: Sage.

Rowntree, B. Seebohm. 1901. *Poverty: A Study of Town Life*. London: Macmillan.

Rubin, David C., and Alan D. Baddeley. 1989. "Telescoping Is Not Time Compression: A Model of the Dating of Autobiographical Events." *Memory and Cognition* 17:653–61.

Rubin, Donald B. 1987. *Multiple Imputation for Nonresponse in Surveys*. New York: Wiley.

Rubin, Donald B., and Nathaniel Schenker. 1986. "Multiple Imputation for Interval Estimation from Single Random Samples with Ignorable Nonresponse." *Journal of the American Statistical Association* 81:366–74.

Rubin, Donald B., and Elaine Zanutto. 2002. "Using Matched Substitutes to Adjust for Nonignorable Nonresponse through Multiple Imputations." In *Survey Nonresponse,* ed. Robert M. Groves et al. New York: Wiley.

Rugg, D. 1941. "Experiments in Wording Questions II." *Public Opinion Quarterly* 5:91–92.

Rugg, D., and Hadley Cantril. 1944. "The Wording of Questions." In *Gauging Public Opinion.* ed. Hadley Cantril. Princeton, NJ: Princeton University Press.

Salmon, C., and J. S. Nichols. 1983. "The Next-Birthday Method of Respondent Selection." *Public Opinion Quarterly* 47:270–76.

Sanbonmatsu, D., and R. Fazio. 1990. "The Role of Attitudes in Memory-Based Decision-Making." *Journal of Personality and Social Psychology* 59:614–22.

Sanchez, Maria Elena. 1992. "Effects of Questionnaire Design on the Quality of Survey Data." *Public Opinion Quarterly* 56:206–17.

Scammon, Richard M., and Ben J. Wattenberg. 1970. *The Real Majority.* New York: Coward-McCann.

Schaefer, David R., and Don A. Dillman. 1998. "Development of a Standard Email Methodology: Results of an Experiment." *Public Opinion Quarterly* 62:378–97.

Schaeffer, Nora Cate. 1991. "Hardly Ever or Constantly? Group Comparisons Using Vague Quantifiers." *Public Opinion Quarterly* 55:395–423.

Schafer, Joseph L. 1997. *Analysis of Incomplete Multivariate Data.* London: Chapman and Hall.

Schober, Michael F., and Fredrick G. Conrad. 1997. "Conversational Interviewing." *Public Opinion Quarterly* 61:576–602.

Schonlau, Matthias, Ronald D. Fricker Jr., and Marc N. Elliott. 2002. *Conducting Research Surveys via E-mail and the Web.* Santa Monica, CA: RAND Corporation.

Schuman, Howard. 1990. "3 Different Pens Help Tell the Story." *New York Times,* March 7.

———. 1992. "Context Effects: State of the Past/State of the Art." In *Context Effects in Social and Psychological Research,* ed. Norbert Schwarz and Seymour Sudman. New York: Springer-Verlag.

Schuman, Howard, and Jean M. Converse. 1971. "The Effects of Black and White Interviewers on Black Responses in 1968." *Public Opinion Quarterly* 35:44–68.

Schuman, Howard, and Otis Dudley Duncan. 1974. "Questions about Attitude Survey Questions." In *Sociological Methodology, 1973–1974,* ed. H. L. Costner. San Francisco: Jossey-Bass.

Schuman, Howard, Jacob Ludwig, and Jon A. Krosnick. 1986. "The Perceived Threat of Nuclear War, Salience, and Open Questions." *Public Opinion Quarterly* 50:519–36.

Schuman, Howard, and Stanley Presser. 1981. *Questions and Answers in Attitude Surveys.* New York: Academic.

Schuman, Howard, Charlotte Steeh, and Lawrence Bobo. 1985. *Racial Attitudes in America: Trends and Interpretations.* Cambridge, MA: Harvard University Press.

Schwarz, Norbert. 2003. "Culture-Sensitive Context Effects: A Challenge for Cross-Cultural Surveys." In *Cross-Cultural Survey Methods,* ed. Janet A. Harkness, Fons J. R. van de Vijver, and Peter Ph. Mohler, 93–100. New York: Wiley.

Schwarz, Norbert, et al. 1985. "Response Scales: Effects of Category Range on Reported Behavior and Comparative Judgments." *Public Opinion Quarterly* 49:388–95.

Schwarz, Norbert, and H. Bless. 1992. "Construction Reality and Its Alternatives: Assimilation and Contrasts Effects in Social Judgment." In *The Construction of Social Judgment*, ed. L. Martin and A. Tesser, 217–45. Hillsdale, NJ: Erlbaum.

Schwarz, Norbert, Bärbel Knäuper, Hans-J. Hippler, Elisabeth Noelle-Neumann, and Leslie Clark. 1991. "Rating Scales: Numerical Values May Change the Meaning of Scale Labels." *Public Opinion Quarterly* 55:570–82.

Schwarz, Norbert, Fritz Strack, and Hans-Peter Mai. 1991. "Assimilation and Contrast Effects in Part-Whole Question Sequences: A Conversational Logic Analysis." *Public Opinion Quarterly* 55:3–23.

Schwarz, Norbert, and Seymour Sudman, eds. 1992. *Context Effects in Social and Psychological Research*. New York: Springer-Verlag.

———, eds. 1994. *Autobiographical Memory and the Validity of Retrospective Reports*. New York: Springer-Verlag.

———, eds. 1996. *Answering Questions: Methodology for Determining Cognitive and Communicative Processes in Survey Research*. San Francisco: Jossey-Bass.

Sebestik, J., et al., 1988. "Initial Experiences with CAPI." *Proceedings of the Bureau of the Census' Annual Research Conference* (U.S. Bureau of the Census, Washington, DC), 357–65.

Sebold, Janice. 1988. "Survey Period Length, Unanswered Numbers, and Nonresponse in Telephone Surveys." In *Telephone Survey Methodology*, ed. Robert M. Groves et al. New York: Wiley.

Shea, Christopher. 2000. "Don't Talk to Humans: The Crackdown on Social Science Research." *Lingua Franca* 10, no. 6: 1–17.

Sheppard, Jane. 2001. "Promoting and Advocating Survey Research, or Is It 'Just Say No'? Refusal Rates over Past Year." http://www.cmor.org (Council for Marketing and Opinion Research).

Sherman, Robert. 2000. "Tests of Certain Types of Ignorable Nonresponse in Surveys Subject to Item Nonresponse or Attrition." *American Journal of Political Science* 44:362–74.

Sigelman, Lee, and Richard G. Niemi. 2001. "Innumeracy about Minority Populations: African Americans and Whites Compared." *Public Opinion Quarterly* 65:86–94.

Simon, Herbert A. 1957. *Models of Men*. New York: Wiley.

Singer, Eleanor. 1978. "Informed Consent: Consequences for Response Rate and Response Quality in Social Surveys." *American Sociological Review* 43:144–62.

Singer, Eleanor, Nancy A. Mathiowetz, and Mick P. Couper. 1993. "The Impact of Privacy and Confidentiality Concerns on Survey Participation: The Case of the 1990 U.S. Census." *Public Opinion Quarterly* 57:465–82.

Singer, Eleanor, John Van Hoewyk, and Mary P. Maher. 1998. "Does the Payment of Incentives Create Expectation Effects?" *Public Opinion Quarterly* 62:152–64.

———. 2000. "Experiments with Incentives in Telephone Survey." *Public Opinion Quarterly* 64:171–88.

Singer, Eleanor, John Van Hoewyk, and Randall J. Neugebauer. 2003. "Attitudes and Behavior: The Impact of Privacy and Confidentiality Concerns on Participation in the 2000 Census." *Public Opinion Quarterly* 67:368–84.

Singer, Eleanor, Dawn R. Von Thurn, and Esther R. Miller. 1995. "Confidentiality Assurances and Responses: A Quantitative Review of the Experimental Literature." *Public Opinion Quarterly* 59:66–77.

Sirken, Monroe G., et al., eds. 1999. *Cognition and Survey Research*. New York: Wiley.

Skogan, Wesley G. 1990. "The National Crime Survey Redesign." *Public Opinion Quarterly* 54:256–72.

Smith, A. F., J. B. Jobe, and D. Mingay. 1991. "Retrieval from Memory of Dietary Information." *Applied Cognitive Psychology* 5:269–96.

Smith, Charles E., Jr., Peter M. Radcliffe, and John H. Kessel. 1999. "The Partisan Choice: Bill Clinton or Bob Dole?" In *Reelection 1996: How Americans Voted*, ed. Herbert F. Weisberg and Janet M. Box-Steffensmeier, 70–87. Chappaqua, NY: Chatham House.

Smith, Eric R. A. N., and Peverill Squire. 1990. "The Effects of Prestige Names in Question Wording." *Public Opinion Quarterly* 54:97–116.

Smith, Tom W. 1978. "In Search of House Effects: A Comparison of Responses to Various Questions by Different Survey Organizations." *Public Opinion Quarterly* 42:443–63.

———. 1982. "House Effects and the Reproducibility of Survey Measurements: A Comparison of the 1980 GSS and the 1980 American National Election Study." *Public Opinion Quarterly* 46:54–68.

———. 1983. "The Missing 25 Percent: An Analysis of Nonresponse on the 1980 General Social Survey." *Public Opinion Quarterly* 47:386–404.

———. 1987. "That Which We Call Welfare by Any Other Name Would Smell Sweeter: An Analysis of the Impact of Question Wording on Response Patterns." *Public Opinion Quarterly* 51:75–83.

———. 1990a. "The First Straw? A Study of the Origins of Election Polls." *Public Opinion Quarterly* 54:21–36.

———. 1990b. "Phone Home? An Analysis of Household Telephone Ownership." *International Journal of Public Opinion Research* 2:369–90.

———. 1991. "Context Effects in the General Social Survey." In *Measurement Error in Surveys*, ed. Paul Biemer et al. New York: Wiley.

———. 1996. "The Impact of Presence of Others on a Respondent's Answers to Questions." *International Journal of Public Opinion Research* 9:33–47.

———. 1997. "Improving Cross-National Survey Response by Measuring the Intensity of Response Categories," General Social Survey (GSS) Cross-National Report 197. Chicago: National Opinion Research Center.

Squire, Peverill. 1988. "Why the 1936 *Literary Digest* Poll Failed." *Public Opinion Quarterly* 52:125–33.

Stimson, James A. 1999. *Public Opinion in America: Moods, Cycles, and Swings*. 2nd ed. Boulder, CO: Westview.

Stokes, Donald E., and Warren E. Miller. 1962. "Party Government and the Saliency of Congress." *Public Opinion Quarterly* 26:531–46.

Stouffer, Samuel, et al. 1949. *The American Soldier*. New York: Wiley.

Strack, F., and L. Martin. 1987. "Thinking, Judging, and Communicating: A Process Account of Context Effects in Attitude Surveys." In *Social Information Processing and Survey Methodology*, ed. H. Hippler, Norbert Schwarz, and Seymour Sudman, 123–48. New York: Springer-Verlag.

Suchman, Lucy, and Brigitte Jordan. 1990. "Interactional Troubles in Face-to-Face Survey Interviews." *Journal of the American Statistical Association* 85:232–41.

Sudman, Seymour. 1980. "Improving the Quality of Shopping Center Sampling." *Journal of Marketing Research* 17:423–31.

Sudman, Seymour, and Norman Bradburn. 1973. "Effects of Time and Memory Factors on Response in Surveys." *Journal of the American Statistical Association* 68:805–15.

Sudman, Seymour, Norman Bradburn, and Norbert Schwarz. 1996. *Thinking about Answers: The Application of Cognitive Processes to Survey Methodology*. San Francisco: Jossey-Bass.

Sudman, Seymour, A. Finn, and L. Lannom. 1984. "The Use of Bounded Recall Procedures in Single Interviews." *Public Opinion Quarterly* 48:520–24.

Taylor, Humphrey. 2000. "Does Internet Research Work? Comparing Online Survey Results with Telephone Survey." *International Journal of Market Research* 41: 51–63.

Taylor, Shelley E. 1981. "A Categorization Approach to Stereotyping." In *Cognitive Processes in Stereotyping and Intergroup Behavior*, ed. David Hamilton. Hillsdale, NJ: Erlbaum.

Taylor, Shelley E., and Susan Fiske. 1978. "Salience, Attention, and Attribution: Top of the Head Phenomena." In *Advances in Social Psychology*, ed. L. Berkowitz. New York: Academic.

Thornberry, Owen T., Jr., and James T. Massey. 1988. "Trends in United States Telephone Coverage across Time and Subgroups." In *Telephone Survey Methodology*, ed. Robert M. Groves et al. New York: Wiley.

Thurstone, Louis. 1928. "Attitudes Can Be Measured." *American Journal of Sociology* 33:529–54.

Tourangeau, Roger. 1984. "Cognitive Science and Survey Methods." In *Cognitive Aspects of Survey Design: Building a Bridge between Disciplines*, ed. T. Jabine et al. Washington, DC: National Academy Press.

———. 1999. "Context Effects on Answers to Attitude Questions." In *Cognition and Survey Research*, ed. Monroe Sirken et al. New York: Wiley.

Tourangeau, Roger, Mick P. Couper, and Frederick Conrad. 2004. "Spacing, Position, and Order: Interpretative Heuristics for Visual Features of Survey Questions." *Public Opinion Quarterly* 68:368–93.

Tourangeau, Roger, and Kenneth A. Rasinski. 1988. "Cognitive Processes Underlying Context Effects in Attitude Measurement." *Psychological Bulletin* 103: 299–314.

Tourangeau, Roger, Kenneth A. Rasinski, and Norman Bradburn. 1991. "Measuring Happiness in Surveys: A Test of the Subtraction Hypothesis." *Public Opinion Quarterly* 55:255–66.

Tourangeau, Roger, Kenneth A. Rasinski, Norman Bradburn, and Roy D'Andrade. 1989. "Carryover Effects in Attitude Surveys." *Public Opinion Quarterly* 53: 495–524.

Tourangeau, Roger, Lance Rips, and Kenneth A. Rasinski. 2000. *The Psychology of Survey Response*. Cambridge: Cambridge University Press.

Tourangeau, Roger, G. Shapiro, A. Kearney, and L. Ernst. 1997. "Who Lives Here? Survey Undercoverage and Household Roster Questions." *Journal of Official Statistics* 13:1–18.

Tourangeau, Roger, and Tom W. Smith. 1996. "Asking Sensitive Questions: The Impact of Data Collection Mode, Question Format, and Question Context." *Public Opinion Quarterly* 60:275–304.

———. 1998. "Collecting Sensitive Information with Different Modes of Data Collection." In *Computer Assisted Survey Information Collection*, ed. Mick P. Couper et al. New York: Wiley.

Tourangeau, Roger, Darby Miller Steiger, and David Wilson. 2002. "Self-Administered Questions by Telephone: Evaluating Interactive Voice Response." *Public Opinion Quarterly* 66:265–78.

Traugott, Michael W. 1987. "The Importance of Persistence in Respondent Selection for Preelection Surveys." *Public Opinion Quarterly* 51:48–57.

Traugott, Michael W., Robert M. Groves, and James Lepkowski. 1987. "Using Dual Frames to Reduce Nonresponse in Telephone Surveys." *Public Opinion Quarterly* 51:522–39.

Traugott, Michael W., and Mee-Eun Kang. 2000. "Push Polls as Negative Persuasive Strategies." In *Election Polls, the News Media, and Democracy*, ed. Paul Lavrakas and Michael Traugott. New York: Chatham House.

Traugott, Michael W., and Elizabeth C. Powers. 2000. "Did Public Opinion Support the Contract with America?" In *Election Polls, the News Media, and Democracy*, ed. Paul Lavrakas and Michael Traugott. New York: Chatham House.

Troldahl, V., and R. Carter. 1964. "Random Selection of Respondents within Households in Phone Surveys." *Journal of Marketing Research* 1:71–76.

Trussell, Norm, and Paul J. Lavrakas. 2004. "The Influence of Incremental Increases in Token Cash Incentives on Mail Survey Response." *Public Opinion Quarterly* 68:349–67.

Tuchfarber, Alfred, and William Klecka. 1976. *Random Digit Dialing: Lowering the Cost of Victim Surveys*. Washington, DC: Police Foundation.

Tuckel, Peter S., and Barry M. Feinberg. 1991. "The Answering Machine Poses Many Questions for Telephone Survey Researchers." *Public Opinion Quarterly* 55:200–17.

Tuckel, Peter S., and Harry O'Neill. 2001. "The Vanishing Respondent in Telephone Surveys." Paper presented at the annual AAPOR conference, Montreal.

Tucker, Clyde, James M. Lepkowski, and Linda Piekarski. 2002. "The Current Efficiency of List-Assisted Telephone Sampling Designs." *Public Opinion Quarterly* 66:321–38.

Turnbull, William. 1947. "Secret vs. Nonsecret Ballots." In *Gauging Public Opinion*, ed. Hadley Cantril et al. Princeton, NJ: Princeton University Press.

Turner, Charles F., et al., 1998. "Automated Self-Interviewing and the Survey Measurement of Sensitive Behaviors." In *Computer Assisted Survey Information Collection*, ed. Mick P. Couper et al. New York: Wiley.

Tversky, Amos, and Daniel Kahneman. 1973. "Availability: A Heuristic for Judging Frequency and Probability." *Cognitive Psychology* 5:207–32.

———. 1974. "Judgment under Uncertainty: Heuristics and Biases." *Science* 185:1124–31.

———. 1979. "Prospect Theory: An Analysis of Decision under Risk." *Econometrica* 47:263–91.

United States Bureau of the Census. 1984. *Statistical Abstracts of the United States, 1984*. Washington, DC: Government Printing Office.

———. 1993. "Memorandum for Thomas C. Walsh from John H. Thompson, Subject: 1990 Decennial Census—Long Form (Sample Write-In) Keying Assurance Evaluations." Washington, DC: U.S. Bureau of the Census.

United States Federal Committee on Statistical Methodology. 1990. *Data Editing in Federal Statistical Agencies*, Statistical Policy Working Paper 15. Washington, DC: U.S. Office of Management and Budget.

Utah v. Evans. 2002. 536 U.S. 452.

van de Vijver, Fons J. R. 2003. "Bias and Equivalence: Cross-Cultural Perspectives." In *Cross-Cultural Survey Methods*, ed. Janet A. Harkness, Fons J. R. van de Vijver, and Peter Ph. Mohler, 143–55. New York: Wiley.

Vehovar, Vasja, et al. 2002. "Nonresponse in Web Surveys." In *Survey Nonresponse*, ed. Robert M. Groves et al. New York: Wiley.

Visser, Penny S., Jon A. Krosnick, Jesse Marquette, and Michael Curtin. 1996. "Mail Surveys for Election Forecasting? An Evaluation of the *Columbus Dispatch* Poll." *Public Opinion Quarterly* 60:181–227.

Viterna, Jocelyn S., and Douglas W. Maynard. 2002. "How Uniform Is Standardization? Variation within and across Survey Research Centers Regarding Protocols for Interviewing." In *Standardization and Tacit Knowledge: Interaction and Practice in the Survey Interview*, ed. Douglas Maynard et al. New York: Wiley.

Voss, D. Stephen, Andrew Gelman, and Gary King. 1995. "Review: Preelection Survey Methodology: Details from Eight Polling Organizations, 1988 and 1992." *Public Opinion Quarterly* 59:98–132.

Wagenaar, W. A. 1986. "My Memory: A Study of Autobiographical Memory over Six Years." *Cognitive Psychology* 18:225–52.

Waksberg, Joseph. 1978. "Sampling Methods for Random Digit Dialing." *Journal of the American Statistical Association* 73:40–46.

Walsh, John P., Sara Kiesler, Lee S. Sproull, and Bradford W. Hesse. 1992. "Self-Selected and Randomly Selected Respondents in a Computer Network Survey." *Public Opinion Quarterly* 54:241–44.

Wänke, Michaela, Norbert Schwarz, and Elisabeth Noelle-Neumann. 1995. "Asking Comparative Questions: The Impact of the Direction of Comparison." *Public Opinion Quarterly* 59:347–72.

Warner, S. 1965. "Randomized Response: A Survey Technique for Eliminating Evasive Answer Bias." *Journal of the American Statistical Society* 60:63–69.

Warriner, Keith, John Goyder, Heidi Gjertsen, Paula Hohner, and Kathleen McSpurren. 1996. "Charities, No; Lotteries, No; Cash, Yes: Main Effects and Interactions in a Canadian Incentives Experiment." *Public Opinion Quarterly* 60:542–62.

Weisberg, Herbert F. 1987. "The Demographics of a New Voting Gap: Marital Differences in American Voting." *Public Opinion Quarterly* 51:335–43.

Weisberg, Herbert F., Jon A. Krosnick, and Bruce D. Bowen. 1996. *An Introduction to Survey Research, Polling, and Data Analysis*. 3rd ed. Thousand Oaks, CA: Sage.

Weisberg, Herbert F., and Stephen T. Mockabee. 1999. "Attitudinal Correlates of the 1996 Presidential Vote." In *Reelection 1996: How Americans Voted*, ed. Herbert F. Weisberg and Janet M. Box-Steffensmeier, 45–69. Chappaqua, NY: Chatham House.

Wentland, Ellen J., with Kent W. Smith. 1993. *Survey Responses: An Evaluation of Their Validity*. San Diego: Academic Press.

Werner, O., and D. T. Campbell. 1970. "Translating, Working through Interpreters, and the Problem of Decentering." In *A Handbook of Cultural Anthropology*, ed. R. Naroll and R. Cohen. New York: American Museum of Natural History.

Wessel, Christina, Wendy Rahn, and Tom Rudolph. 2000. "An Analysis of the 1998 NES Mixed-Mode Design." NES Technical Report 57. http://www.umich.edu/~nes.

Wilcox, Clyde, Lee Sigelman, and Elizabeth Cook. 1989. "Some Like It Hot: Individual Differences in Reponses to Group Feeling Thermometers." *Public Opinion Quarterly* 53:246–57.

Williams, Margaret. 2004. "Women Judges: Accession at the State Court Level." Ph.D. diss., Ohio State University, Columbus.

Willimack, Diane K., Elizabeth Nichols, and Seymour Sudman. 2002. "Understanding Unit and Item Nonresponse in Business Surveys." In *Survey Nonresponse*, ed. Robert M. Groves et al. New York: Wiley.

Willimack, Diane K., Howard Schuman, Beth-Ellen Pennell, and James M. Lepkowski. 1995. "Effects of a Prepaid Nonmonetary Incentive on Response Rates and Response Quality in a Face-to-Face Survey." *Public Opinion Quarterly* 59:78–92.

Willis, Gordon B., Theresa J. DeMaio, and Brian Harris-Kojetin. 1999. "Is the Bandwagon Headed to the Methodological Promised Land? Evaluating the Validity of Cognitive Interviewing Techniques." In *Cognition and Survey Research*, ed. Monroe Sirken et al. New York: Wiley.

Wilson, Timothy D., and Sara D. Hodges. 1992. "Attitudes as Temporary Constructions." In *The Construction of Social Judgments*, ed. Leonard Martin and Abraham Tesser. New York: Springer-Verlag.

Winkels, Jeroen W., and Suzanne Davies Withers. 2000. "Panel Attrition." In *Researching Social and Economic Change: The Uses of Household Panel Studies*, ed. David Rose. London: Routledge.

Wolf, Frederic M. 1986. *Meta Analysis: Quantitative Methods for Research Synthesis.* Beverly Hills, CA: Sage.

Wright, Debra L., William S. Aquilino, and Andrew J. Supple. 1998. "A Comparison of Computer-Assisted and Paper-and-Pencil Self-Administered Questionnaires in a Survey on Smoking, Alcohol, and Drug Use." *Public Opinion Quarterly* 62:331–53.

Wright, Tommy. 2001. "Selected Moments in the Development of Probability Sampling." *Survey Research Methods Section Newsletter* 13:1–6.

Xu, Minghua, Benjamin J. Bates, and John C. Schweitzer. 1993. "The Impact of Messages on Survey Participation in Answering Machine Households." *Public Opinion Quarterly* 57:232–37.

Zaller, John R. 1992. *The Nature and Origins of Mass Opinion.* Cambridge: Cambridge University Press.

Zaller, John R., and Stanley Feldman. 1992. "A Simple Theory of the Survey Response: Answering Questions versus Revealing Preferences." *American Journal of Political Science* 36:579–616.

Zdep, S. M., et al. 1979. "The Validity of the Randomized Response Technique." *Public Opinion Quarterly* 43:544–49.

Zipp, John F., and Joann Toth. 2002. "She Said, He Said, They Said: The Impact of Spousal Presence in Survey Research." *Public Opinion Quarterly* 66:171–208.

Name Index

Subject Index